THE HOT UNIVERSE

INTERNATIONAL ASTRONOMICAL UNION

UNION ASTRONOMIQUE INTERNATIONALE

THE HOT UNIVERSE

PROCEEDINGS OF THE 188TH SYMPOSIUM OF THE
INTERNATIONAL ASTRONOMICAL UNION
HELD IN KYOTO, JAPAN, AUGUST 26–30, 1997

EDITED BY

KATSUJI KOYAMA
Kyoto University, Japan

SHUNJI KITAMOTO
Osaka University, Japan

and

MASAYUKI ITOH
Kobe University, Japan

KLUWER ACADEMIC PUBLISHERS
DORDRECHT / BOSTON / LONDON

A C.I.P. Catalogue record for this book is available from the Library of Congress.

ISBN 0-7923-5058-8 (HB)

---

*Published on behalf of
the International Astronomical Union
by
Kluwer Academic Publishers, P.O. Box 17, 3300 AA Dordrecht, The Netherlands.*

*Sold and distributed in the North, Central and South America
by Kluwer Academic Publishers,
101 Philip Drive, Norwell, MA 02061, U.S.A.*

*In all other countries, sold and distributed
by Kluwer Academic Publishers,
P.O. Box 322, 3300 AH Dordrecht, The Netherlands.*

Printed on acid-free paper

*All Rights Reserved
©1998 International Astronomical Union*

*No part of the material protected by this copyright notice may be reproduced or utilized
in any form or by any means, electronic or mechanical, including photocopying,
recording or by any information storage and retrieval system, without written permission
from the publisher.*

Printed in the Netherlands.

Contents

Preface xv
The Hot Universe: An Overview 3
    Y. Tanaka

Session 1: Plasma and Fresh Nucleosynthesis Phenomena

1-1. Sun and Stars

Theory of Flares and MHD Jets 9
    K. Shibata
X-Ray Coronae from Stars 13
    R. Pallavicini
The Quest for X-Rays from Protostars 17
    T. Montmerle
Solar Arcade Flarings as Magnetodynamic Process of Mass and Energy
Shedding from Energized Magnetic Regions 21
    Y. Uchida, S. Hirose, S. Cable, S. Uemura, K. Fujisaki, M. Torii,
    and S. Morita

1-2. Supernovae, Supernova Remnants and Galactic Hot Plasma

X-Rays from Supernova 1993J and Ejecta Instabilities 27
    K. Nomoto and T. Suzuki
Galactic 1.8 MeV Emission from ^{26}Al 31
    N. Prantzos
Gamma-Ray Line Spectroscopy Results from COMPTEL 35
    V. Schönfelder
Non-Equilibrium Condition in the SNR 39
    H. Tsunemi
Astrophysical Plasmas and Atomic Processes 43
    J.S. Kaastra
Hot Plasma in the Galaxy 47
    S. Yamauchi

1-3. Galaxies and Their Clusters

Metal Abundances in the Hot ISM of Elliptical Galaxies 53
    M. Loewenstein and R.F. Mushotzky
Hot Gaseous Halo in the Elliptical and Spiral Galaxies 57
    H. Awaki
Hot Gas in Groups and Their Galaxies 61
    T.J. Ponman and A. Finoguenov
ASCA Observations of Distant Clusters of Galaxies 65
    T.G. Tsuru

Session 2: Future Space Programs

Early Results from the Low-Energy Concentrator Spectrometer on-board
BeppoSAX 71
    A.N. Parmar, M. Bavdaz, F. Favata, T. Oosterbroek,
    A. Orr, A. Owens, U. Lammers, and D. Martin
The Astro-E Mission 75
    Y. Ogawara

AXAF in Context: A Revolution 79
   M. Elvis
ABRIXAS 83
   G. Hasinger, J. Trümper, and R. Staubert
The INTEGRAL Mission 87
   W. Hermsen and C. Winkler

Session 3: Diagnostics of High Gravity Objects with X- and Gamma Rays

3-1. White Dwarfs and Neutron Stars

Super-Eddington Sources in Galaxies 93
   G. Fabbiano
Mass Measurement of Accreting Magnetic White Dwarfs with Hard X-Ray Spectroscopy 97
   M. Ishida and R. Fujimoto
Photoionized Plasmas in X-Ray Binary Pulsars: ASCA Observations ... 101
   F. Nagase
Binary Structure of Accreting Neutron Stars 105
   D.A. Leahy
Millisecond Time Variations of X-Ray Binaries 107
   J.H. Swank
GRO J1744-28 and the Rapid Burster; Bizarre Objects! 111
   W.H.G. Lewin
X-Ray and Gamma-Ray Observations of Isolated Neutron Stars 113
   D.J. Thompson
Search for Nonthermal X-Rays from Supernova Remnant Shells 117
   R. Petre, J. Keohane, U. Hwang, G. Allen, and E. Gotthelf
Detection of TeV Gamma Rays from SN1006 121
   T. Tanimori
Very High Energy Gamma Rays from Plerions: CANGAROO Results ... 125
   T. Kifune

3-2. Black Hole Binaries

Galactic Radio-Jet Sources: Multiwavelength Observations 131
   Ph. Durouchoux and D. Hannikainen
Accretion Disks in Black Hole Candidates Observed with ASCA 135
   T. Dotani

3-3. AGNs

Evidence for Strong Gravity in the AGN Plasmas 141
   K. Iwasawa
X-Ray Aspects of the IRAS Galaxies 145
   E.J.A. Meurs
The Properties of Blazers Detected by EGRET 149
   J.R. Mattox
High Energy Phenomena in AGN Jets 153
   F. Takahara

3-4. Gamma-Ray Bursts

Gamma-Ray Burst Observations with BATSE 159

G.J. Fishman

Discovery of X-Ray Counterparts to Gamma Ray Bursts by BeppoSAX . 163
L. Piro

A Review of GRB Counterpart Searches 167
C. Kouveliotou

Quick Observations of the Fading X-Rays from Gamma-Ray Bursts with
ASCA 171
T. Murakami

Session 4: Large Scale Hot Plasmas and Their Relation with Dark Matter

X-Ray Large Scale Structure and XMM 177
M. Pierre

Hierarchical Structure in Dark Matter Distributions 181
K. Makishima

Diffuse EUV Emission from Clusters of Galaxies 185
S. Bowyer, R. Lieu, and J.P. Mittaz

Hot Electrons and Cold Photons: Galaxy Clusters and the Sunyaev-Zel'dovich
Effect 189
J.P. Hughes

X-Ray Observations of the Hot Intergalactic Medium 193
Q.D. Wang

ASCA Sky Survey Observations and the Cosmic X-Ray Background
in 2-10 keV 197
H. Inoue, T. Takahashi, Y. Ueda, A. Yamashita, Y. Ishisaki, and Y. Ogasaka

Contributed Papers

Session 1: Plasma and Fresh Nucleosynthesis Phenomena
1-1. Sun and Stars

An Attempt to Classify Solar Microwave-Bursts by Source Localization
Characteristics and Dynamics of Flare-Energy Release 205
A. Krüger, B. Kliem, J. Hildebrandt,
V.P. Nefedev, B.V. Agalakov, and G.Ya. Smolkov

Resistive Processes in the Preflare Phase of Eruptive Flares 207
T. Magara and K. Shibata

Three-Dimensional Simulation Study of Plasmoid Injection into Magnetized
Plasma... 209
Y. Suzuki, T.-H. Watanabe, A. Kageyama, T. Sato, and T. Hayashi

Impulsive Flare Plasma Energization in the Light of Yohkoh Discoveries 211
Y.M. Voitenko

MHD Simulation of Chromospheric Evaporation in a Solar Flare Based on
Magnetic Reconnection Model 213
T. Yokoyama and K. Shibata

Coronal Variability and Flaring of the RS CVn Binaries σ^2 CrB and
HR1099 215
A. Brown, R.A. Osten, S.A. Drake, K.L. Jones, and R.A. Stern

HgMn Stars as Apparent X-Ray Emitters 217
   S. Hubrig and T.W. Berghöfer
Variable High Velocity Jets in the Symbiotic Star CH Cygni 219
   T. Iijima
X-Ray Emission from Distant Stellar Clusters 220
   N.S. Schulz and J.H. Kastner
X-Ray Coronae from Single Late-Type Dwarf Stars 222
   K.P. Singh, S.A. Drake, and N.E. White
Chemical Abundances of Early Type Stars 224
   S. Tanaka, S. Kitamoto, T. Suzuki, K. Torii, M.F. Corcoran,
   and W. Waldron
ASCA Observations of 44I Bootis and VW Cephei 226
   C.S. Choi and T. Dotani
ASCA Observation of NGC1333 Star Forming Region 228
   M. Itoh, H. Fukunaga, K. Koyama, Y. Tsuboi, S. Yamauchi,
   N. Kobayashi, M. Hayashi, and S. Ueno
X-Ray Observations of Orion OB1b Association 230
   M. Nakano
Flares and MHD Jets in Protostar 232
   M. Hayashi, K. Shibata, and R. Matsumoto
X-Ray Valiabilities from Protostars in the R CrA Molecular Cloud 234
   K. Hamaguchi, H. Murakami, K. Koyama, and S. Ueno
ASCA Observations of Class I Protostars in the Rho Oph Dark Cloud .. 236
   Y. Tsuboi, K. Koyama, Y. Kamata, and S. Yamauchi

1-2. Supernovae, Supernova Remnants and Galactic Hot Plasma

The Half-Life of Titanium 44 and SN 1987A 241
   Y.S. Mochizuki and S. Kumagai
Black Hole Disk Accretion in Supernovae 243
   H. Nomura, S. Mineshige, M. Hirose, K. Nomoto, and T. Suzuki
The X-Ray Spectrum of Supernova SN1993J 245
   S. Uno
Neutrino Transport in Type II Supernovae: Boltzmann Solver vs Monte Carlo
Method 247
   S. Yamada, H.T. Janka, and H. Suzuki
Radio Emission from Extended Shell-Like SNRs 249
   A.I. Asvarov
Discovery of X-Ray Emission from the Radio SNR G352.7-0.1 251
   K. Kinugasa, K. Torii, H. Tsunemi, S. Yamauchi, K. Koyama, and T. Dotani
The Effect of Dust Sputtering in Supernova Remnant 253
   T. Kurino, M. Fujimoto, and A. Habe
ASCA Observation of the Cygnus Loop Supernova Remnant 254
   E. Miyata and H. Tsunemi
Thermal and Non-Thermal X-Rays from SN 1006 and IC 443 256
   M. Ozaki and K. Koyama
X-Ray Study of Crab-Like and Composite SNRs 258
   K. Torii, H. Tsunemi, and P. Slane

The Morphologies of SNRs and Their ISM and CSM 260
    Z.-R. Wang
The AD 393 Guest Star and the SNR RX J1713.7-3946 262
    Z.-R. Wang, Q.Y. Qu, and Y. Chen
Multifrequency Spectral Studies of SNRs 263
    X.Z. Zhang, X.J. Wu, L.A. Higgs, T.L. Landecker, D.A. Green,
    and D.A. Leahy
New Detection of X-Ray Pulsar Nebulae by ASCA 265
    N. Kawai, K. Tamura, and S. Shibata
Pulsar Nebulae 267
    S. Shibata, N. Kawai, and K. Tamura
Magnetic Reconnection as the Origin of Galactic Ridge X-Ray Emission 269
    S. Tanuma, T. Yokoyama, T. Kudoh, K. Shibata, R. Matsumoto,
    and K. Makishima
Observations of Interstellar O VI Absorption at 3 km/s Resolution 271
    E.B. Jenkins, U.J. Sofia, and G. Sonneborn
Can Galactic γ-Ray Background be due to Superposition of γ-Rays from
Millisecond Pulsars? 273
    V.B. Bhatia, S. Mishra, and N. Panchapakesan
Transition Probabilities for ^{1}H in Strong Magnetic Fields 275
    J.M. Benkő and K. Balla
Diagnostic of Astrophysical Plasma in Neighborhood of Neutron Stars .. 277
    M. Mijatović and E.A. Solov'ev

1-3. Galaxies and Their Clusters

Evolution of Multiphase Hot Interstellar Medium in Elliptical Galaxies ..281
    Y. Fujita, J. Fukumoto, and K. Okoshi
Solving the Mysteries in Hot ISM in Early-Type Galaxies 283
    K. Matsushita, T. Ohashi, and K. Makishima
X-Ray Observation of the Normal Spiral Galaxies NGC2903 and NGC628
with ASCA 284
    T. Mizuno, H. Ohbayashi, N. Iyomoto, and K. Makishima
Hydrodynamical Model of X-Ray Emitting Gas Around Elliptical
Galaxies 286
    R. Saito and T. Shigeyama
Simulated X-Ray Emission from Starburst Driven Winds 287
    D.K. Strickland, I.R. Stevens, and T.J. Ponman
ASCA X-Ray Observation of the Lobe Dominant Radio Galaxy NGC 612 289
    M. Tashiro, H. Kaneda, and K. Makishima
X-Ray Observations of M32 with ASCA 291
    T. Toneri, K. Hayashida, and M. Loewenstein
A Circulation Hypothesis of Spiral Galaxies 293
    C. Linfei
X-Ray Properties of ASCA Observed 43 Clusters of Galaxies 295
    F. Akimoto, M. Watanabe, A. Furuzawa, Y. Tawara, Y. Kumai,
    S. Satoh, and K. Yamashita
Generation and Implications of Post-Merger Turbulence in Clusters of

Galaxies 297
    I. Goldman
Clusters of Galaxies in a Flat CHDM Universe 299
    A. Habe, C. Hanyu, and S. Yachi
ASCA Study of Shapley Supercluster 300
    H. Hanami
ASCA Observation of "Failed Cluster" of Galaxies Candidate, 0806+20 . 302
    K. Hashimotodani, K. Hayashida, and T.T. Takeuchi
ASCA Observations of the Abell 496 Cluster of Galaxies 304
    I. Hatsukade, J. Ishizaka, M. Yamauchi, and K. Takagishi
The Supply of Magnetic Fields from a cD Galaxy to Intra-Cluster Space 306
    H. Hirashita, S. Mineshige, K. Shibata, and R. Matsumoto
Temperature Structure in Merging Coma Cluster of Galaxies 308
    H. Honda, M. Hirayama, H. Ezawa, K. Kikuchi, T. Ohashi,
    M. Watanabe, H. Kunieda, and K. Yamashita
The Fate of Intra-Group Medium 310
    Y. Ishimaru
Development of New Analysis Method for Mapping Observations of
Clusters of Galaxies 312
    K. Kikuchi, T. Ohashi, H. Ezawa, M. Hirayama, H. Honda, and R. Shibata
Constraints on the Matter Fluctuation Spectrum from X-Ray Cluster
Number Counts..314
    T. Kitayama and Y. Suto
Magnetic Field Amplification and Intergalactic Plasma Heating through
Magnetic Twist Injection from Rotating Galaxies 315
    R. Matsumoto, A. Valinia, T. Tajima, S. Mineshige, and K. Shibata
Mapping the Virgo Cluster of Galaxies with ASCA 317
    T. Ohashi, K. Kikuchi, K. Matsushita, N.Y. Yamasaki,
    A. Kushino, and ASCA Virgo Project Team
ASCA Observations of Three Gravitational Lensing Clusters of Galaxies;
CL0500-24, CL2244-02, and A370 319
    N. Ota, K. Mitsuda, and Y. Fukazawa
Evolution of X-Ray Clusters of Galaxies with Active Protogalaxies 321
    H. Saga, S. Yachi, and A. Habe
ASCA Observation of A1674 323
    T. Takai, K. Hayashida, K. Hashimotodani, and W. Kawasaki
Uncertainty in Mass Determination of Galaxy Clusters due to the Bulk
Motion of Intracluster Medium 325
    M. Takizawa and S. Mineshige
ASCA Observation of Groups of Galaxies 327
    Y. Tawara, S. Sato, A. Furuzawa, K. Yamashita, K. Isobe, and Y. Kumai
Matter Distribution in the Galaxy Clusters A539 and A2319 329
    D. Trèvese, G. Cirimele, and M.de Simone
Sunyaev-Zel'dovich Effect Observation Project with the Nobeyama 45-m
Telescope...330
    M. Tsuboi, T. Ohno, A. Miyazaki, T. Kasuga, A. Sakamoto, and T. Noguchi

Session 2: Future Space Programs

Development of STJ as a New X-Ray Detector 335
    N.Y. Yamasaki, T. Ohashi, K. Kikuchi, H. Miyazaki,
    E. Rokutanda, A. Kushino, and M. Kurakado

Multilayer X-Ray Optics for Future Missions 337
    K. Yamashita, H. Kunieda, Y. Tawara, and K. Tamura

Direct Measurement of the CCD with a Sub-Pixel Resolution by Using a
New Technique 339
    K. Yoshita, S. Kitamoto, E. Miyata, H. Tsunemi, and K.C. Gendreau

Final Performances of the X–Ray Mirrors of the JET–X Telescope 341
    E. Poretti

Session 3: Diagnostics of High Gravity Objects with X- and Gamma Rays

3-1. White Dwarfs and Neutron Stars

Spin-Down of Neutron Stars and Compositional Transitions in the Cold
Crustal Matter 345
    K. Iida and K. Sato

The Equations of Motion for Binary Systems with Relativistic
Quadrupole-Quadrupole Moments Interaction 346
    C. Xu and X. Wu

Radiative Winds from Accretion Disks in CVs 348
    M. Hachiya, Y. Tajima, and J. Fukue

Supersoft X-Ray Source RX J0019.8+2156 350
    K. Matsumoto, M. Okugami, and J. Fukue

ASCA Observations of the SgrA Region 352
    Y. Maeda, K. Koyama, H. Murakami, M. Sakano, K. Ebisawa,
    T. Takeshima, and S. Yamauchi

Energy Spectra of X 1636–536 Observed with ASCA 354
    K. Asai, T. Dotani, K. Mitsuda, H. Inoue, Y. Tanaka, and W.H.G. Lewin

The Circumstellar Matter of Cygnus X-3 356
    S. Kitamoto and K. Kawashima

The ASCA Observation Campaign of SS433 358
    T. Kotani, N. Kawai, M. Matsuoka, and W. Brinkmann

The ASCA Results of GRO J1744-28 360
    M. Nishiuchi

Structure and Time-Dependent Behavior of Be Star Disks in Be/X-Ray
Binaries 362
    A.T. Okazaki

VRI Light Curve Analysis of SS 433 364
    M. Okugami, Y. Obana, and J. Fukue

Bright X-Ray Stars near the Galactic Center 366
    M. Sakano, M. Nishiuchi, Y. Maeda, K. Koyama, and J. Yokogawa

Discovery of QPO from X Persei with RXTE 368
    T. Takeshima

Non-Detection of kHz QPOs in GX 9+1 and GX 9+9 370
    R. Wijnands, M. van der Klis, and J. van Paradijs

Spectral Evolution During Dipping of the Low Mass X-Ray Binary
XBT 0748-676 372
   M.J. Church, M. Balucińska-Church, K. Mitsuda, T. Dotani, and K. Asai
Two-Dimensional Accretion Disk Models of a Neutron Star 374
   M. Fujita and T. Okuda

3-2. Black Hole Binaries

A Model for a Thin Magnetised Disc in LMC X-3 379
   P. Hoyng
Disk Oscillations as the Origin of Quasi-Periodic Oscillations in Black
Hole Candidates 381
   T. Yamasaki, S. Mineshige, and S. Kato
Rapid X-Ray Variability of Cyg X-1 382
   P.C. Agrawal, B. Paul, A.R. Rao, M.N. Vahia, J.S. Yadav,
   T.M.K. Marar, S. Seetha, and K. Kasturirangan
Optical Flares of Cygnus X-1 384
   N.G. Bochkarev, E.A. Karitskaya, and V.M. Lyuty
Cygnus X-1: X-Ray Emission Mechanism and Geometry 386
   W. Cui, S.N. Zhang, J.H. Swank, X.-M. Hua, K. Ebisawa, and T. Dotani
X-Ray Determination of the Black-Hole Mass in Cygnus X-1 388
   A. Kubota
Spectral Evolution of the Thermal Component in Cygnus X-1 during
Intensity Variations in the Soft State 390
   M. Balucinska-Church, M.J. Church, K. Mitsuda, T. Dotani, H. Inoue,
   and F. Nagase
ASCA Observations of the Superluminal Jet Source GRS 1915+105 392
   K. Ebisawa
X-Ray Timing Studies of GRS 1915+105 394
   B. Paul, P.C. Agrawal, A.R. Rao, M.N. Vahia, J.S. Yadav,
   T.M.K. Marar, S. Seetha, and K. Kasturirangan
Multiwavelength Temporal Behavior of GRS 1915+105 396
   D. Hannikainen and Ph. Durouchoux
Rapid Hard X-Ray Variability in GRO J0422+32 398
   F. van der Hooft
Three-Dimensional Local MHD Simulations of High States and Low States
in Magnetic Accretion Disks 400
   T. Matsuzaki, R. Matsumoto, T. Tajima, and K. Shibata
Electron-Positron Pair Winds from Central Luminous Accretion Disks
with Radiation Drag 402
   Y. Tajima and J. Fukue
X-Ray Fluctuations from the Advection-Dominated Accretion Disk with
a Critical Behavior 404
   M. Takeuchi

3-3. AGNs

Compton Reflection in the Vicinity of a Rotating Black Hole 409
   A. Maciolek-Niedźwiecki and P. Magdziarz
Two-Temperature Transonic Accretion Disks 411

K.E. Nakamura, M. Kusunose, R. Matsumoto, and S. Kato
Enhancement of Turbulent Viscosity by Global Magnetic Fields in
Accretion Disks 412
    Y. Nakao
Accretion–Disk Corona Advected by External Radiation Drag 413
    Y. Watanabe and J. Fukue
Numerical Simulation of Relativistic Jet Formation in Black Hole
Magnetosphere 415
    S. Koide, K. Shibata, and T. Kudoh
Radiation from Advection-Dominated Flows 417
    T. Manmoto
Black-Hole Accretion Corona Immersed in the Disk Radiation Fields 419
    T. Miwa, Y. Watanabe, and J. Fukue
Determination of NGC 4151 Nucleus Mass from Parameters of Narrow
Satellites of Broad Lines 421
    N.G. Bochkarev
The X-Ray Spectrum and Variability of NGC 4151 422
    K.M. Leighly, M. Matsuoka, M. Cappi, and T. Mihara
Reverberation Mapping Analysis of the Broad-Line Region in Seyfert
Galaxy NGC 4151 424
    S.J. Xue and F.Z. Cheng
X-Ray Properties of High-Z Radio-Quiet Quasars: ASCA Observations . 426
    M. Cappi, M. Matsuoka, C. Vignali, A. Comastri, T. Mihara, C. Otani,
    G.G.C. Palumbo, and S.J. Xue
ASCA Observations of Narrow Line Seyfert 1 Galaxies 428
    K. Hayashida
A Model of the Broad-Band Continuum of NGC 5548 430
    P. Magdziarz and O. Blaes
How Unbiased is a [OIII]λ5007-Bright Sample? 432
    S. Ueno, J.D. Law-Green, H. Awaki, and K. Koyama
BeppoSAX Observation of the Quasar 3C273 434
    T. Mineo, G. Cusumano, M. Guainazzi, and P. Grandi
Change in the Warm Absorber in MR 2251-178 436
    C. Otani, T. Kii, and K. Miya
ASCA Observations of the Quasar Concentration 1338+27 438
    T. Yamada, Y. Ueda, T. Takahashi, T. Mihara, N. Kawai, and Y. Ishisaki
Search for 10 TeV Gamma-Rays from the Nearby AGNs with the Tibet
Air Shower Array 440
    L.K. Ding, T. Kobayashi, K. Mizutani, A. Shiomi, Y.H. Tan,
    T. Yuda, and the Tibet ASγ Collaboration
Microvariability of S5 0716+714, a γ-Ray Blazar 442
    R. Nesci, E. Massaro, M. Maesano, F. Montagni, F. D'Alessio,
    G. Tosti, and M. Luciani
X-Ray Properties of LINERs and Low Luminosity Seyfert Galaxies
Observed with ASCA 444
    Y. Terashima, H. Kunieda, P.J. Serlemitsos, and A. Ptak

Evidence for a Dramatic Activity Decline in the Nucleus of the Radio
Galaxy Fornax A 446
    N. Iyomoto, K. Makishima, M. Tashiro, K. Matsushita,
    Y. Fukazawa, H. Kaneda, and S. Osone
ROSAT X-Ray Observations of the Radio Galaxy 4C46.09 447
    D.A. Leahy
The Black Hole Grazer 449
    Y. Taniguchi and O. Kaburaki
Optical Variability in AGNs; Disk Instability or Starbursts? 451
    T. Kawaguchi, S. Mineshige, M. Umemura, and E.L. Turner
Testing the Central Engines in AGNs 453
    N. Yamazaki, Y. Taniguchi, and O. Kaburaki
Dynamical Evolution of Accretion Flow onto a Black Hole 455
    M. Yokosawa

3-4. Gamma-Ray Bursts

Possible X-Ray Counterparts to Gamma-Ray Bursts, GRB930131 and
GRB940217 459
    R. Shibata, T. Murakami, Y. Ueda, A. Yoshida, F. Tokanai,
    C. Otani, N. Kawai, and K. Hurley
Importance of the High Energy Channel for the Gamma-Ray Burst Data 461
    A. Mészáros, Z. Bagoly, L.G. Balázs, I. Horváth, and P. Mészáros

Session 4: Large Scale Hot Plasmas and Their Relation with Dark Matter

Optical Follow-Up Observations of the ASCA Large Sky Survey 465
    M. Akiyama, K. Ohta, T. Yamada, Y. Ueda, T. Takahashi,
    M. Sakano, T. Tsuru, and B. Boyle
Recent Report on the ASCA GIS Source Catalog Project 467
    Y. Ishisaki, T. Ohashi, Y. Ueda, T. Takahashi, K. Makishima,
    and the GIS Team
ASCA Deep Sky Survey 469
    Y. Ogasaka
Optical Follow-Up Observations of ASCA Lynx Deep Survey 471
    K. Ohta, M. Akiyama, K. Nakanishi, T. Yamada, K. Hayashida,
    T. Kii, and Y. Ogasaka
The NEP ROSAT Survey 473
    C.R. Mullis, I.M. Gioia, and J.P. Henry

Preface

The present decade is opening new frontiers in high-energy astrophysics. After the X-ray satellites in the 1980's, including Einstein, Tenma, EXOSAT and Ginga, several satellites are, or will soon be, simultaneously in orbit offering spectacular advances in X-ray imaging at low energies (ROSAT; Yohkoh) as well as at high energies (GRANAT), in spectroscopy with increased bandwidth (ASCA; SAX), and in timing (XTE). While these satellites allow us to study atomic radiation from hot plasmas or energetic electrons, other satellites study nuclear radiation at gamma-ray energies (CGRO) associated with radioactivity or spallation reactions. These experiments show that the whole universe is emitting radiation at high energies, hence we call it the "hot universe." The hot universe, preferentially emitting X- and gamma-rays, provides us with many surprises and much information.

A symposium "The Hot Universe" was held in conjunction with the XXIIIrd General Assembly of the International Astronomical Union, at Kyoto on August 26-30 in 1997. The proceedings are organized as follows. Synthetic view of "the hot universe" is discussed in Section 1, "Plasma and Fresh Nucleosynthesis Phenomena". Timely discussions on the strategy for future missions "Future Space Program" are found in Section 2. Then the contents are divided into two major subjects: the compact objects and thin hot diffuse plasmas. Section 3 is devoted to the category of compact objects which includes white dwarfs, neutron stars, and gravitationally collapsed objects: stellar mass black holes or active galactic nuclei. The second subject, thin hot diffuse plasma, is discussed in Section 4. Here, new results on the large scale hot plasmas and the relationship with dark matter are brought together.

Excellent suggestions on the scientific program are due to the Scientific Organizing Committee consisting of Drs A. Fabian, N. Geherels, G. Hasinger, K. Koyama (chair), D. Leahy, E. Meurs, T. Montmerle, F. Pacini, R. Pallavicini, A. Parmar, R. Sunyaev and J. Swank.

Kyoto, October 30, 1997

Katsuji Koyama, Shunji Kitamoto, and Masayuki Itoh

IAU Symposium No. 188
The Hot Universe

The Hot Universe: An Overview

THE HOT UNIVERSE: AN OVERVIEW

Y. TANAKA
Astronomical Institute, University of Amsterdam
Kruislaan 403, 1098 SJ Amsterdam, the Netherlands, and
Institute of Space and Astronautical Science
Sagamihara, Kanagawa-ken 229, Japan

1. Introduction

The universe contains an extremely wide variety of temperature structures from 3K to 1 billion K and even beyond. This symposium focuses on the hot part of the universe. The "hot universe" is by far the best place to study high-energy astrophysics. In this overview, I shall be based mainly on the results in the X-ray band that best manifests the hot universe. However, needless to say that multi-wavelength investigations, from radio, infrared through gamma-rays, are essential for comprehensive understanding.

The universe consists of hierarchical structures. Regradless of their scales, we find hot objects wherever the gravitational potential is sufficiently deep. However, the observed phenomena and the physics are various, depending on the nature of the potential.

2. Extended Objects

2.1. CLUSTERS OF GALAXIES

Clusters of galaxies contain a huge amount of hot gas (intracluster medium: ICM), far exceeding the total stellar mass. This is a great discovery of X-ray astronomy. X-ray study of clusters has rapidly advanced with the Einstein Observatory and more recently ROSAT and ASCA. Observation of clusters provides crucial tests for the cosmological models. X-ray measurements allow determination of the total gas mass, and also, when hydrostatic equilibrium can be justified, the total gravitational mass that is dominated by dark matter. Thus obtained baryon fraction strongly suggests a low-Ω_0 universe. By observing clusters at various redshifts, we can study the cluster evolution. No obvious evolutionary effects observed after $z \sim 0.5$ also confronts $\Omega_0=1$ universe. Search for more distant clusters is clearly important.

The ICM is polluted with a large amount of heavy elements believed to originate from early type galaxies. The recent ASCA results of abundances indicate a major contribution from type II supernovae in the early phase of galaxy evolution. Since clusters will retain most of the enriched material, study of the heavy element abundances allows "archaeology" of the history of nucleosynthesis in the universe.

2.2. GALAXIES

Hot halos are commonly present in early type galaxies, as established with the Einstein Observatory. The ASCA results on ellipticals invariably show low abundances of their hot gas, 0.3 to utmost 1 solar. This contradicts seriously the expected high abundances of 2–6. This discrepancy is not easily resolved, and clearly involves some fundamental problems.

Hot gas is not only in early type galaxies. There is increasing evidence for extensive hot gas in the host galaxies of AGN. Could it be related to the central activities? The case of our own Galaxy is suggestive of it. The presence of a 2.6 million $M_\odot$ black hole at our galactic center, though currently inactive, is quite convincing. We find $\sim 10^8$ K plasma as much as several thousand $M_\odot$ within ~ 100 pc of the center, most probably produced in the past activities of the central black hole. History of its X-ray activities is recorded in the 6.4-keV iron fluorescence line from molecular clouds.

Yet unresolved thermal emission in the temperature range $3 \cdot 10^7 - 10^8$ K extends along the plane of our Galaxy. If it is diffuse emission as the recent ASCA results support, it presents real puzzles. The energy density of the plasma is more than an order of magnitude larger than in other modes, i.e. H I and H II gas, magnetic fields, or cosmic rays. Such a hot plasma cannot be gravitationally bound, therefore must be continuously replenished. The energy rate of the mass outflow is enormous, 10^{42-43} erg s^{-1}. What are the energy source and the production mechanism of such a plasma? Other spiral galaxies may also have such hot plasmas.

2.3. SUPERNOVAE AND SUPERNOVA REMNANTS

A wealth of observational results shows us extreme richness of supernova physics, but on the other hand enormous complexity. (1) A wide variety in morphology; shells, center-filled, and even high-speed shrapnels (Vela SNR). (2) Local abundance anomalies. (3) Non-thermal cases; pulsar-powered synchrotron nebulae, and in some SNR due probably to very efficient particle acceleration in the expanding shell. What causes such distinct varieties?

The nucleosynthesis products of type Ia and II supernovae are the key to understand chemical evolution of galaxies and the ICM. High-quality X-ray spectra of SNR have become available with ASCA. Yet, the abundance determination is not straightforward and still requires a breakthrough.

3. Compact Objects

3.1. ACTIVE GALACTIC NUCLEI

Convincing evidence for supermassive black holes in AGN has finally become available recently. Among others, the broadened iron line from Seyfert nuclei represents the relativistic effects of the direct vicinity of black holes. Finding the mass and spin of the black holes would be the next challenge.

X-ray results on Seyfert 2 galaxies lend strong support to the unified scheme of Seyferts in terms of the viewing angle with respect to a surrounding torus. Could this scheme be extended to QSOs? In fact, observations of some infrared-luminous yet X-ray faint galaxies have shown their intrinsic luminosities as high as a QSO. Many hidden-nucleus QSOs may well exist.

Recent discoveries of low-luminosity AGN in normal galaxies, much lower than typical Seyferts, and a supermassive black hole in our own Galaxy suggest a possibility that every normal galaxy hosts a supermassive black hole. Search for more low-luminosity AGN might find a link to the mystery, "where did QSOs go?".

The long debate on CXB is converging. The ROSAT deep survey shows no lack of sources to account for the CXB flux. The spectral paradox below 20 keV is resolved at least qualitatively. The remaining issue is the high-energy cut-off. While luminous QSOs tend not show a high-energy cut-off, most AGN must show a proper cut-off to form the observed CXB spectrum.

X-ray emission of AGN is powered by mass accretion, and probably comes from an accretion disk, much the same as X-ray binaries, except for Blazars. In fact, there are striking similarities in X-ray properties between AGN and black-hole binaries. Despite huge differences in mass and scale, the accretion phenomena in both systems seem to be essentially the same. The same is also true for relativistic jets. They are observed from both systems. Apart from these, we still need to understand what causes the distinction between radio-loud and radio-quiet AGN.

3.2. STELLAR-MASS BLACK HOLES, NEUTRON STARS, WHITE DWARFS AND STARS

Many black-holes have been discovered among X-ray binaries. So far, ten of them were established from the optical mass functions, giving a dynamical mass exceeding $3M_\odot$. In addition, the characteristic ultrasoft X-ray spectrum (in eight of the above ten) is most probably a black-hole signature.

Most black-hole sources are low-mass X-ray binaries, and remarkably all the low-mass black-hole binaries are transients. Transient outbursts occur frequently, and half as many have turned out to be black-hole binaries. RXTE and BeppoSAX will discover many more transients, hence we expect many more black holes. Transients are also valuable for studying accretion phenomena, since the accretion rate sweeps many orders of magnitude. The bizarre objects, such as the Rapid Burster, GRO J1744-28, happen to be transient sources. Is this merely accidental, or giving an important hint?

Are the relativistic jets a signature of black holes? We will know the answer from radio outbursts of future X-ray transients. Galactic transient jets are also precious for studying the condition for jet formation, because of closer distances, frequent occurrence, and short dynamical time scales.

An abrupt transition between a soft thermal and a hard power-law spectral state occurs regardless of whether the compact object is a weakly-magnetized neutron star or a black hole. This is a fundamental property of an accretion disk, and has been suspected as due to a radical change in the disk structure around a certain accretion rate. Yet another change occurs at a further low accretion rate below which sources are turned down to a quiescent state. Whether this is due to a choked flow or a transition into an advection-dominated disk is a subject of current debate.

Concerning short-time variabilities, the recently discovered kHz QPO strongly suggest fast spin of neutron stars in low-mass X-ray binaries.

Physics of neutron star interior is extremely important. Mass and radius of neutron stars that provide a strong constraint on the equation of state can in principle be obtained from X-ray bursts, since they occur on the neutron star surface. For instance, the absorption line at 4.1 keV observed in some bursts from several bursters is most probably a gravitationally redshifted atomic line. If the responsible element is identified, M/r is readily determined. It is worth emphasizing that the static nuclear force that determines the equation of state cannot be measured by ground-based experiments. Neutron star cooling is also an important probe of the interior.

White dwarfs and stars are also part of the hot universe. Among recent advances, I will mention three. (1) New method of mass estimation of white dwarfs from X-ray spectrum of cataclysmic variables, (2) the ROSAT discovery of many more T Tauri stars, and (3) the ASCA discovery of strong X-rays from pre-T Tauri protostars. These discoveries will bring a new scope to physics of star formation.

4. Still Higher Energies

The energy range above 100 keV is dominated by non-thermal processes. In addition to CGRO, RXTE and BeppoSAX, next-generation missions with still higher angular and/or spectral resolution in this energy range will discover new wonders of "superhot" universe.

Last but not least, the gamma-ray burst is still a most enigmatic phenomenon in the universe. Soft X-ray afterglow established with BeppoSAX allows follow-up observations to pin-point the source location. Identification of the optical counterpart is a real crucial step towards the goal of understanding the phenomenon.

Session 1: Plasma and Fresh Nucleosynthesis Phenomena

1-1. Sun and Stars

THEORY OF FLARES AND MHD JETS

KAZUNARI SHIBATA
National Astronomical Observatory
Mitaka, Tokyo 181, Japan

Abstract: Recent development on the theory and numerical modeling of solar flares and jets is reviewed with emphasis on the magnetic reconnection model. Application to protostellar flares and jets is also discussed.

1. Introduction: Why do we need magnetic reconnection ?

Hot plasmas have very high electrical conductivity so that their magnetic Reynolds number R_m is very large; for example, we find

$$R_m = \frac{t_D}{t_A} = \frac{V_A L}{\eta_{Spitzer}} \sim 10^{13} \left(\frac{B}{10G}\right)\left(\frac{L}{10^9 \text{cm}}\right)\left(\frac{n}{10^9 \text{cm}^{-3}}\right)\left(\frac{T}{10^6 \text{K}}\right) \gg 1,$$

for solar coronal plasma, where the diffusion time is $t_D \sim 3 \times 10^6$ years (!), and the Alfven time $t_A = L/V_A \sim 10 - 100$ sec. Since the observed time scale of solar flares is a few min − a few hours and thus $10 - 100 t_A$, the simple Ohmic dissipation cannot exlain solar flares. This is a universal property of various hot plasmas in our universe from magnetically confined fusion plasmas to intergalactic plasmas, and gives us fundamental difficulty in understanding explosive energy release such as solar flares.

If we consider very small scale (e.g., 0.01 cm), R_m becomes small (~ 100) so that we can explain time scale. However, in this case, we cannot explain total flare energy ($\sim 10^{29} - 10^{32}$ erg) since we need a large volume to explain total energy of flares whose size is $\sim 10^9 - 10^{10}$ cm. Hence we need coupling between micro-scale physics (resistivity) and macro-scale physics (flow dynamics), which is an essential element of magnetic reconnection.

Recently, Yohkoh discovered a number of evidence of magnetic reconnection in solar flares (e.g., Tsuneta 1998), such as cusps, X-ray plasmoids, X-ray jets, and so on. Although the origin of resistivity has not yet been fully understood at present, it has been established that the non-uniform resistivity (such as anomalous resistivity) can lead to fast reconnection at time scale of $10 - 100 t_A$ (e.g., Ugai 1989, Yokoyama and Shibata 1994) and hence the numerical modeling of solar flares and related mass ejections (jets and plasmoids) based on the reconnection model has greatly been advanced, which will be reviewed in this article.

2. Magnetic Reconnection Model for Solar Flares and Jets

Cusp-Shaped Flares: Yokoyama and Shibata (1998) succeeded to construct a fully self-consistent reconnection model of cusp-shaped flares discovered by Yohkoh (Tsuneta et al. 1992, Tsuneta 1998 for a review), taking into account the effect of heat conduction (Yokoyama and Shibata 1997) and evaporation. According to their results (see Fig. 1 of Yokoyama and Shibata 1998), the adiabatic slow shock dissociates into *isothermal slow shock* and *conduction front* (Forbes et al. 1989), and the fast shock becomes nearly isothermal and post-fast-shock plasmas become much denser than in the adiabatic case. This fast shock structure explains the loop top hard X-ray source discovered by Masuda et al. (1994). Yokoyama and Shibata (1998) further found that the temperature of flare loops scales with the field strength B as

$$T_{flare} \simeq 2 \times 10^7 \left(\frac{B}{30 G}\right)^{6/7} \left(\frac{L}{10^9 \text{cm}}\right)^{2/7} \left(\frac{n}{10^9 \text{cm}^{-3}}\right)^{-1/7} \text{ K},$$

where L is the half length of the loop, n is the electron number density. This formula is applicable not only to solar flares but also to stellar flares (and even to galactic flares) as far as the Spitzer conductivity is applicable.

X-ray Plasmoid Ejections: Magara et al. (1997) developed a reconnection model of X-ray plasmoid ejections discovered by Yohkoh (Shibata et al. 1995, Tsuneta 1998), assuming initially a sheared force free arcade (see Kusano et al. 1995 and Choe and Lee 1996 for the formation and evolution of such a sheared arcade), and succeeded to explain the observation (Ohyama and Shibata 1997) that acceleration of a plasmoid (to a few 100 km/s) precedes a hard X-ray impulsive peak (if the latter is a measure of electric field strength at the neutral point).

X-ray Jets: X-ray jets are also discovered by Yohkoh (Shibata et al. 1992, Shimojo et al. 1996). These jets occur in association with microflares or subflares and their apparrent velocity is 10 – 1000 km/s. Yokoyama and Shibata (1995,1996) developed 2D numericlal simulation model of X-ray jets, in which magnetic reconnection occurs in the current sheet between emerging flux and pre-existing coronal field (Fig. 1). They found; (1) reconnection proceeds with formation of magnetic islands (plasmoids) by tearing instability, coalescence of islands, and their ejections, (2) reconnection jets

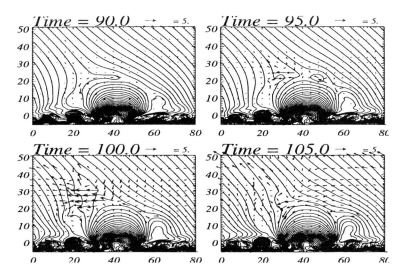

Figure 1. Numerical simulation of X-ray jets based on emerging flux reconnection model (Yokoyama and Shibata 1995, 1996). Note that plasmoids (magnetic islands) are repeatedly created and ejected from the current sheet, and disappear due to secondary reconnection.

collide with the ambient field to form fast shock at the colliding point, (3) both hot and cool jets are formed simultaneously, which were actually observed by Canfield et al. (1996) as X-ray jets and Hα surges.

Unified Model: Yohkoh has revealed that various (apparently different) flares show common properties such as ejection of plasma. Numerical simulations of reconnection show also ejection of plasmoids (or flux rope in 3D space). Hence Shibata (1996) proposed a unified model, *plasmoid-induced-reconnection model*, to explain various flares. In this model, if the current sheet is long, ejected plasmoids (helical loop) are directly observed, while if it is short, plasmoids collide with ambient field to disappear as a result of secondary reconnection (Fig. 1), though the helical field and mass are released into open flux tubes to accelerate spinning jets along them. The physics of the spinning jet is basically the same as that of magnetically driven jets from accretion disks (e.g., Uchida and Shibata 1985, Shibata and Uchida 1986).

3. Astrophysical Application: Protostellar Flares and Jets

Hayashi et al. (1996) studied the interaction between a protostellar magnetosphere and an accretion disk surrounding it, using 2D axi-symmetric numerical simulations. They found that once the accretion disk plasma is coupled with stellar field, the stellar field is enormously sheared by the rotation of the disk to expand outward, like solar coronal mass ejections. The expanding loop finally detatches from the stellar field due to reconnection to form a plasmoid or a helically twisted toroid. During this stage, a very hot plasma ($\sim 10^8$ K with total energy $\sim 10^{36}$ erg) is created, which explains protostellar flares observed by ASCA and ROSAT (Koyama et al. 1996, Tsuboi et al. 1998, Monmerle 1998). The velocity of the ejected plasmoid is 200 – 400 km/s, which corresponds to optical jets. This model also shows the ejection of the cold gas from the disk (e.g., Uchida and Shibata 1985, Shibata and Uchida 1986), which corresponds to high velocity neutral winds.

Finally it should be stressed that the processes discussed in this article could occur in various astrophysical situations (Makishima 1997), not only in solar and stellar magnetosphere but also in our Galaxy (Tanuma et al. 1998), and even in cluster of galaxies (Makishima 1997, Matsumoto et al. 1998, Hirashita et al. 1998).

The author would like to thank T. Yokoyama, M. Ohyama, T. Magara for their help in preparing his talk.

References

Canfield, R., et al. 1996, ApJ ApJ, 464, 1016.
Forbes, T. G., Malherbe, J. M., and Priest, E. R. 1989, Sol. Phys. 120, 258.
Hayashi, M. R., Shibata, K., Matsumoto, R. 1996, ApJ 468, L38.
Hirashita, H. et al. 1998, in these proceedings.
Koyama, K., et al. 1996, PASJ, 48, L87.
Kusano, K., Suzuki, Y., Nishikawa, K. 1995, ApJ, 441, 942.
Choe, G. S., and Lee, L. C. 1996, ApJ 472, 360.
Magara, T., Shibata, K., Yokoyama, T. 1997, ApJ, 487, 437.
Makishima, K. 1996, in Proc. X-ray Imaging and Spectroscopy of Cosmic Hot Plasmas, Makino, F. et al. (eds.), Universal Academy Press, p. 137.
Masuda, S., et al. 1994, Nature, 371, 495.
Matsumoto, R., et al. 1998, in these proceedings.
Montmerle, T. 1998, in these proceedings.
Ohyama, M., and Shibata, K. 1997, PASJ, 49, 249.
Shibata, K., and Uchida, Y. 1986, PASJ, 38, 631.
Shibata, K., et al. 1992, PASJ, 44, L173.
Shibata, K., et al. 1995, ApJ, 451, L83.
Shibata, K. 1996, Adv. Space Res., 17, (4/5)9.
Shimojo, M., et al. 1996, PASJ, 48, 123.
Tanuma, S., et al. 1998, in these proceedings.
Tsuboi, Y., et al. 1998, in these proceedings.
Tsuneta, S., et al. 1992, PASJ, 44, L63.
Tsuneta, S. 1998, in these proceedings.
Uchida, Y., and Shibata, K. 1985, PASJ, 37, 515.
Ugai, M. 1989, Phys. Fluids B, 1, 942.
Yokoyama, T., and Shibata, K. 1994, ApJ, 436, L197.
Yokoyama, T., and Shibata, K. 1995, Nature, 375, 42.
Yokoyama, T., and Shibata, K. 1996, PASJ, 48, 353.
Yokoyama, T., and Shibata, K. 1997, ApJ, 474, L61.
Yokoyama, T., and Shibata, K. 1998, in these proceedings.

X-RAY CORONAE FROM STARS

R. PALLAVICINI
Osservatorio Astronomico di Palermo
Palazzo dei Normanni, 90134 Palermo, Italy

1. Introduction

A number of major advances in stellar coronal physics have occurred since 1990 mainly as a consequence of imaging observations by ROSAT and spectroscopic observations by ASCA. These can be summarised as follows:

1. an all-sky survey has been performed by ROSAT at a sensitivity of $\sim 2 \times 10^{-13}$ erg cm^{-2} s^{-1}, complemented by pointed observations an order of magnitude deeper;
2. complete mapping and deeper pointings have been obtained for virtually all open clusters closer than ~ 500 pc, and covering the age range from ~ 30 Myr to ~ 700 Myr;
3. complete mapping and deeper pointings have been obtained for several Star Forming Regions (SFRs) covering the age range ~ 1 to ~ 10 Myr;
4. spectroscopic observations of bright coronal sources have been obtained with EUVE and ASCA allowing the derivation of the temperature structure and elemental abundances.

In this paper I will discuss briefly the second and fourth topic in the above list, i.e. imaging observations of open clusters by ROSAT and spectroscopy of bright coronal sources by ASCA. X-ray observations of young stellar objects in SFRs are discussed by Montmerle elsewhere in this volume.

2. Imaging observations of open clusters

Open clusters form homogeneous samples of stars with approximately the same age and chemical composition, but different masses. Prior to ROSAT, only two clusters (the Hyades and the Pleiades) had beed observed in sufficient detail at X-ray wavelengths. ROSAT has observed a dozen nearby open clusters covering the age range from ~ 30 Myr to ~ 700 Myr. These observations allow investigating the evolution of stellar angular momentum with age in stars of different masses and convection zone depths.

A comparison of the X-ray luminosity functions of different clusters shows that X-ray emission steadily decreases from younger clusters (like IC2602 and IC2391 at ~ 30 Myr or α Per at ~ 50 Myr) to intermediate age clusters (like the Pleiades and NGC2516 at $\sim 70\text{-}100$ Myr), to older clusters (like the Hyades, Coma and Praesepe at an age of $\sim 500\text{-}700$ Myr). While this is generally true for solar-type stars (cf. Randich 1997 and references therein), there are significant differences with regard to the age dependence of X-ray luminosity functions for stars of different spectral types. For instance, while G and K stars in α Per (~ 50 Myr) are brighter in X-rays than stars of the same spectral type in the Pleiades (age ~ 70 Myr), this is not true anymore for M dwarfs, which show virtually the same X-ray luminosity function in α Per and in the Pleiades (Randich et al. 1996). This can be understood as a consequence of magnetic braking by stellar winds, the braking time scale being determined by the depth of the convection zone. For G and K stars, the convective zone is sufficiently shallow to allow a significant braking in the time interval from the age of α Per to that of the Pleiades, while there is no significant angular momentum loss in the same time interval for M stars with deeper convective zones.

Although the general picture is quite consistent, there are a number of disturbing facts, indicating that the current scenario is at best only a first order approximation. For instance, late-type stars in the Praesepe cluster (which has the same age and chemical composition as the Hyades) are much weaker in X-rays than similar stars in the Hyades (Randich & Schmitt 1995). On the contrary, F and G stars in the Coma cluster (at an age of ~ 500 Myr) are in good agreement with the Hyades. Jeffries et al. (1997) have found similar discrepancies between NGC2516 (a southern "twin" of the Pleiades) and the general pattern shown by the Pleiades, NGC6475 (~ 200 Myr) and the Hyades. They have attributed it to the subsolar metallicity ([Fe/H]$= -0.3$) of NGC 2516, suggesting that metallicity may be an additional parameter (through its effects of the depth of the convection zone) affecting angular momentum evolution.

3. Spectroscopy of coronal sources

Stellar coronae have thermal spectra rich in emission lines. $1-T$ and $2-T$ models and solar abundances were usually assumed in the analysis of early observations with low spectral resolution, such as those obtained with the IPC detector on *Einstein* and the PSPC detector on ROSAT. Only recently, with the EUVE and ASCA missions, it has become possible to exploit the diagnostic potential of X-ray coronal spectra, to derive the temperature distribution and elemental abundances.

Somewhat surprisingly, most (albeit not all) coronal sources observed with ASCA (and EUVE as well) showed reduced metal abundances (by typical factors of 3 to 5) with respect to *solar* abundances (White 1996, and references therein). In many cases, the best fit to the ASCA CCD spectra over the range 0.5 to 10 keV is obtained with a $2-T$ thermal model with variable non-solar abundances, i.e. by leaving individual elements free to vary. In other cases, a sufficiently good fit is obtained by leaving only the overall metallicity free to vary, i.e. by varying the individual abundances in the same proportion with respect to solar abundances. When the individual abundances are left free to vary, they show little evidence for a FIP effect, although evidence for a FIP effect has been reported for some of the stars observed with EUVE. At any rate, if a FIP effect is present, it is the opposite to that observed in the solar corona, where ions with low First Ionization Potential (FIP) like Iron are enhanced with respect to ions with high FIP.

The fact that the coronal abundances measured by ASCA and EUVE in a given star are typically lower than solar photospheric abundances does not imply necessarily that they are also lower than the photospheric abundance of the same star. Most stellar coronal sources for which medium resolution spectroscopy has been obtained with ASCA are bright RS CVn and Algol-type binaries, which are known to have often subsolar photospheric abundances (e.g. Randich et al. 1994) although it is yet unclear whether the observed weakness of the optical absorpion lines is due to lower metallicity or rather to the effect of surface activity. For instance, the RS CVn binary λ And observed by ASCA (Ortolani et al. 1997) was found to have a coronal abundance a factor 5 lower than solar but perfectly consistent with its measured photospheric abundance.

While the subsolar coronal abundances found by ASCA in several RS CVn and Algol-type binaries may not be inconsistent with the photospheric abundances of these stars as a class, there are at least a few cases in which there is a clear discrepancy between the measured coronal and photospheric abundances. One case is the young rapidly rotating star AB Dor observed simultaneously by ASCA and EUVE (Mewe et al. 1996). The coronal metallicity of this star is ~ 0.3 solar in spite of the fact that AB Dor is a ZAMS star with a measured photospheric abundance that is solar. Another case is the young active star HD35850 (Tagliaferri et al. 1997): again the coronal metallicity measured by ASCA is ~ 0.3 while the measured photospheric metallicity is solar. On the contrary, the active giant β Cet observed by both ASCA and SAX shows a solar photospheric abundance (Maggio et al. 1997), while the analysis of a recent SAX observation of Capella gives a coronal metallicity ~ 0.8 solar (Favata et al. 1997) consistent with both the EUVE value and the photospheric abundance, but at variance with previously reported results from ASCA (e.g. White et al. 1996).

4. Conclusions

The observations carried out over the past few years have allowed substantial progress to be made in our understanding of stellar coronal emission. As an example, I have illustrated here only two problems, that of late-type stars in open clusters and of elemental abundances in bright coronal sources. There are several other areas in which the new observations have provided crucial information. For instance, ROSAT data have given further support to the shock heating model for early-type stars; on the other hand, cool dwarfs offer the best opportunity to probe convection, angular momentum evolution and dynamos. The existence of a coronal/wind dividing line (DL) among cool giants has been confirmed by ROSAT, but the origin of this DL remains speculative. Moreover, the DL seems to disappear among supergiants and all hybrid stars are in fact detected in X-rays. PMS stars in the age range $\sim$ 1 to 10 Myr have been confirmed to be strong X-ray sources, extending to younger ages the dependence of coronal emission on age already shown by cluster data. More importantly, ROSAT and ASCA have detected X-rays from very young ($\sim$ 0.1 Myr) embedded objects in several SFRs (Koyama et al. 1996) showing that X-rays are a new powerful tool to study star formation and early stellar evolution. Finally, the puzzling results obtained by ASCA and EUVE on coronal abundances of active stars point at the existence of photospheric/coronal abundances anomalies or at problems with the atomic physics entering the spectral codes used in the analysis.

References

Favata, F., Mewe, R., Brickhouse, N.S., Pallavicini, R., Micela, G., 1997, A&A 324, L37
Jeffries, R.D., Thrurston, M.R., Pye, J.P., 1997, MNRAS, in press
Koyama, K., Hamaguchi, K., Ueno, S., Kobayashi, N., Feigelson, E.D., 1996, PASJ 48, L87
Maggio, A., Favata, F., Peres, G., Sciortino, S., Rosner, R., 1997, A&A , in press
Mewe, R., Kaastra, J.S., White, S.M., Pallavicini, R., 1996, A&A 315, 170
Ortolani, A., Maggio, A., Pallavicini, R., Sciortino, S., Drake, J.J., Drake, S.A., 1987, A&A 325, 664
Randich, S., 1997, in G. Micela, R. Pallavicini and S. Sciortino (eds.), *Cool Stars in Clusters and Associations: Magnetic Activity and Age Indicators*, in press
Randich, S., Giampapa, M.S., Pallavicini, R., 1994, A&A 283, 893
Randich, S., Schmitt, J.H.M.M., 1995, A&A 298, 115
Randich, S., Schmitt, J.M.M.M., Prosser, C.F., Stauffer, J.R., 1996, A&A 305, 785
Tagliaferri, G., Covino, S., Fleming, T.A., Gagné, M., Pallavicini, R., Haardt, F., Uchida, Y., 1997, A&A 321, 850
White, N.E., 1996, in R. Pallavicini and A.K. Dupree (eds.), *Ninth Cambridge Workshop on Cool Stars, Stellar Systems, and the Sun*, ASP Conference Series 109, 193

THE QUEST FOR X-RAYS FROM PROTOSTARS

THIERRY MONTMERLE
Service d'Astrophysique, Centre d'Etudes de Saclay
CEA/DSM/DAPNIA/SAp
91191 Gif-sur-Yvette Cedex, France

1. What is a protostar ?

The field of low-mass star formation and early evolution has made rapid progress in recent years, thanks in particular to observations in the IR and mm ranges. The current evolutionary scheme calls for two main stages, themselves divided into two substages (e.g., André & Montmerle 1994): (*i*) *protostars*, comprizing the newly discovered so-called "Class 0 sources", detected mostly or only in the mm range, which are young protostars with estimated ages $\sim 10^4$ yrs, and "Class I sources", visible in the near- to mid-IR, which are evolved protostars with estimated ages $\sim 10^5$ yrs; (*ii*) *T Tauri stars*, which are visible in the IR but also in the optical, the younger being the "classical" T Tauri stars (called "Class II" in the IR), and the "weak-line" T Tauri stars ("Class III" in the IR), with a large age spread of $\sim 10^6 - 10^7$ yrs. According to current models (e.g., Shu et al. 1987), protostars consist of a forming star surrounded by an extended envelope (up to $\sim 10,000$ AU in radius); the star forms via an accretion disk inside a cavity $\sim$ several 100 AU in radius. The disk probably plays an important role in generating molecular outflows, running through the envelope. Classical T Tauri stars are only surrounded by a disk, which disappears at the weak-line T Tauri stage.

The important difference between a "young" (i.e., Class 0) protostar and an "evolved" (i.e., Class I) protostar is not so much of structure, but rather of content. In Class 0 protostars the envelope contains most of the (future stellar) mass : $M_{env} \gg M_\star$, while the opposite holds for Class I protostars, where the envelope is comparatively tenuous. Translated into extinction in the visible, a Class I envelope has $A_V \sim$ several 10, while for Class 0, $A_V > 100$ or more, i.e., $N_H \sim 10^{22} - 10^{23}$ cm^{-2}.

2. Why look for X-rays ?

As amply demonstrated in the case of T Tauri stars, and more generally in all late-type stars, including the Sun (as vividly illustrated by the *Yohkoh* results), X-ray emission is closely linked with magnetic fields, via reconnection of field lines of opposite polarity, while the magnetic fields themselves are believed to originate in the dynamo effect resulting from convective motions (e.g., Montmerle et al. 1993).

In "ordinary" stars, however, the X-rays come from a hot ($\gtrsim 10^7$ K) plasma trapped in magnetic loops anchored to the photosphere. T Tauri stars and protostars offer more exotic possibilities, with possible combinations of magnetic field lines linking the star, the accretion disk, and the envelope. In T Tauri stars, although the situation likely exists, the comparison between the X-ray properties of classical vs. weak-line T Tauri stars shows no evidence so far for a significant star-disk contribution to the X-ray emission. In protostars, only recently has some X-ray evidence been found for star-disk or star-envelope interactions (see below). The X-rays thus appear as a unique probe of the magnetic structure of very inner regions of protostars. Also, a high X-ray flux may significantly affect the physical state of the circumstellar material, for instance via ionization (accretion, Glassgold et al., 1997, or coupling of matter with the magnetic fields), heating, and also induce various irradiation effects on dust grains, gas chemistry, etc. (e.g., Casanova et al., 1995; Maloney et al., 1996).

There are difficulties, however, because of the high extinction associated with protostars. This extinction results not only of the amount of envelope material along the line-of-sight, but also from the fact that protostars are generally embedded in dense interstellar matter, typically "cores" of molecular clouds, where gas densities exceed 10^5 cm^{-3}. As a result, reliable estimates of A_V are not easy to obtain. Since the X-ray photoelectric cross-section varies roughly as E_X^{-3} (e.g., Ryter 1996), X-rays are a priori easier to detect in the higher-energy range for a given sensitivity.

3. Success of the quest

In the course of observations of dense molecular clouds, which are the seat of low-mass star formation, and which were already known to harbor X-ray emitting T Tauri stars, *ROSAT* and *ASCA* were able to detect X-ray emission from Class I protostars.

The first detections were obtained on nearby molecular clouds, ρ Oph and R CrA, at distances $\gtrsim 150$ pc from the Sun. In a deep (~ 30 ksec) *ROSAT* PSPC exposure, Casanova et al. (1995) found Class I sources in the error boxes (typically $\approx 10-20''$ in radius) of X-ray sources in the ρ Oph cloud, but confusion with other IR sources prevented secure identifications. With *ASCA*, Koyama et al. (1996) detected a cluster of Class I sources in R CrA, but here again the identifications were not secure, since the typical *ASCA* error box is $\gtrsim 1'$ in radius. A reliable identification could be obtained with the *ROSAT* HRI, when Grosso et al. (1997) detected a variable X-ray source within $\sim 2''$ of an IR source associated with a well-documented Class I source in the ρ Oph cloud, YLW15. More examples are now available, as Table 1 shows.

One way to estimate the efficiency of protostars to convert energy into X-rays is to compute their bolometric-to-X-ray luminosity ratios L_X/L_{bol}. As can be seen from Table 1, most protostars have $L_X/L_{bol} \sim 10^{-3} - 10^{-2}$, which is on the high side of the values obtained for T Tauri stars. The younger protostars, as estimated from their IR and mm properties (see, e.g., André & Montmerle 1994), tend to have the highest ratios. For these objects, the interpretation of X-ray emission in terms of magnetic activity at the surface of the forming star (especially flares) is fully consistent with that observed on T Tauri stars.

Table 1 – X-ray detected protostars

Source name	Cluster	A_V [mag]	L_X (10^{30} erg s^{-1}) (0.1 – 2.4 keV)	$\log(L_X/L_{bol})$	Flare ?	Ref.
IRS1/TS2.6	RCrA	41	~1.7	−4.3	No	1
			2.5		No	2
IRS2/TS13.1	RCrA	32	~0.9	−3.6	No	1
			9.3		No	2
IRS5/TS2.4	RCrA	39	~1.4	−3.4	No	1
			9.3		No	2
IRS7/R1	RCrA	35	~6.3	−3.4	Yes	1
			<1.0		-	2
IRS9/R2	RCrA	32	~1.0	−3.4	Yes	1
			<1.2		-	2
YLW15	ρ Oph	33	160	−2.1	No	3
YLW15	ρ Oph	33	>10^3	>−1.6	Yes	4
			$L_{X,tot} > 10^4$ †	>1		
EL29	ρ Oph	57	7.3	−4.4	Yes	5
WL6	ρ Oph	57	4.7	−3.3	Periodic	4
SVS16	NGC1333	28	280	−2	No	6

Refs.: (1) Koyama et al. (1996); (2) Neuhaüser & Preibisch (1997); (3) Casanova et al. (1995); (4) Grosso et al. (1997); (5) Kamata et al. (1997); (6) Preibisch (1997)

† Integrated over all X-ray energies.

In one remarkable case, however, the L_X/L_{bol} ratio has been found to be > to ≫ 1. This value has been reached by YLW15 during a "superflare" (Grosso et al. 1997) lasting a few hours. It cannot be well determined because the value of the extinction, as well as of the temperature (since the HRI has no spectral resolution), is uncertain, but as shown in Table 2, the X-ray luminosity must have been very high, perhaps as high as ~ 10^{36} erg s^{-1}. The characteristic cooling time allows to compute the plasma density and the size of the emitting region, as well as the value of the equipartition magnetic field (see Grosso et al. 1997 for details). In the case of YLW15, the plasma density and the magnetic field are more or less solar, but the size of the emitting region (assumed to be in the form of a tube with an aspect ratio a = length/diameter = 10) can be of the order of 1 AU. The large L_X/L_{bol} ratio, as well as the large size of the emitting region, strongly suggest that the X-ray activity, while still magnetic in origin, cannot be due to the star alone, but rather is the result of star-disk and/or star-envelope interactions (see, e.g., Hayashi et al. 1996). (Another possibility is that the central star is a binary, and then there can be star-star interactions like in the very X-ray-active RS CVn systems.)

4. Conclusions

The above results show that large numbers of hard photons are likely to be present in the protostellar systems. Not only are the effects mentioned in Sect. 2 on the circumstellar gas and the dust important, but it is also likely that the disk would be bombarded by shocks and energetic protons associated with the X-ray flares. As the early Sun should have passed through the flaring Class I pro-

Table 2 – X-ray properties of the YLW15 superflare

	T_X (10^7 K)	1	3	6	12
$A_V=20$	log $L_{X,tot}^{\S}$	33.7	34.0	34.3	34.5
	length (AU) †	0.32	0.24	0.24	0.20
$A_V=30^*$	log $L_{X,tot}$	34.6	35.1	35.2	35.5
	length (AU)	0.64	0.52	0.48	0.44
$A_V=40$	log $L_{X,tot}$	35.0	35.9	36.1	36.4
	length (AU)	1.2	1.2	0.92	0.84

§ Corrected for extinction. † Length of a magnetically confined plasma tube (aspect ratio = 10). * Best guess for the total visual extinction: $A_V = A_{V,ISM} + A_{V,envelope}$.

tostellar phase, the existence of such shocks, as well as X-ray and particle irradiation phenomena, may resolve several long-standing mysteries (isotopic anomalies, chondrule melting, excess particle irradiation) in ancient meteorites (e.g., Feigelson 1982, Cameron 1995; Lee et al., 1997). Also, the study of protostellar X-rays may hold clues to the origin of outflows, generally thought to be driven by MHD mechanisms.

The next step now is the search for X-rays from the youngest protostars, namely the Class 0 sources, allowing to push the "age barrier" for X-ray emission down to $\sim 10^4$ yrs. To do this, an extended energy range, as well as a good sensitivity, are required to overcome the large extinction of these objects. Future X-ray satellites, *AXAF*, *XMM* and *Astro-E*, have the required characteristics. In addition, their spectroscopic capabilities should allow to directly measure the extinction, leading to more accurate determinations of the X-ray luminosities. If X-rays are indeed found from these very young stars, then this will show that X-rays are a lifetime companion to all stages of stellar evolution.

References
André, P., & Montmerle, T., 1994, Ap.J., **420**, 837
Cameron, A.G.W. 1995, Meteoritics, **30**, 133
Casanova, S., Montmerle, T., Feigelson, E.D. & André, P. 1995, Ap.J., **439**, 752
Feigelson, E.D. 1982, Icarus, **51**, 155
Glassgold, A.E., Najita, J., & Igea, J. 1997, Ap.J., **480**, 344
Grosso, N., Montmerle, T., Feigelson, E.D., André, P., Casanova, S., & Gregorio-Hetem, J. 1997, Nature, **387**, 56
Hayashi, M.R., Shibata, K., & Matsumoto, R. 1996, Ap.J., **468**, L37
Kamata, Y., et al., PASJ, in press
Koyama, K., Hamaguchi, K., Ueno, S., Kobayashi, N. & Feigelson, E.D. 1996, PASJ, **48**, L87
Lee, T., Shu, F.H., Shang, H. & Glassgold, A.E. 1997, Science, in press
Maloney, P.R., Hollenbach, D.J., & Tielens, A.G.G.M. 1996, Ap.J., **466**, 561
Montmerle, T., Feigelson, E.D., Bouvier, J., & André, P. 1993, in *Protostars & Planets III*, U. of Arizona Press, p. 689
Neuhäuser, R. & Preibisch, Th. 1997, Astr.Ap., **322**, L37
Preibisch, T. 1997, Astr.Ap., **324**, 690
Ryter, C.E. 1996, Astr.Sp.Sci., **236**, 285
Shu, F.H., Adams, F.C., & Lizano, S. 1987, Ann.Rev.Astr.Ap., **25**, 23

SOLAR ARCADE FLARINGS AS MAGNETODYNAMIC PROCESS OF MASS AND ENERGY SHEDDING FROM ENERGIZED MAGNETIC REGIONS

Y. UCHIDA, S. HIROSE, S. CABLE, S. UEMURA, K. FUJISAKI, M. TORII AND S. MORITA
Physics Department, Science University of Tokyo, Tokyo

Abstract. Prototypes of magnetic actions in producing and shedding the X-ray-emitting high temperature plasmas in various astrophysical objects are witnessed in the spatially-resolved form on the Sun by the Solar X-ray Satellite "Yohkoh". The most prominent of those are arcade flarings seen as powerful arcade flares in active regions with strong magnetic field. Larger scale fainter X-ray arcade formation observed at high latitudes, shedding a large amount of mass and energy as CME's (coronal mass ejections) also belongs to this category. Since some features found by the new observation by Yohkoh are incompatible with the so-called "classical model of arcade flarings", we advance an alternative model based on the quadruple magnetic sources in the photosphere.

1. Introduction

Solar flares are sudden releases of energy stored in the stressed magnetic field. A superhot plasma region is created due to the fast release of magnetic energy in the Alfven-transit time, much faster than the energy loss times by radiation or by heat conduction in the rarefied plasma in the corona. Coronal magetic field having a plasma frozen to it can store energy in the form of magnetic stress if it is distorted by the motion of the footpoints in the high β ($= p_{gas}/p_{mag}$) gas in the photosphere. If the magnetic field configuration has a neutral point (or an array of neutral points) in it, the stressed magnetic fluxes with the loaded mass and helicity can be transferred to other regions of lower stress , due to reconnections in the neutral point region. and can reduce the accumulated magnetic stress energy in the system.

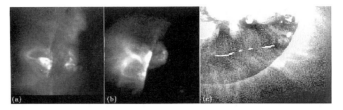

Figure 1. Quadrupolarity of Arcade Flares and Arcade Formations: High connections to both sides of the flare core and a brightening at the axis of the dark tunnel below in (a) Feb 21, 1992 flare, and in (b) Dec 2, 1991 flare. (c) Pre-event coronal structure on Jan 18, 1992, of the "Giant Cusp" event of Jan 25, 1992.

2. Pre-event Corona around the Site of Arcade Flarings

Recent observations by Yohkoh have revealed essential new information about the arcade flaring processes (Tsuneta et al. 1992). Detailed analyses of the observations of the still faint pre-event coronal structures of arcade flarings have revealed that there exist in the fainter pre-event corona, structures like the high connections from the top of the flare core back to the photosphere on both sides, or like the brightening at the axis of the dark tunnel below the cusped arcade (Uchida 1996), or the "overlapped dual arcades" in the case of high latitude arcade formation events (Uchida et al. 1997)(Fig 1). Those strongly suggest that the bipolar source picture that the "classical model" was based upon may not be valid. Corresponding to the "overlapped dual arcades", it was found that the photospheric magnetic field below them *is not* a simple bipolar region, but many patches of wrong polarities exist (cf., Martin 1990) on the other side of the previously called "polarity-reversal line", and the situation may correspond to quadruple arrays of magnetic sources, explaining the "overlapped dual arcades" as its separatric surface (Uchida 1997).

Also, the findings about the process of X-ray arcade formation itself, like the "spine" rising from below the X-ray arcade *sometime after* the dark filament rose, forming ultimately a feature like the "Giant Cusp" (Fujisaki et al. 1997), are also something not expected in the "classical model".

The so-called "classical model" (Sturrock 1966, and others) has a scenario that the magnetic arcade connecting the bipolar regions supports the dark filament on the top part of it, and the arcade is opened up by the rise of the dark filament, and a magnetic neutral point is *created* between the legs of the pulled out arcade. The reclosing of the once opened-up arcade through magnetic reconnection is thought to provide the difference in the energies of the opened and reclosed fields to the flare. Uchida (1980) noted that this model has a difficulty in energy that the energy of the dark filament rise should then be larger than the flare energy itself. One should then explain the energy of the dark filament rise, instead, but that is never

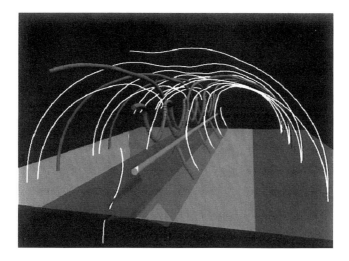

Figure 2. Magnetic Structure in the MHD Simulations of Hirose et al.(1998): The observed photospheric field with a large number of opposite-polarity patches on the other sides are modelized as quadruple arrays. The separatric surface in the field is the model's counterpart of the observed "overlapped dual arcades".

addressed. Also a question was raised about the suspension of the dark filament mass at the top of the otherwise convex arcade, especially initially when no dip has developed yet.

3. The Process in the Quadruple Source Model

Uchida et al. (1980, 1996,1997) advanced a model having quadruple magnetic sources in the photosphere, in place of the "classical model". In this model, there exist an array of neutral points in the field components $B_\perp$ (field perpendicular to the polarity-reversal line in the middle of the arrays of +, -, +, - sources) from the beginning. The longitudinal field $B_\parallel$ dominating in this neutral sheet of $B_\perp$ suspends the dark filament gas in the form of a thin partition, held by the squeezed antiparallel part of $B_\perp$ above the central polarity reversal line (Fig 2). The gravity acting on the dark filament mass stabilizes this.

We have performed a 2.5D MHD simulations with the destabilization of the dark filament by a current injection from a small emerging magnetic region. We found in the simulations (Hirose et al. 1997) that the thin partition type dark filament can exist in equilibrium first, separating the oppositely-directed vertical $B_\perp$ on both sides. When the dark filament is squeezed out, the strong antiparallel vertical parts of $B_\perp$ pushed from both sides (the right and left quadrants) can start reconnecting in the part of the neutral sheet from which the dark filament has been squeezed out. A considerable amount of mass and helicity loaded on the reconnecting flux can be transferred directly across the separatric surface, as a part of the flux tube undergoes the reconnection at the center. The dark filament is dynamically passive here, and pushed up by the fluxes swelling up across the separatric surface, rather than actively pulling the arcade into a tipped structure as supposed in the "classical model".

The reconnection in this case produces the flare core, heated cusp-shaped region, just like the observed cusps in arcade flares, *but here*, the heated part also has the upper connections connecting back the top of the heated cusp to the photosphere on both sides just as observed in Fig. 1(a),(b) (Uchida 1996). There is also a feature pressed down (S-shaped structure if seen from above), explaining the observed brightening feature near the axis of the dark tunnel. There are no counterparts for these in the "classical model".

The released part of the rising flux tubes will further be accelerated by the magnetic buoyancy, rolling down the slope of the magnetic potential gradient outwards (melon-seed effect). This final process in the mass and energy shedding, allowed *by* the occurrence of magnetic reconnection, may explain the large energy and mass in CME (coronal mass ejection)'s.

4. Conclusion

Observations by Yohkoh brought us some essential information about the global model of arcade flarings that was hidden in the still faint pre-event structures, supporting the quadruple magnetic source model, rather than the classical "opening up – reclosing" model.

These findings and new interpretations about energy and mass shedding from arcade flares and arcade formations will provide prototypes for their much greater versions in other active stars like dMe and RSCVn.

5. References

Fujisaki,K., et al. 1997, submitted to *Publ.Astron.Soc.Japan*.
Hirose, S., et al., 1997, in preparation.
Martin, S., in *Dynamics of Quiescent Prominences*, (Springer Lecture Note on Physics) **363**, pp1-44.
Sturrock, P.A., 1966, *Nature*, **221**, 695.
Tsuneta, S., et al., 1992, *Publ. Astron. Soc. Japan*, **44**, L63-69.
Uchida, Y., 1980, in *Skylab Workshop, Solar Flares*, ed. P.A. Sturrock (University of Colorado Press), p67, and p110.
Uchida, Y., 1996, *Adv. Space Res.*, **17**, (4/5)19-28
Uchida,Y., et al., 1997, submitted to *Publ.Astron.Soc.Japan*.

Session 1: Plasma and Fresh Nucleosynthesis Phenomena

1-2. Supernovae, Supernova Remnants and Galactic Hot Plasma

X-RAYS FROM SUPERNOVA 1993J AND EJECTA INSTABILITIES

K. NOMOTO AND T. SUZUKI

Department of Astronomy & Research Center for the Early Universe, University of Tokyo, Bunkyo-ku, Tokyo 113, Japan, and Institute for Theoretical Physics, University of California, Santa Barbara, CA 93106-4030, USA

Abstract.

Among the mechanisms of X-ray emissions from supernovae, we focus on the circumstellar interaction. In particular, we relate the new X-ray features of SN1993J to the hydrodynamical instabilities in the ejecta. First, we model the early-time two component spectral feature by invoking instabilities in the dense cooling shell in the ejecta. Second, we model the gradual increase in the X-ray light curve as X-rays emitted from the reverse shocked ejecta; here the model requires a large scale change in the density distribution due to Rayleigh-Taylor instabilities around the core-envelope interface.

1. Introduction

Supernova explosions are important sources of cosmic X-ray and gamma-ray emissions. They produce various heavy elements whose line X-ray emissions provide a unique tool to trace the chemical evolution of galaxies and the intra-cluster medium. Supernova explosions synthesize radioactive elements whose decays are significant sources of line gamma-rays and positrons. Some supernovae undergo circumstellar interactions which cause intense X-ray emissions.

More specifically, X-ray emissions from supernovae are expected from i) the shock breakout when the color temperature exceeds $\sim 10^6$ K, ii) Compton degradation of line γ-rays from the radioactive decays, iii) pulsar emissions, and iv) circumstellar interaction. Among these mechanisms, we focus on circumstellar interaction in SN 1993J.

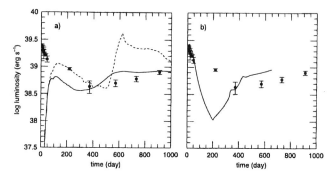

Figure 1. X-ray light curves for various models as compared with the observations of SN 1993J with ROSAT (Zimmermann *et al.* 1996). a) The dashed line shows the X-rays emitted from the reverse shocked ejecta, where the density structure is the same as original 3H11 (solid line in Fig. 2). The solid line shows the X-rays from the reverse shocked ejecta, where the density structure is modified as seen from the dashed line in Figure 2. b) The dotted line shows the X-rays emitted from the CSM which has a large clump in the outer region. The blast wave hits the clump around day 200 causing the X-ray enhancement.

Recent topics on SN 1987A are described in my contribution to the IAU/Joint Discussion 8 (speciall issue of New Astronomy 1998).

2. Early Spectra and Instabilities at the Cooling Shell

SN 1993J has been identified as a Type IIb supernova (SN IIb), whose spectral and light curve features can be well reproduced as the explosion of a red-supergiant whose hydrogen-rich envelope is as small as $\lesssim 1\ M_\odot$ (Nomoto *et al.* 1993; Podsiadlowski *et al.* 1993). The thin-envelope model implies that the progenitor of SN 1993J had lost most of its H-rich envelope and was surrounded with CSM. Then X-ray emissions from the interaction between the ejecta and circumstellar matter are expected, as actually has been observed with ROSAT (Zimmermann *et al.* 1994) and ASCA (Kohmura *et al.* 1994) as seen in Figure 2.

Following Suzuki & Nomoto (1995; see also Fransson *et al.* 1996), we have studied circumstellar interactions using a realistic ejecta model 3H11 (Nomoto *et al.* 1995; Iwamoto *et al.* 1997) and a CSM model with a parameterized density distribution of $\rho_{\rm CSM} = \rho_1 (r/r_1)^{-s}$. Here $s = s_{\rm in}$ at $r < r_1 = 4.9 \times 10^{15}$ cm, while $s = s_{\rm out}$ at $r \geq r_1$. The solid line in Figure 2 shows the initial density structure of the ejecta 3H11, where the jump is the interface between the He core and the H-rich envelope.

The collision between the ejecta and circumstellar matter creates a reverse shock which is radiative to form a cooling dense shell in the ejecta. X-rays emitted from the reverse shock are mostly absorbed by this shell at early times (Fig. 2a). Instead, early hard X-rays up to $\sim$ day 50 are well

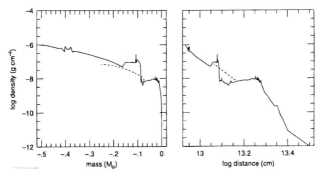

Figure 2. The density profile of the outer ejecta of 3H11 (solid). The dashed line is the profile adopted to account for the gradual increase in the X-ray observations.

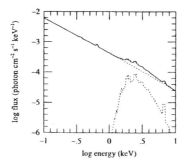

Figure 3. The calculated spectrum at day 20. The upper dashed line is the X-ray from the shocked circumstellar matter, and the lower one is the X-ray leaked through the cooling dense shell. The total emission is shown by the solid line.

modeled as thermal emissions from the shocked CSM (Fig. 2b).

Re-analysis of the ASCA observations has shown that the X-ray spectrum at day 20 consists of two components: (1) a high-temperature component with low absorption and (2) a low-temperature component with high absorption (Uno *et al.* 1997b). These two components are likely originate from (1) the forward shocked CSM and (2) the reverse shocked ejecta, respectively. However, the hydrodynamical model shows that the column-density of the cooling shell at day 20 is too high for X-rays from the reverse shocked layer to be observed.

We resolve this problem as follows: the cooling dense shell is Rayleigh-Taylor unstable, thus being clumpy. Its column density must be less than in the spherical model, so that X-rays from the ejecta can be leaked out through the cooling shell. We calculate the same hydrodynamical model but with the column density of the cooling shell reduced by a factor of 4, and obtain the spectrum in Figure 3. The upper dashed line is the X-rays from the shocked CSM and the lower one is the X-rays leaked through the cooling shell, which well reproduces the observed feature.

3. X-Ray Light Curve and Instabilities at the Core Edge

After day ~ 400, the ROSAT X-ray light curve shows a gradual increase (Figure 2; Zimmermann et al. 1996), as also reported by ASCA observations (Uno et al. 1997a). Here we explore several models for the light curve.

1. The dashed line in Figure 2a shows the X-rays emerged from the reverse shocked ejecta of original 3H11. The CSM has $s_{in} = 1.7$ and $s_{out} = 2.0$. The X-ray flare has been predicted to occur around day 500 when the reverse shock hits the density jump at the H/He interface.
2. The solid line in Figure 2 also shows the X-rays from the reverse shocked ejecta, where the density profile is modified as shown by the dashed line in Figure 2. The CSM has $s_{in} = 1.7$ and $s_{out} = 2.5$. The X-ray luminosity increases more gradually because the reverse shock arrives at the interface as early as day 300 - 400 and propagates against the power-law density profile rather than the jump.
3. The solid line in Figure 2b shows alternative possibility that a collision between the blast wave and a large clump in the CSM (around day 200) enhances the X-ray emission from the CSM.

To reproduce the gradual increase in the X-ray luminosity, we assume that the Rayleigh-Taylor instabilities have made the density distribution at the core-envelope shallower, from the sharp jump to the power-law (r^{-10}) as shown by the dashed line in Figure 2. The instabilities in the H-rich envelope should be more extensive than our original model 3H11. An asymmetric structure due to the spiral-in of the companion star into the envelope (Nomoto et al. 1995) would cause more extensive mixing.

This work has been supported in part by the grant-in-Aid for Scientific Research (05242102, 06233101) and COE research (07CE2002) of the Ministry of Education, Science, and Culture in Japan, and by the National Science Foundation in US under Grant No. PHY94-07194.

References

Fransson, C., Lundqvist, P., & Chevalier, R.A., 1996, ApJ, 461, 993
Iwamoto, K., Young, T.R., Nakasato, N., Shigeyama, T., Nomoto, K., Hachisu, I., Saio, H., 1997, ApJ, 477, 865
Kohmura, Y. et al., 1994, PASJ, 46, L157
Nomoto, K., Iwamoto, K., & Suzuki, T., 1995, Phys. Rep., 256, 173
Nomoto, K., Suzuki, T., Shigeyama, T., et al., 1993, Nature, 364, 507
Podsiadlowski, Ph., Hsu, J.J.L., Joss, P.C., & Ross, R.R., 1993, Nature, 364, 509
Suzuki, T., & Nomoto, K., 1995, ApJ, 455, 658
Uno, S., et al., 1997a, in X-Ray Imaging and Spectroscopy of Cosmic Hot Plasmas, ed. F. Makino and K. Mitsuda (Universal Academy Press, Tokyo), p.399
Uno, S., et al., 1997b, in this volume
Zimmermann, H.-U., Lewin, W.H.G., & Aschenbach, B., 1996, MPE report, 263, 289
Zimmermann, H.-U., et al., 1994, Nature, 367, 621

GALACTIC 1.8 MEV EMISSION FROM AL-26

N. PRANTZOS
Institut d'Astrophysique de Paris
98bis, Bd. Arago, 75014 Paris

1. Introduction

The detection of the relatively short-lived ($\sim 10^6$ yr) radioactive ^{26}Al in our Galaxy, more than 12 years ago, convincingly showed that nucleosynthesis is still active today on a large scale. Still, despite considerable efforts in the past decade, the site of this nucleosynthetic activity cannot be identified from theory alone; several potential sites are still on the list, been able to provide the 2-3 $M_\odot$ of ^{26}Al per Myr, implied by the detected 1.8 MeV flux of $\sim 4\ 10^{-4}$ cm^{-2} s^{-1} from the galactic plane (see [7] for a review).

The recent observations of the Compton Gamma-Ray Observatory allowed for the first time to establish a map of the distribution of ^{26}Al in the Galaxy [2], suggesting that massive stars are its most probable sources. On the other hand, the GRIS instrument detected an unexpectedly broad line [4], with potentially important implications for our understanding of the fate of nucleosynthesis ejecta [1]. Future observations with the INTEGRAL satellite are expected to clarify this point and to reveal details of the 1.8 MeV map, corresponding to major sites of galactic nucleosynthesis.

2. Production sites of Al-26

Several astrophysical sites may produce important amounts of ^{26}Al, by hydrostatic or explosive nucleosynthesis [7].

- Wolf-Rayet (WR) stars, with initial mass M>30 $M_\odot$ (for solar metallicity) synthesize ^{26}Al in their H-burning cores and eject it in the ISM by their strong stellar winds. The most recent models of WR stars [3], with improved physical ingredients (mass loss and nuclear reaction rates, opacities) account better than previous ones for observed WR properties (luminosities, composition, statistics) and show that galactic WR stars could produce up to ~1 $M_\odot$ of ^{26}Al in the past Myr; uncertainties in stellar models may reduce this figure by a factor ~2-3.

- Models of 10-30 $M_\odot$ stars (with no mass loss) show that they synthesize ^{26}Al hydrostatically in their Ne layers and (to a smaller extent) explosively in their C-Ne layers, during the SNII explosion; in the latter case, neutrino induced nucleosynthesis may also produce additional ^{26}Al. The amount of hydrostatically produced ^{26}Al is subject to important uncertainties, related to the treatment of convection in the corresponding stellar layers, and so is the result of ν-nucleosynthesis (due to its sensitivity to the poorly known ν spectra). Massive stars may produce up to 2 $M_\odot$ of ^{26}Al per Myr [8], but 3-5 times smaller amounts cannot be excluded.

-Intermediate mass stars (~4-9 $M_\odot$) can produce important amounts of ^{26}Al during their AGB phase, by hydrostatic H-burning in the bottom of their convective envelopes, *if* the temperature in that zone is T>7 10^7 K. Although some Hot Bottom Burning may indeed take place in those stars, the amount of ^{26}Al produce is highly uncertain, due to our current inability to evaluate the depth of the convective envelope.

- Novae may also synthesize interesting amounts of ^{26}Al by explosive H-burning, but the uncertainties involved in the current calculations (related mainly to the treatment of time-dependent convection, even in hydrodynamic models) do not allow any definite predictions.

In summary, several nucleosynthesis sites may be able to provide ~2 $M_\odot$ of ^{26}Al per Myr, with theoretical uncertainties been much larger in the case of of novae and AGB stars than in the case of SNII or WR stars.

3. Diffuse galactic 1.8 MeV emission from ^{26}Al

In view of the difficulty to decide on purely theoretical grounds on the most plausible source of galactic ^{26}Al, it was suggested that a clear signature could be provided by the spatial profile of the 1.8 MeV emission in the galactic plane: an asymmetric profile with prominent features in the tangent directions to the spiral arms, would favour massive stars (WR, SNII), while a smooth profile would favour an old population (low mass AGB stars and novae).

The COMPTEL instrument aboard the Compton GRO performed the first mapping of the 1.8 MeV line in the early 90ies [2]. The results of the 5-year survey (1991-1996) appear on Fig. 1. The emission profile is clearly assymetric and, besides the inner Galaxy, several prominent features appear, three of which (at galactic longitude l~ 32°, -50° and -74°) can be tentatively identified with spiral arms. This is a strong argument for a massive star origin of the galactic ^{26}Al although, in principle, massive AGB stars with M>5 $M_\odot$ or 0NeMg rich novae from progenitors of some-

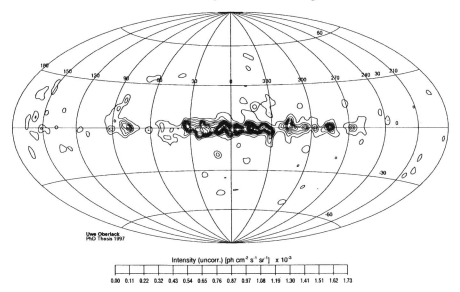

Figure 1. Sky-map of 1.8 MeV emission, from the 5-year survey of COMPTEL (1991-1996). The emission is concentrated in the galactic plane, and mostly in the inner Galaxy (-30° to +30° of galactic longitude), but also in several hot-spots; some of them (at l∼ 32°, -50°, -74°) can be tentatively identified with spiral arms, while others with more localised nearby sources like Vela (∼-90°) or Cygnus (∼80°).
The overall picture (emission assymetric w.r.t. the Galactic Center, and several hot-spots identified with spiral arms), points to massive stars at the origine of Al-26. (*Courtesy: R. Diehl and the COMPTEL team*).

what higher mass cannot be excluded. A composite picture, with a smooth background from low mass sites and a superimposed massive stellar population concentrated in the spiral arms cannot be excluded either, but the overall picture loses in simplicity.

Two of the remaining hotspots (l∼ 80°, -90°) are certainly not related to spiral arms. The former can be identified with the Cygnus region, where the activity of several tens of SN in the past Myr can account for the observed emissivity. The latter coincides with the Vela region and originates probably from the Vela remnant (age: ∼11 000 yr). This detection, the first of a point source of ^{26}Al, would allow to calibrate the yield of individual SNII, once the still uncertain distance to Vela (200-400 pc) is more precisely evaluated.

In a recent balloon flight, the GRIS experiment (with a large field of view, of ∼100°, but with a unique spectroscopic capability in this energy range, due to its Ge detector) detected the 1.8 MeV emission from the Galactic Center direction at the ∼7 σ level. The line profile is surprisingly broad, with an intrinsic FWHM W=5.4±1.4 keV [4]. This corrsponds to a Doppler velocity of 540±140 (W/5.4) km/s for a spherical expansion or to an effective temperature $kT \sim 39\pm20$ keV ($T \sim 4.5\pm2.3 \; 10^8$ K) for a thermal origin. Indeed, the line was expected to be rather narrow, its width reflecting only galactic differential rotation or random motions of the ISM, in which ^{26}Al was supposed to be at rest (thermalised in a short time $<<1$ Myr after its ejection). But galactic rotation alone can account only for a FWHM ~ 1.7 keV only, i.e. 3-4 times smaller than observed.

If confirmed, this result will certainly have important implications for our understanding of the fate of nucleosynthesis ejecta. Some of these implications have been scetched in [1]; the most interesting concerns probably the possibility for the ejected metals (from either WR, SNII or novae, but not from AGB stars, their winds been very slow) to condense in grains and travel undeccelerated for timescales of ~ 1 Myr. More refined calculations and a better understanding of the grain properties are necessary to find whether this idea is realistic or not.

4. Conclusions

A major step in our understanding of the nucleosynthesis of ^{26}Al in our Galaxy was achieved in the 90ies, with the first mapping of the galactic 1.8 MeV emission by COMPTEL. The assymetric profile of the emission and the (quite probable) identification of several prominent features with spiral arms clearly favour massive stars (WR and/or SNII) as the sources of ^{26}Al (although a contribution from lower mass objects, like AGB stars and novae, cannot be excluded).

Future observations with the INTEGRAL satellite, to be launched in 2001-2002, are expected to reveal details of that map (corresponding to major sites of galactic nucleosynthesis), to help establishing the poorly known spiral pattern in the inner Galaxy, to allow calibration of the ^{26}Al yields of individual sites (through confirmation of the Vela emission and detection of other point sources) and to improve our understanding of the fate of nucleosynthesis ejecta (in case the reported width of the 1.8 MeV line is confirmed).

References

1. Chen W. et al., 1997, in *The Transparent Universe*, Eds. T. Courvoisier & Ph. Durouchoux (ESA), in press
2. Diehl R. et al., 1995, A&A, 298, 445
3. Meynet G. et al., 1997, A&A, 320, 460
4. Naya J. et al., 1996, Nature, 384, 44
5. Oberlack et al., 1994, ApJS, 92, 433
6. Prantzos N., 1993, ApJ, 405, L55
7. Prantzos N. & Diehl R., 1996, Phys. Rep., 267, 1
8. Timmes F. et al., 1995, ApJ, 449, 204

GAMMA-RAY LINE SPECTROSCOPY RESULTS FROM COMPTEL

V. SCHÖNFELDER

Max-Planck-Institut für extraterrestrische Physik
P.O. Box 1603, D-85740 Garching, Germany
E-mail: vos@mpe-garching.mpg.de

Abstract. COMPTEL aboard NASA's Compton Observatory has led to a major progress in the field of astronomical gamma-ray line spectroscopy. Highlights are the all-sky map of the 1.8 MeV line from radioactive ^{26}Al, the first detection of the 1.156 MeV line from radioactive ^{44}Ti from a Supernova remnant (Cas A), and the detection of excessive MeV emission from the Orion complex that may be ascribed to excitation of ^{12}C and ^{16}O nuclei.

1. The 1.8 MeV All-sky Map

As described in the paper of N. Prantzos (these proceedings), radioactive ^{26}Al is supposed to be produced in core collapse supernovae, by hydrostatic nuclear burning in massive stars with strong stellar winds, by novae, by AGB stars or a mixture of all four. Due to its long decay time of about 10^6 years, ^{26}Al traces nucleosynthesis sites over the past million years. The 1.8 MeV line was discovered by HEAO-C [1]. COMPTEL has now produced an all-sky map in the light of this line (see Fig. 1 [2], [3]).

The following conclusions can be derived from the map: First, the galactic plane stands out clearly. Therefore, the bulk of the observable ^{26}Al has to be of Galactic, not local origin. Second, the emission *along* the plane is remarkably irregular and asymmetric around the Galactic Center. There is more emission on the right side of the Galactic Center than on the left side. Substantial emission comes from regions outside the Inner Galaxy. Along the plane there are hot spots and empty regions. The reason for the clumpiness may be that nearby sources contribute significantly. Possible source candidates are: Vela, Carina, the Cygnus region, and tangential projections to the spiral arms. These findings all support our understanding that ^{26}Al is procuded in massive stars, whether these are only core-collapsed supernovae or other massive stars, is not yet clear.

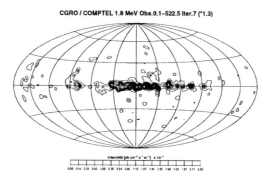

Figure 1. All-sky map in the light of the 1.809 MeV line from radioactive ^{26}Al [2].

2. Detections of 1.156 MeV ^{44}Ti Line from Cas A

Cas A is the youngest known supernova in our Galaxy. A COMPTEL maximum-likelihood map at 1.156 MeV from radiactive ^{44}Ti is shown in Fig. 2. If the flux measured is converted into the mass of ^{44}Ti, produced at the time of the explosion (about 1670), $(1 \text{ to } 3) \cdot 10^{-4}$ $M_\odot$ are obtained, depending on the assumed decay-time ($\tau = 78$ to 96 years) and the distance (~ 3 kpc), which both are not accurately known [4]. Theoretical predictions of the ^{44}Ti-yield of this core-collapse SN range from a few 10^{-5} to 10^{-4} $M_\odot$. Since the production of 10^{-4} $M_\odot$ of ^{44}Ti should have been accompanied by about 0.1 $M_\odot$ of ^{56}Ni, its decay products should have made it an optically bright shining SN. If at all, only a weak SN was seen. Flamstedt reported the detection of a 6th magnitude star in 1680, which might have been Cas A. It now seems that Cas A was surrounded by a thick dust shell before and at the time of the explosion, which absorbed the optical light. Though the supernova shock after the event destroyed the dust, the evaporated atoms are still there, and they have been recently detected by ROSAT through their X-ray absorption [5]. This first detection of a SN-remnant in the light of the ^{44}Ti-line illustrates the potential for finding previously undetected young Galactic SN by conducting a systematic line search. No convincing evidence for such events has been found in the COMPTEL data, yet [6].

3. Gamma-ray Observations from ORION

COMPTEL has detected excessive gamma-ray emission from the Orion complex in the 3 to 7 MeV gamma-ray band [7]. This excess has been tentatively identified with nuclear interaction lines from excited Carbon and Oxygen nuclei. The gamma-ray emission essentially extends over the entire Orion cloud complex (consistent on a coarse scale with its CO-distribution).

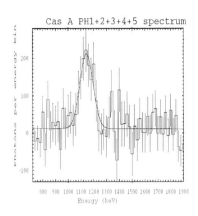
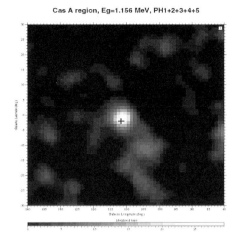

Figure 2. Detection of the 1.156 MeV line from radioactive ^{44}Ti from Cas A. Left side: background-subtracted energy spectrum; right side: maximum-likelihood map. Data are from the sum of phase 1 to 5 from COMPTEL [4]. From the width of the measured line an expansion velocity of the remnant of 7200 ± 2900 km can be inferred [4].

The 3 to 7 MeV luminosity exceeds that at lower (< 3 MeV) and higher (> 7 MeV) energies by a factor of about 10. The origin of the 3 to 7 MeV excess remains unclear. The tentative identification with nuclear interaction lines is based on its spectral shape. The problems of the interpretation are manyfold: e.g. the absence of line emission below 3 MeV, which puts constraints on the chemical composition and the energy spectrum of the energetic particles producing these lines, or the high ionisation rate and the resulting short life-times and pathlengthes of the energetic particles. Recently, Parizot [8] has pointed out that it may be easier to explain the Orion gamma-ray observations by putting the source of energetic particles *outside*, rather than *inside* the Orion cloud, e.g. into the Orion-Eridanus Superbubble. Another complication of the gamma-ray line hypothesis has been pointed out by Dogiel et al. [9]. The same energetic nuclei which produce the gamma-ray lines should also produce knock-on electrons, which then may be visible via their bremsstrahlung at X-ray energies. A search in ROSAT data, performed over the entire Orion region seen in gamma-rays, showed no correlation between X- and gamma-ray line emission. The ROSAT upper limit to the 0.5 to 2 keV X-ray luminosity is about 10^{33} erg/sec. This limit is close to the expected luminosity. After Ramaty et al. [10] became aware of these investigations, they pointed out that there should be two additional processes, which produce even higher X-ray fluxes. These are inverse bremsstrahlung of the energetic particles with ambient electrons, and K_α-line emission, stimulated by interactions with the en-

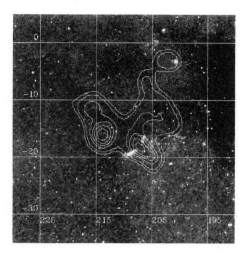

Figure 3. Superposition of ROSAT 0.5 – 2 keV X-ray map of the Orion region onto the 3 to 7 MeV intensity profiles.

ergetic particles. The sum of all three components, indeed, exceeds the ROSAT upper limit, and this might be a real problem for the gamma-ray line model [10]. One way out of this possible discrepancy might be to assume that the gamma-ray line production in Orion is not diffuse in origin, but comes from a few localised emission regions, which cannot be resolved by COMPTEL. In this case, much larger X-ray absorbing column densities might exist around these source, which would explain the non-visibility of the X-radiation [9]. Finally, it should be pointed out that the COMPTEL team considers the present Orion-results as preliminary [7]. We definitely need more data (not only at gamma-ray energies, but also at X-ray energies above 2 keV) and better analysis methods to substantiate the line hypothesis.

References

1. Mahoney, W.A. et al., 1984, A&AS **120**, p. 311-314
2. Oberlack, U. et al., 1996, A&AS **120**, p. 311-314
3. Oberlack, U., 1997, Thesis: *Über die Natur der Galaktischen* ^{26}Al *Quellen: Untersuchung des 1.8 MeV Himmels mit COMPTEL*, TU München
4. Iyudin, A. et al., 1997, Proc. of 2nd INTEGRAL Workshop, ESA-SP 382, 37-42
5. Hartmann, D. et al., 1997, Proc. of 4th Int. Conf. on Nucl. in the Cosmos Notre Dame, USA, Nucl. Phys. A.
6. Dupraz, C. et al., 1997, A&A **324**, 683
7. Bloemen, H. et al., 1997, Ap.J Letters **475**, L25
8. Parizot, E.M.G., 1997, submitted to A&A, and Proc. of 4th Compton Symp., AIP, in press
9. Dogiel, V. et al., 1997, 25th Int. Cosmic Ray Conf., Vol. **3**, 133-136
10. Ramaty, R. et al., 1997, Conf. Proc. of 4th Compton Symp., AIP, in press

NON-EQUILIBRIUM CONDITION IN THE SNR

H. TSUNEMI
Graduate School of Science, Osaka University
1-1 Machikaneyama-cho, Toyonaka, Osaka 560, Japan

1. Introduction

SNe are one of the major source of the heavy elements in Galaxy. They produced high temperature plasma heavily contaminated with heavy elements mixing with the interstellar matter. Due to the temperature and the density, the shock heated plasma does not reach the collisional ionization condition in young SNR, like Cassiopeia A, Tycho and Kepler. The emission is well represented with a non-equilibrium condition plasma.

ASCA observed these young SNR with its high energy resolving power. We could detect three types of emission lines from Si, S: $K\alpha$ from He-like ion, $K\beta$ from He-like ion and $K\alpha$ from H-like ion. The intensity ratios between these lines, we could determine the non ionization degree which is characterized by the ionization parameter, τ, the product of the electron density and the time elapsed after the shock heating. With taking into account the density and the age of the young SNR, it makes sense that they do not reach the collisional equilibrium condition.

ASCA observed middle-aged SNR. The plasma in them are thought to reach the CIE condition with taking into account their ages. Furthermore, they gathered relatively large amount of interstellar matter which diluted the ejecta. The metal abundance there should have represented the average abundance of the interstellar matter. We report there the observational results of them.

2. The Cygnus Loop

The Cygnus Loop is a typical middle aged ($\sim 18000 years$) SNR well studied in various wavelength. The X-ray spectrum in the north-east rim clearly shows that the plasma does not reach the CIE condition (Miyata et al., 1994). The X-ray emitting plasma in the rim is always gathering the in-

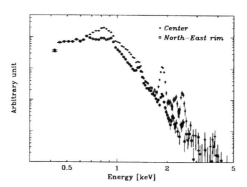

Figure 1. The SIS spectra for the north-east rim and for the center region of the Cygnus Loop. They are plotted by equalizing O lines.

terstellar material. The newly coming interstellar materials are recently shock heated. The elapsed time after the shock heating has almost nothing to do with the SNR age. Therefore, it makes sense that they are fresh plasma showing NEI condition (Miyata 1995). The metal abundances in the north east rim are smaller than those of cosmic values. This fact, it is still puzzling, clearly supports that the X-ray emitting plasma in the rim is dominated not by the ejecta but by the interstellar matter.

The X-ray spectrum from the center portion of the Cygnus Loop consists of two components (Miyata et al., 1997). One is relatively low temperature showing clear O-K lines. Line intensity ratio between O-VII and O-VIII shows almost the same value to that observed in the north-east rim. The other is relatively high temperature showing clear Si and S lines. The low temperature component must come from the rim region while Si and S lines come from the core region of the Loop which can be considered to be superposed to each other due to the projection effect. The plasma conditions in the core are metal rich, high temperature and low ionization parameter. The abundance of Si and S in the core region are 4 and 6 times those of cosmic values which are about 20-40 times higher than those observed in the north east rim. Figure 1 shows the X-ray spectrum of the core component. The apparent line widths of Si and S show 4 and 6 times those of cosmic values with a electron temperature of a few keV.

The metal rich plasma in the core region of the Loop is relatively concentrated in a compact region. What is the origin of this metal rich plasma ? The straightforward interpretation is to understand it as fossil of the ejecta from the SN. The plasma component in the core region depends on the progenitor type. Based on the theoretical model (Thielemann et al., 1996), we can estimate the metal abundance. By comparing the mass of Si, S and Fe, Miyata et al. (1997) claimed that the Cygnus Loop is originated from the type II SN. This is the first time that the SN type of the Cygnus Loop is determined from the metal abundance.

3. The Vela SNR

The Vela SNR is also a typical middle aged ($\sim 10000 years$) SNR. There is an X-ray pulsar in its center confirming the type II origin. Far from the so called plerion type SNR, it has a relatively large shell showing a thermal emission.

ROSAT observation revealed that the Vela SNR is rather circular structure. Furthermore, it discovered that there are several cusp-shape structures outside the X-ray emitting shell. Aschenbach et al. (1994) claimed that they are the fragments produced in the SN explosion.

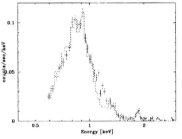

Figure 2. X-ray spectrum of the fragment A.

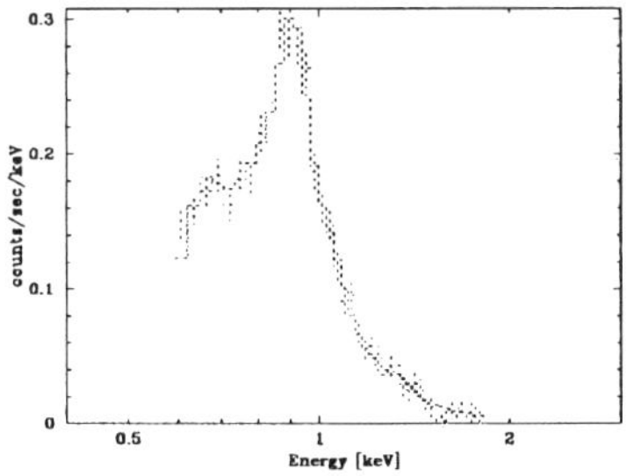

Figure 3. X-ray spectrum of the fragment B.

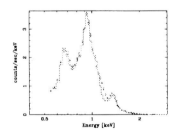

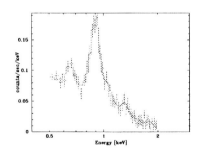

Figure 4. X-ray spectrum of the fragment D.

Figure 5. X-ray spectrum of the fragment E.

ASCA observed so far four fragments out of six reported (A through F). They are the fragment, A, B, D and E. The other two are too dim to be detected by Asca. The X-ray spectra for them are shown in figures 2-5. Some are bigger than the FOV of the SIS which makes difficult to estimate the background level partly due to the radiation damage effect on the CCD.

TABLE 1. The best-fit parameters by the NEI model.

Fragment	kT(keV)	O^*	Ne^*	Mg^*	Si^*	Fe^*
A	0.31±0.02	0.03	0.15	0.11	1.0	0.04
B	0.33±0.01	0.09	0.49	0.14	0.03	0.05
D	0.36±0.02	1.2	2.4	2.3	0.0	0.15
E	0.32±0.02	0.08	0.33	0.15	0.11	0.17
rim	0.24±0.03	0.02	0.07	0.11	0.08	0.01

* Theses are the best fit values. The relative values are determined by ±25% whereas the absolute values have large uncertainties.

Table 1 shows the result of metal abundance for them. Most of the spectra come from line emissions. Therefore, it is quite difficult to reach the absolute values for metal abundance. Whereas, the relative values among various elements are quite reliable. Keeping this point in mind, we notice that each fragment is characterized by one or two prominent elements. We also observed the shell region where we found relatively poor metal abundance. This reminds us the poor metal abundance in the shell region of the Cygnus Loop. Therefore, the metal abundance in the interstellar matter near the Vela SNR is much lower than those of the the fragments.

If they are the mixture of metal rich plasma with the interstellar matter, the metal rich plasma must be a debris of the progenitor star. It should have a so-called onion skin structure: each layer contains individual rich element. Therefore, the fragments may represent the constituent of some layer. It is quite interesting how they passed the shock front without being destroyed, since the shock front region is so dense.

The ASCA observation confirmed the NEI condition for the plasma condition in the young SNR. It also revealed that the plasma in the middle aged SNR does not reach the CIE condition. Furthermore, we detected possible fossils from them which might represent the metal abundance of the progenitor star of the SN.

References

Aschenbach, B., Egger, R. & Trumper, J. 1995, *Nature*, **373**, 587
Miyata, E., Tsunemi, H., Pisarski, R., & Kissel, S. E. 1994, *PASJ*, 46, L101
Miyata, E., PhD thesis of Osaka university 1995
Miyata, E., Tsunemi, H., Kohmura, T., Suzuki, S., & Kumagai, S. 1997, submitted
Thielemann, F.-K., Nomoto, K., & Hashimoto, M. 1996, *ApJ*, **460**, 408

ASTROPHYSICAL PLASMAS AND ATOMIC PROCESSES

J.S. KAASTRA
SRON
Sorbonnelaan2, 3584 CA Utrecht, The Netherlands

1. Introduction

Several plasma codes are available for the analysis of hot astrophysical plasmas. Among the oldest are the Raymond-Smith model (Raymond & Smith 1977) and Mewe-Gronenschild model (Mewe et al. 1985, 1986). Minor updates to this last code have resulted in the *meka* model (Kaastra 1992); major updates (most importantly the ionization balance and the treatment of the Fe-L complex) resulted in the *mekal* code (Mewe, Kaastra & Liedahl 1995). Here also the plasma codes of Masai (1984) and Landini & Monsignori Fossi (1990) should be mentioned. The RS, meka and mekal codes are included in the XSPEC fitting package, the latest mekal code is incorporated in the SPEX package (Kaastra et al. 1996). Both Masai's code and SPEX contain non-equilibrium ionization (NEI) modes. All these codes differ in details, see Brickhouse et al. (1995) for an overview. Most important for the analysis of X-ray data is the ionization balance that is used for iron and the treatment of the Fe-L complex.

Masai (1997) has shown the role of the ionization balance. A simulated ASCA spectrum yielded best-fit temperatures of 0.85, 1 and 1 keV and iron abundances of 0.7, 0.9 and 1.0 times solar using the ionization balance for iron of Arnaud & Rothenflug (1985), Arnaud & Raymond (1992) and Masai, respectively. Thus, the use of different ionization balances may lead to rather large differences in the derived parameters.

In a similar way, the treatment of the Fe-L complex can give large differences. For example, Matsushita (this conference) has fitted the spectrum of the elliptical galaxy NGC 4636 with different plasma codes. Comparing the RS and mekal codes, she finds for O-S abundances of 0.42 and 0.73 times solar, respectively, and for iron of 0.44 and 0.70. The temperatures derived from both codes is 0.76 and 0.66 keV, respectively. In general, we

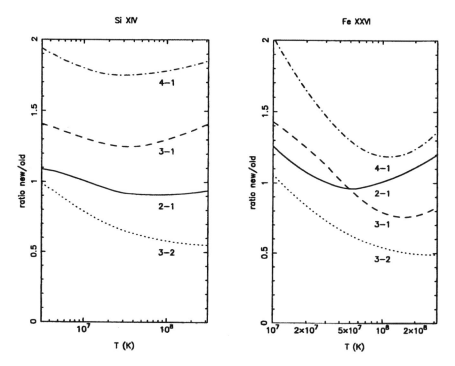

Figure 1. Ratio of new line calculations over old calculations as described in the text. Transitions shown are for Si XIV: 2-1 (2.00 keV), 3-1 (2.38 keV), 4-1 (2.50 keV) and 3-2 (0.37 keV); for Fe XXVI: 2-1 (6.97 keV), 3-1 (8.25 keV), 4-1 (8.70 keV) and 3-2 (1.29 keV)

recommend presently the mekal code for the analysis of hot plasmas, although there are certainly still large uncertainties in several aspects of that code, as we show below.

2. Excitation rates

Triggered by the major improvements obtained for the Fe-L complex by using the latest excitation calculations available, and also by the need of higher accuracy and more details for the analysis of coming high-resolution missions such as AXAF, XMM and ASTRO-E, we (R. Mewe, J.S. Kaastra et al.) have started a project to revise and extend all atomic data used in the mekal-code. This includes a systematic search through the literature and a critical evaluation of the data used. Recently we finished the hydrogen iso-electronic sequence. Although it is generally assumed that the data for the H- and He-sequence are rather accurate, our investigation showed that this holds only for the 2-1 transitions in the H-sequence. We have used the most recent atomic data available for the excitation cross sections in order to determine the line strength for all transitions up to $n = 5$ in

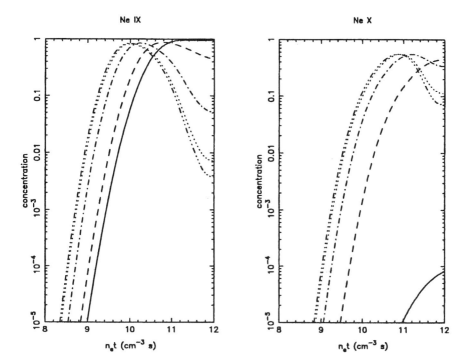

Figure 2. Ion fraction as a function of $n_e t$ for Ne IX and Ne X in an ionizing plasma with a temperature of 1 keV with an additional cool electron component with a temperature 0.1 keV and a relative density as compared to the 1 keV component of 0 (-...), 0.1 (...), 1 (-.-), 9 (—) and ∞ (solid line).

the H-sequence. These data include the effects of resonances at low energies and temperatures. In fig. 1 we plot the ratio of some of these newly calculated line strengths to the line strengths used in the mekal code. The data for Si XIV are from Aggarwal & Kingston (1992) and for Fe XXVI from Kisielius et al. (1996). It is seen that for the most important temperature ranges the Lymanα (2-1) transitions are accurate to $\sim$10%, but for Lymanβ (3-1) and Lymanγ (4-1) the differences between the old and new calculations can be up to 40-70%. A similar difference exists for Hα (3-2). It is evident that it is very important to have the best available rates for these ions in the plasma codes used for X-ray spectral modelling, and we intend to release an update of the SPEX code when the most important rates (including the He- and Li-sequence and others) have been included.

3. Non-equilibrium ionization

Non-equilibrium ionization (NEI) plays a dominant role in supernova remnants. In most cases spectral fits are made based upon the assumption of a plasma that is instanteneously heated to the shock temperature and ionizes subsequently by collisions with thermal electrons. A Maxwellian distribution is not always the best description of the plasma, however. For example, Itoh (1984) showed that in the early stages of ionization ($n_e t < 3\,10^9$ cm^{-3}s) the electron population consists of hot shock-heated electrons and cooler electrons ejected during the ionization process. He showed the effects of this on the plasma temperature evolution. However it also affects the NEI ionization process, as shown here. An additional cooler electron component has insufficient energy to ionize but gives mainly rise to an enhanced recombination rate, thereby slowing down the ionization process. This is illustrated in fig. 2 for H-like and He-like Ne. As a result, the ionization time scale t as derived from the measured ion concentrations is over-estimated by a factor which in the worst case can be an order of magnitude. It is evident that also the abundance estimates depend upon the presence or absence of such an additional cool electron component. An example where this process may play a role is RCW 86 (Vink et al 1997). ASCA spectra of this supernova remnant taken at different locii showed $n_e t$ values of $\sim 2\,10^9$ cm^{-3}s, which combined with an electron density of 0.2 cm^{-3} leads to a typical plasma age of 300 year, much smaller than the age of the remnant (at least 2000 year). Moreover, the remnant shows very low abundances, not only for elements like O, Si and Fe, but also for Ne, which is hard to reconcile with e.g. dust sputtering models, since Ne cannot be bound in dust. Given also the low $n_e t$ value, Vink et al suggested that the above mentioned effect plays a role in RCW 86.

References

Aggarwal, K.M., Kingston, A.E. (1992): Phys.Scr., 46, 193
Arnaud, M., Raymond, J. (1992): ApJ 398, 394
Arnaud, M., Rothenflug, R. (1985): A&AS 60, 425
Brickhouse, N., Edgar, R., Kaastra, J. et al (1995): Legacy 6, 4
Itoh, H. (1984): ApJ 285, 601
Kaastra, J.S. (1992): An X-ray spectral code for optically thin plasmas, Internal SRON-Leiden report, version 2.0.
Kaastra, J.S., Mewe, R., Nieuwenhuijzen, H. (1996): in UV and X-ray spectroscopy of astrophysical and laboratory plasmas, p. 411, eds. K. Yamashita and T. Watanabe, Univ. Acad. Press
Kisielius, R., Berrington, K.A., Norrington, P.H. (1996): A&AS 118, 157
Landini, M., Monsignori Fossi, B.C. (1990): A&AS 82, 229
Masai, K. (1984): Ap&SS 98, 367
Masai, K. (1997): A&A 324, 410
Mewe, R., Gronenschild, E.H.B.M., Van den Oord, G.H.J. (1985): A&AS 62, 197
Mewe, R., Kaastra, J.S., Liedahl, D.A. (1995): Legacy 6, 16
Mewe, R., Lemen, J.R., Van den Oord, G.H.J. (1986): A&AS 65, 511
Raymond, J.C., Smith, B.W. (1977): ApJS 35, 419
Vink, J., Kaastra, J.S., Bleeker, J.A.M. (1997): A&A, in press.

HOT PLASMA IN THE GALAXY

S. YAMAUCHI
Faculty of Humanities and Social Sciences, Iwate University,
3-18-34 Ueda, Morioka, Iwate 020, Japan

1. Introduction

In the X-ray band, we can see weak and extended X-rays along the Galactic plane and near the Galactic Bulge region, although these regions are dominated by many point sources (e.g., Warwick et al. 1985). The Tenma satellite discovered conspicuous emission lines from selected regions near the Galactic plane (Koyama et al. 1986). These lines are identified with K-shell line from He-like Fe, hence the extended emission is attributable to optically thin hot plasmas with temperatures of several keV. The origin of the thin hot plasmas, however, have been debatable, because no class of X-ray objects shows such high temperature plasma emissions. To investigate the origin of the extended X-rays, we are currently observing the Galactic plane regions with the ASCA satellite. In this paper, we report on the ASCA results: the hard X-ray imaging and spectroscopy of the hot plasma in the Galaxy.

2. Galactic Ridge Plasma

At first, I present the results of the hot plasma along the Galactic plane, the Galactic Ridge emission. Figure 1 (a) shows an X-ray mosaic image in the $l\sim28°$ region, the direction of the Galactic arm, called the Scutum arm. Several weak discrete sources are found, but most of the X-rays are unresolved. Therefore, with the ASCA spatial resolution of about $1'$, we confirmed that the Galactic Ridge X-rays are mainly due to diffuse hot plasmas.

Figure 1 (b) shows a typical X-ray spectrum of the Galactic Ridge emission. We see emission lines from highly ionized Si, S, and Fe. In fact, the X-ray spectra from all the observed regions are well represented by 2-temperature plasma models. The temperatures of the soft and the hard

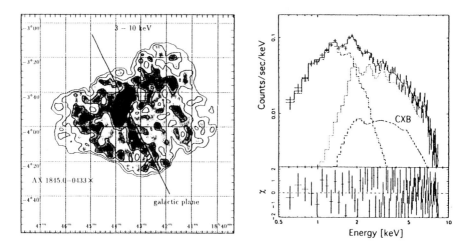

Figure 1. X-ray mosaic image of the Galactic Ridge emission in the 3–10 keV energy band (left) and a typical X-ray spectrum (right).

components are found to be 0.6–1 keV and 5–10 keV, respectively. The soft component can be explained by an integrated emission of supernova remnants or superbubbles, but the hard component is puzzling. Therefore, in this paper we will concentrate on the hard component, which is dominated above 3 keV.

The Galactic plane distribution of the hard component is found to be globally symmetric with respect to the Galactic Center, which agrees well with results of the 6.7 keV iron line mapping performed with Ginga (Koyama et al. 1989; Yamauchi and Koyama 1993). Further details of the Galactic Ridge emission are presented in Kaneda et al. (1997).

3. Galactic Center Plasma

Since the 6.7 keV iron line is originated from a thin hot plasma, we can know the location of the hot plasma using this line. With the Ginga satellite, we discovered an enhancement of the line flux near the Galactic Center (Koyama et al. 1989; Yamauchi et al. 1990), which means that a large amount of hot plasma is located in the Galactic Center region. The surface brightness distribution of the hot plasma is elliptical with the major and minor axes of $1.8°$ and $1.0°$, respectively.

In order to investigate the Galactic Center plasma in further detail, we have made the ASCA 6.7 keV iron line map in the inner $1° \times 1°$ region, as is given in figure 2. We see two diffuse plasma components near the Galactic

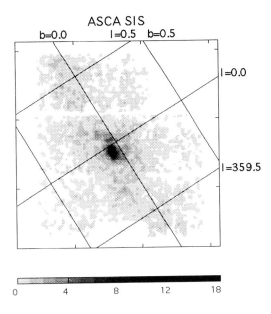

Figure 2. The 6.7 keV iron line map in the Galactic Center region.

Center. One is the extended emission which was already discovered with Ginga. The other is newly found component, and is more compact at the Galactic Center, with the size of about $3' \times 2'$.

We have made X-ray spectra from 8 regions of the Galactic Center, and fitted with thin thermal models. Then we found that the intrinsic spectra are very similar with each others at the plasma temperature of about 10 keV, while the N_H value is variable from position to position. Therefore, we infer that the Galactic Center plasma is in the same physical conditions over the entire region.

4. Galactic Bulge Plasma

We also found diffuse excess X-rays above the background level in the Galactic Bulge region. The X-ray flux decreases as the Galactic latitude increases from $|b| = 0°$ to $|b| = 3°$. However, the spectra above 3 keV are similar to that of the Galactic Center. In fact, by model fitting, we obtained the plasma temperatures of the Galactic Bulge region to be 5–10 keV.

5. Summary and Discussion

We summarize physical parameters of the hot plasma in table 1, where the dynamical time scales may indicate the maximum age of the hot plasma. We found that the hot plasmas with temperatures of 5–10 keV and essentially the physical parameters, are widely distributed in the Galaxy.

TABLE 1. Physical parameters of the hot plasma in the Galaxy

Parameter	Ridge	Center ($1.8°\times1.0°$)	Center ($3'\times2'$)	Bulge
Temperature (keV)	5–10	~ 10	~ 10	5–10
Thermal energy (erg)	$\sim 10^{56}$	$\sim 10^{54}$	$\sim 3\times 10^{50}$	$\sim 10^{55}$
Dynamical time scale (yr)	$\sim 10^{5}$	$\sim 10^{5}$	$\sim 4\times 10^{3}$	$\sim 10^{6}$

We discovered two hot gas components in the Galactic Center region, with the temperatures much higher than that bounded by the Galactic gravity. Together with the two different dynamical age for each plasma, we infer that the Galactic Center region has exhibited intermittent activities, which would produced the Galactic Center plasmas.

Another hint of the intermittent activities is found in the neutral iron line at 6.4 keV. The peak position of the 6.4 keV line map agrees with the position of the molecular cloud Sgr B2 (see figure 3 in Koyama et al. 1996). The X-ray spectrum from the Sgr B2 region shows, not only the strong 6.4 keV line, but also a deep absorption K-edge of iron, which is typically found in X-rays reflected by cool gas. However, in the Galactic Center region, we see no X-ray source to be reflected. This puzzle can be solved if the Galactic Center was bright 300 years ago, the light traveling time from the Galactic Center to Sgr B2. This would be the most recent activity in the central region of the Galaxy.

The author thanks Prof. K. Koyama for his helpful discussion and Dr. H. Kaneda and Mr. Y. Maeda for their kind supports to this work.

References

Kaneda, H. et al. (1997) Complex Spectra of the Galactic Ridge X-rays Observed with ASCA, *Astrophys. J.*, in press

Koyama, K. et al. (1986) Thermal X-Ray Emission with Intense 6.7-keV Iron Line from the Galactic Ridge, *PASJ*, **38**, 121

Koyama, K. et al. (1989) Intense 6.7-keV Iron Line Emission from the Galactic Centre, *Nature*, **339**, 603

Koyama, K. et al. (1996) ASCA View of Our Galactic Center: Remains of Past Activities in X-rays?, *PASJ*, **48**, 249

Warwick, R. S. et al. (1985) The Galactic Ridge Observed by Exosat, *Nature*, **317**, 218

Yamauchi, S. et al. (1990) Optically Thin Hot Plasma near the Galactic Center: Mapping Observations of the 6.7 keV Iron Line, *Astrophys. J.*, **365**, 532

Session 1: Plasma and Fresh Nucleosynthesis Phenomena

1-3. Galaxies and Their Clusters

METAL ABUNDANCES IN THE HOT ISM OF ELLIPTICAL GALAXIES

M. LOEWENSTEIN AND R. F. MUSHOTZKY
NASA/GSFC
Code 662, Greenbelt, MD 20771, USA

1. Introduction

In elliptical galaxies, where most of the stars were formed at an early epoch, the total mass, spatial distribution, and relative abundances of metals are intimately connected to the galaxy formation process. Determinations of the hot interstellar medium metallicity from X-ray spectral analysis are more direct, less model-dependent, and more radially extensive than optical estimates based on broad-band colors or line indices, and provide a view into the nucleosynthetic histories of elliptical galaxies.

2. Abundances in Gas-Rich Ellipticals

ASCA spectra can generally be decomposed into soft and hard components (Matsumoto *et al.* 1997). The soft component consists of emission from hot interstellar gas originally ejected from evolved stars, and shows a wide range of X-ray-to-optical flux ratios. The hard component generally scales linearly with optical luminosity, with a relative normalization and spectrum consistent with measurements of the integrated emission from X-ray binaries in spiral galaxy bulges. In this brief review, we focus on the 18 gas-rich ellipticals that have the most reliable abundance estimates, drawing upon the results presented in K. Matsushita's thesis, as well as on our own analysis.

The metallicities derived from fitting thermal plasma models are driven primarily by the Fe abundance, and range from 0.1 to 0.7 solar. Since it has generally been assumed that abundances of the mass-losing stars are supersolar, and since Type Ia SN are expected to further enrich the hot gas to Fe abundances of at least three times solar, X-ray abundances of elliptical galaxies are 3-30 times lower than what might naively be expected.

3. Reconciling Optical and X-ray Abundances

Clearly, either the SNIa rate is much lower than estimated, and/or SNIa ejecta is not efficiently mixed into the hot ISM. Can optical and X-ray abundances be reconciled, even if SNIa enrichment is neglected?

Hot gas abundances can be diluted by accretion of primordial or intergalactic material (Brighenti & Mathews 1997); however, since metallicities are not lower in more gas-rich systems this cannot be a dominant effect.

Plasma code inadequacies are an extra source of uncertainty for X-ray abundances; however, excluding the questionable Fe L region in spectral fits does not systematically raise the metallicity (Buote & Fabian 1997). Higher abundances can also be accommodated in more complex spectral models. Buote and Fabian have recently found that the best fit to *ASCA* data often consists of a two-temperature plasma, with $kT \sim 1.5$ keV in the secondary component, and that the abundances in such fits are systematically higher by about a factor of 2 compared to models with a single gas phase plus X-ray binaries. However, we have found that the temperature obtained from the He-like to H-like Si line ratios is in excellent agreement with the single-phase model, and is not consistent with the presence of hotter gas in the amounts suggested by Buote and Fabian.

The conventional wisdom that elliptical galaxies have supersolar abundances is based on measurements of the nuclear Mg2 index. However, when one accounts for the Mg overabundance with respect to Fe and aperture effects, it becomes clear that the global Fe abundances of the stars and hot gas are not grossly discordant. We have compared recent estimates of the global Fe abundance using optical data (Trager 1997) with the X-ray measurements. Comparing the 8 galaxies present in both samples reveals generally minor discrepancies, although a few galaxies still have unaccountably low X-ray abundances (Figure 1). The average optical Fe abundances are about 0.45 solar compared to 0.3 solar for the hot gas.

4. Silicon-to-Iron Ratios

Elemental abundance ratios provide constraints on the primordial IMF and relative numbers of Type Ia and Type II supernovae. In variable abundance fits for 7 ellipticals, we find Si-to-Fe ratios consistent with, or somewhat less than, the solar value (Figure 2). This is lower than the Mg-to-Fe ratio derived from nuclear optical spectra, and is more in line with what has been measured in intergroup media. The Si abundance provides an independent and robust (to plasma code uncertainties) strong upper limit on the effective SNIa rate that is consistent with what is derived using Fe – about 0.03 SNU.

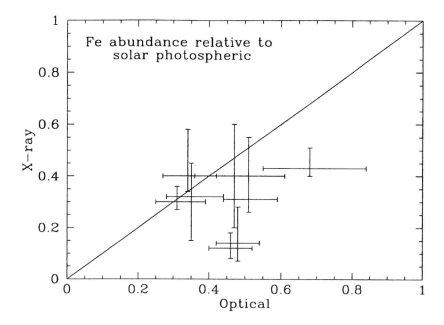

Figure 1. X-ray versus optical global iron abundance.

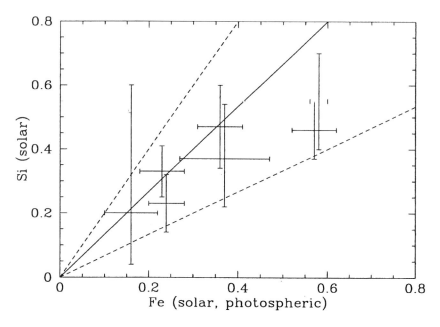

Figure 2. Si versus Fe abundance in the hot X-ray emitting gas. The solid line denotes Si:Fe in the ratio 1:1, while the broken lines denote the ratios 3:2 and 1:2 with respect to the (meteoritic) solar ratio.

5. Implications of Low Abundances for ICM Enrichment

The mass-averaged Fe abundance in an elliptical galaxy, based on X-ray and optical data, is about one-half solar – only slightly higher than what is measured in the intracluster medium. Since the gas-to-galaxies mass ratio is 2-10, there is several times more Fe in the ICM than is locked up in stars. If all cluster metals come from the same SNII-enriched proto-elliptical galaxy gas, then 50-90% of the original galaxy mass has been lost. However, the actual amount of material directly associated with the SNII ejecta is roughly an order of magnitude less: if there is selective mass-loss of nearly pure SNII ejecta, it is possible to lose most of the metals without losing most of the mass. It is also possible that there is another significant source of ICM enrichment, although this fails to explain why the ICM Fe mass correlates so well with total elliptical galaxy luminosity (Arnaud et al. 1992).

6. Concluding Remarks

X-ray spectra of elliptical galaxies are adequately fit by models consisting of gas with subsolar Fe abundance and roughly solar Si-to-Fe ratio, plus a hard X-ray binary component. The consistency of the strength and spectrum of the hard component with that expected from X-ray binaries, along with its more compact spatial distribution supports this model over ones where the hard component is primarily due to a hotter gas phase. Complications in the form of an extra soft continuum or multiple phases can be considered, but the consistency of Si line diagnostic and continuum temperatures demonstrates that the data – at the present level of sensitivity and spectral resolution – do not require these. Optical and X-ray Fe abundance estimates are converging, although there are some cases with anomalously low X-ray values. Problems in the Fe L spectral region remain; however, the main effect of improvements in atomic parameters is likely to be improved spectral fits rather than a radical upward revision in abundances.

Occam's razor would seem to demand that we provisionally accept the reality of low abundances in elliptical galaxies. As a result, we need to seriously reevaluate our notions of elliptical galaxy chemical evolution, intracluster enrichment, and Type Ia supernova rates.

References

Arnaud, M., Rothenflug, R., Boulade, O., Vigroux, L., & Vangioni-Flan, E. 1992, *A & A*, **254**, 49
Brighenti, F. & Mathews, W. G. 1997, *ApJ*, in press
Buote, D. A., & Fabian, A. C. 1997, *MNRAS*, submitted
Matsumoto, H., et al. 1997, *ApJ*, **482**, 133
Matsushita, K. 1997, Ph.D. thesis, University of Tokyo
Trager, S. C., 1997, Ph.D. thesis, University of California, Santa Cruz

HOT GASEOUS HALO IN THE ELLIPTICAL AND SPIRAL GALAXIES

H. AWAKI
Department of Physics, Kyoto University
Kitashirakawa, Sakyo, Kyoto, Japan

1. Introduction

The Einstein observations revealed that starburst and luminous elliptical galaxies had X-ray halo. These galaxies have quite different stellar population. Starburst galaxies contain young massive stars, while elliptical galaxies generally contain an old-metal rich population dominated by K and M giants. Therefore a question is why these two type of galaxies commonly have hot gas, in spite of quite different stellar populations. In order to address this question, we observed these galaxies with ASCA. In this paper, I would like to present observational results, then compare the physical parameters of the hot gas in these galaxies.

2. Hot Gaseous Halo in Spiral Galaxies

2.1. STARBURST GALAXIES

Starburst galaxies are one type of spiral galaxies with violent star formation of a rate > 1 Mo yr^{-1}. M82 is a well-known nearby starburst galaxy. Figure 1a shows an X-ray image of M82, superposed on the optical image. The diffuse X-rays are extending toward the minor axis of the galaxy with the extension of ~ 6 kpc.

We analyzed the ASCA SIS spectrum within a 6 arcmin radius (Figure 1b). The X-ray spectrum was well described by a model consisting of a two-temperature plasma with kT~ 0.31 keV and kT~ 0.95 keV and a power law spectrum. The lower temperature plasma was more extended than the higher temperature plasma (Tsuru et al. 1997), consistent with the temperature gradient found by the ROSAT PSPC (Strickland et al. 1997). We further obtained that the metal abundance in the hot gas is less than 1 solar, in particular, iron abundance is found to be extremely small.

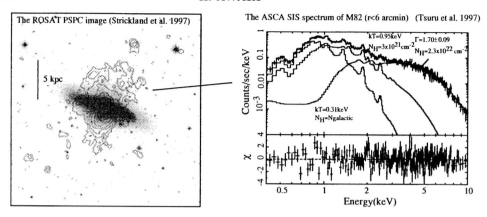

Figure 1. (a) the ROSAT PSPC image and (b) the ASCA SIS spectrum for M82.

TABLE 1. Summary of the hot gas in M82

size of halo	6 kpc
temperature	$(0.3\text{-}1)\times 10^7$ K
abundance	< 1 solar, (Fe/H) < 0.1 solar
total gas mass	10^7 Mo
total energy	10^{56} erg
cooling time	1 Gyr
outflow	yes (dM/dt$\sim$0.7Mo yr^{-1}, v_w=(1-3)$\times 10^3$ km s^{-1})
morphology	extending to the minor axis

The Soft band X-ray luminosity has been found to be well correlate to the far infrared luminosity (e.g. David et al. 1992) and the Bracket γ luminosity (e.g. Ward 1988). Therefore the origin of hot gas would be the starburst activity. Type II SNe frequently occurs in starburst galaxies with the SN rate of 0.1 ($L_{fir}/10^{44}$ erg s^{-1}) yr^{-1}, and total energy of 10^{51} erg. Then the energy input rate by the SN is estimated to be 10^{43} erg s^{-1}, far exceed the observed luminosity of 10^{40} erg s^{-1}. Therefore the soft X-ray can be explained by hot gas heated by type II SNe.

2.2. LINERS AND NORMAL SPIRAL GALAXIES

LINERs are another types of spiral galaxies. ASCA observed several LINERs, and found thin thermal X-rays with kT$\sim$0.5 keV (Terashima et al. 1997). The metal abundance was sub-solar. For NGC 1097 and NGC 5194, iron abundance was extremely small. These parameters are similar to those for starburst galaxies, hence most of LINERs would have similar hot gas

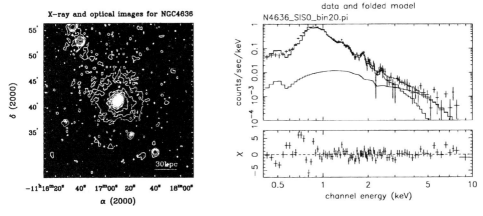

Figure 2. (a) the ROSAT PSPC image and (b) the ASCA SIS spectrum for NGC4636.

with starburst galaxies.

How about normal spiral galaxies? Fabbiano & Trinchieri (1985) pointed out that the X-rays from normal galaxies are described by the superposition of discrete sources. In fact, ROSAT HRI observed M31, and found that most of the X-rays are resolved into point sources (e.g. Primini et al. 1993). Normal galaxy, which have less activity in the nuclear region, would be gas-poor system.

3. Hot Gaseous Halo in Elliptical galaxies

The other famous halo object is an early type galaxy. Figure 2a is the X-ray contour of NGC 4636 overlayed on the optical image. An extended emission with r~50 kpc surrounding the galaxy is clearly seen. The radial profile of the Einstein results was well fitted with a single β model of β=0.45. Using this model, the total mass of X-ray gas was estimated to be $7 \times 10^9 M_\odot$. Recently ASCA performed deep observations on NGC 4636, and found that the radial profile was described with a two-β model (Makishima et al. 1997).

We found that the ASCA SIS spectrum is well fitted by a two-component model; a thin thermal emission (kT~0.76 keV) and a hard component (kT >4 keV) (Awaki et al. 1994). The abundance ratio of the thin thermal component was obtained to be similar to the solar, but smaller than 1 solar.

I note that there are luminous elliptical galaxies with less halo. NGC 4365 has the same optical luminosity as that of NGC4636. The spectrum of NGC 4365 is characterized by a thermal model with kT> 4 keV (Matsumoto et al. 1997).

In early type galaxies, there are many evolved star. Using the mass loss

TABLE 2. Summary of the hot gas in NGC 4636

size of halo	40 kpc
temperature	1×10^7 K
abundance	0.3 solar
total gas mass	10^{10} Mo
total energy	10^{59} erg
cooling time	1 Gyr
wind	no (static halo, binding mass 10^{12} Mo)
morphology	spherical

rate of 0.015 $[L_B/(10^9 Lo)]$ Mo yr^{-1} (Faber & Gallagher 1976), the accumulated mass from evolved stars is estimated to be 10^{10} $[L_B/(7\times10^{10} Lo)]$ [t/10Gyr] Mo. This is comparable to the mass of the hot gas. Therefore, evolved stars can supply all the mass of the hot gas. Unlike to starburst galaxies, type Ia SN frequently occurs in early type galaxies. The energy input rate by the SN is estimated to be 4×10^{41} $(r_{SNIa}/0.22)(E_{SNIa}/6\times10^{50} erg)$ $(L_B/10^{11}$ Lo) erg s^{-1}. This is nearly equal to the luminosity of the hot gas. Accordingly, the origin of the hot gas would be late type stars; red super giants and/or type Ia SN.

4. Summary

Although spiral and elliptical galaxies have hot gas, the physical parameters (e.g. iron abundance, wind, etc) are different with each other. The difference is attributable due to the stellar population in the galaxy. The stellar population depends on evolutional stage of a galaxy. Therefore, the hot gas in starburst galaxies and in elliptical galaxies would be produced in an early stage (<1Gyr) and a late stage after star formation, respectively.

References

Awaki, H. et al. (1993), *PASJ*,**46**,L65–L70
David, L.P. et al. (1992), *ApJ*,**388**,82–92
Faber, S.M., Gallagher, J.S. (1976), *ApJ*,**204**,365–378
Fabianno, G., Trinchieri G. (1985), *ApJ*,**296**,430–446
Makishima, K. et al. (1997), in this volume
Matsumoto, H. et al. (1997), *ApJ*,**482**,133–142
Primini, M.,et al. (1993), *ApJ*,**410**,615–625
Strickland, D.K. et al. (1997), *A&A*,**320**,378–394
Terashima, Y. et al. (1997), *ASP Conference Series*,**113**,54–55
Tsuru, T. et al. (1997), *PASJ*,in press
Ward, M. (1988), *MNRAS*,**231**,1p–5p

HOT GAS IN GROUPS AND THEIR GALAXIES

TREVOR J. PONMAN
School of Physics and Astronomy,
University of Birmingham, Birmingham, B15 2TT, UK

AND

ALEXIS FINOGUENOV
Space Research Institute,
Profsoyuznaya 84/32, 117810 Moscow, Russia

1. Introduction

It is clear that there is an important interplay between galaxies and the group environment. At the velocity dispersions ($\sim 100\,\mathrm{km\,s^{-1}}$) characteristic of groups, the galaxies interact strongly, leading to triggering of star formation, and galaxy merging. We can expect to see evidence of such processes through differences in the properties of galaxies in groups compared to field galaxies. Conversely, the galaxies affect their environment, as is apparent from the presence of heavy elements in the hot intergalactic medium (IGM) in groups, which emit characteristic X-ray lines.

In the present contribution, we will examine the X-ray properties of galaxies within groups, and the distribution of heavy elements in the IGM revealed by imaging X-ray spectrometers.

2. Properties of galaxies in groups

Much of the evidence regarding the peculiarities of galaxies in groups has been derived from the study of *compact* groups. Some of the effects seen are: morphological and dynamical peculiarity (Mendes de Oliveira & Hickson 1994), HI deficiency in spirals (Williams & Rood 1987), lowered velocity dispersion in ellipticals (Zepf & Whitmore 1993) and an anti-correlation of spiral fraction with velocity dispersion (Hickson et al. 1988) and X-ray luminosity (Ponman et al. 1996). However, any enhancement in star formation rates appears to be very limited, and few galaxies look like the remnants of recent mergers (Zepf 1993).

What of the X-ray properties of the galaxies? In the field, many ellipticals have hot gaseous halos (Fabbiano et al. 1992), whilst spirals show complex emission from sources within the body of the galaxy, and where active star formation is present, also from hot coronae or winds (Read et al. 1997). The ROSAT PSPC is an excellent instrument for detecting X-ray emission from the hot IGM, but is not well suited to separating this from emission associated with individual galaxies. In contrast, as can be seen in Fig.1, the ROSAT HRI, with its much higher spatial resolution, is able to study the emission from group galaxies, with little impact from the IGM. We have analysed ROSAT HRI observations of 11 Hickson compact groups

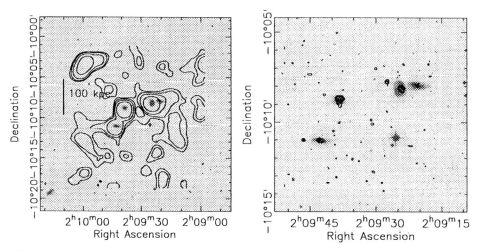

Figure 1. The X-ray emission from the compact group HCG16 (contours) overlaid on an optical image, as viewed (left) by the ROSAT PSPC, and (right) by the HRI.

(HCGs), detecting emission from a total of 27 galaxies and deriving X-ray upper limits for a further 20. The properties of the late type (spiral and irregular) galaxies in this sample are shown in Fig.2. The detected systems, which include almost all of the galaxies with $L_B > 4 \times 10^{10} L_\odot$, all lie above the field spiral line. This provides clear evidence that late type galaxies in these systems are X-ray overluminous. This is *not* a random sample of HCGs, but a set chosen in part because of the presence of disturbed and infrared-bright galaxies. Nonetheless, it is clear that in these systems at least, the galaxies show enhanced X-ray activity.

In the case of early type galaxies, one might expect to see a *reduction* in L_X relative to field galaxies, since dark halos, which are required to retain the hot gas halos of ellipticals, should be stripped by tidal interactions in such tight groups, as shown by simulations. The results (Fig.2) show little sign of any such reduction in L_X, even if one disregards the high values for the dominant central ellipticals. This suggests that dark halos have not

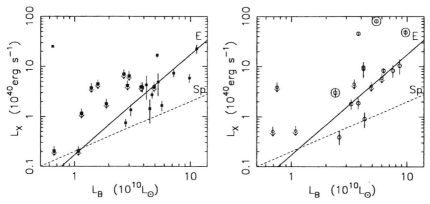

Figure 2. X-ray vs blue luminosity for a sample of (left) late type galaxies and (right) early type galaxies in HCGs (2σ upper limits are denoted with down arrows), compared to the relations (from Fabbiano *et al.* 1992) for field spirals and ellipticals. Elliptical galaxies marked with a double circle are dominant central ellipticals, whose luminosity may well be enhanced by a contribution from the group IGM.

been stripped in many of these galaxies, a result which offers a challenge to theorists.

3. Heavy elements in the intergalactic medium

The ASCA SIS provides an opportunity to map the strength of the emission lines from a number of elements within the IGM in groups and cool clusters. The best determined abundances are those of Fe and Si. Since the Si/Fe ratio is very different in the ejecta from supernovae of types Ia and II, the distribution of these elements gives strong clues to the processes which have contaminated the primordial IGM.

Fukazawa *et al.* (1997) have studied a sample of 40 clusters and groups with ASCA, deriving abundances for Fe and Si in an inner and outer region in each case. Finoguenov & Ponman (in preparation) have studied a much smaller sample, of three groups plus one poor cluster, but have mapped the radial distribution of the elements in detail – allowing for the effects of projection and the blurring of the instrument psf. The results of these two studies are consistent (cf Fig.3) and show that

- Abundance gradients are common in cool systems
- Abundance ratios are ∼solar in the inner regions of both groups and clusters, indicating a mixture of SNIa and SNII ejecta
- In clusters the abundance ratio is tilted towards SNII abundances (high Si/Fe) at larger radii
- It appears that the total iron mass-to-light ratio may be lower in groups than clusters (though observations do not yet extend out to the virial

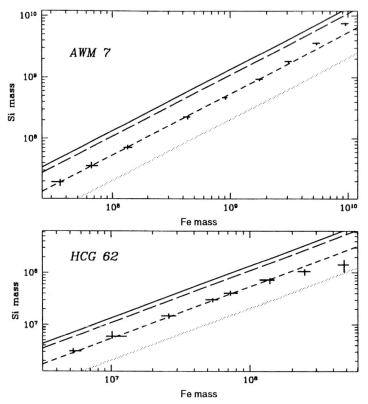

Figure 3. Cumulative mass of Si from the centre of the system is plotted against that of Fe, for the poor cluster AWM7 and the compact galaxy group HCG62. In each plot, the short dashed lines corresponds to solar abundance ratio, the grey line to SNIa (Thielmann et al. 1996), and the two uppermost lines to two models for SNII (Woosley & Weaver 1995, Tsujimoto et al. 1995).

radius in such small systems)

These results support a picture in which SNIa play a larger role in groups than in clusters, due to the loss of much of the SNII ejecta from the relatively shallow potential well in these smaller systems.

References

Fabbiano G., Kim D.W. & Trinchieri G. 1992, ApJS 80, 531
Fukazawa Y. et al. 1997, PASJ submitted
Hickson P., Kindl E. & Huchra J.P. 1988, ApJ 331, 64
Mendes de Oliveira C. & Hickson P. 1994, ApJ 427, 684
Ponman T.J., Bourner P.D.J., Ebeling H. & Böhringer H. 1996, MNRAS 283, 690
Read A.M., Ponman T.J. & Strickland D.K. 1997, MNRAS 286, 626
Williams B.A. & Rood H.J. 1987, ApJS 63, 265
Zepf S.E. 1993, ApJ 407, 448
Zepf S. & Whitmore B.C. 1993, ApJ 418, 72

ASCA OBSERVATIONS OF DISTANT CLUSTERS OF GALAXIES

TAKESHI GO TSURU
Department of Physics, Kyoto University
Kitashirakawa-Oiwake, Sakyo, Kyoto, JAPAN, 606-01
e-mail: tsuru@cr.scphys.kyoto-u.ac.jp

1. Introduction

Investigation of evolution of X-ray properties of clusters of galaxies is a key study of cosmology. The most important result in this field before *ROSAT* and *ASCA* observatories is the detection of the negative evolution of the X-ray luminosity function at red-shift lower than 0.6 (Gioia et al., 1990). Following it, many groups have been investigating evolution of X-ray luminosity function from various surveys with *ROSAT* observatory (eg. Collins et al., 1997). Many of them indicate no negative evolution at the red-shift lower than 0.7, which is against the *Einstein* result.

ASCA added new and key information to the study of the evolution of clusters; temperature and metal abundance of distant clusters of galaxies. I review it and show new results in this report.

2. Results

2.1. TEMPERATURE AND X-RAY LUMINOSITY

Tsuru et al. (1996) and Mushotzky and Scharf (1997) compiled *ASCA* data of distant clusters of galaxies (mostly $0.6 > z > 0.1$). By comparing it with the data of nearby clusters obtained with the previous observatories (eg. David et al., 1993), they suggested no evidence for a change in the temperature and X-ray luminosity relationship. However, it is still doubtful because it is shown without enough cross-calibration among *ASCA* and the other previous observatories. Recently, Fukazawa (1997) compiled *ASCA* data of nearby clusters. Then, I make comparison with the data and show result in this report, which is free from the difficulty of the cross-calibration.

ASCA observations of two very distant clusters, AXJ2019+1127 and MS1054-0321 at the red-shifts of 1.0 and 0.829 respectively, were reported very recently (Hattori et al. 1997a; Donahue et al. 1997). Including the two clusters, I show the temperature and X-ray luminosity relationship in the figure 1. The figure indicates no significant difference among the epochs, which implies no evidence for a strong evolution in this relationship.

2.2. TEMPERATURE AND GAS MASS RELATIONSHIP

Next, I compare the two clusters with nearby clusters in the temperature and gas mass relationship in the figure 2. I also plot data of the clusters at the red-shift of around 0.5. All the temperatures plotted in the figure are obtained with *ASCA*. The gas masses for nearby clusters are calculated from results of imaging analyses of *ASCA* data (Fukazawa 1997). The gas masses of the other (distant) clusters are obtained with *ROSAT* observatory except for that of 3C295 which is determined with *Einstein* observation (Hughes and Birkinshaw 1995; Donahue 1996; Schindler et al. 1997; Henry and Henriksen 1986; Donahue et al. 1997; Hattori et al. 1997a; Hattori et al. 1997b; Hughes 1997). All the gas masses except for AXJ2019+1127 are defined as those within 1.0 Mpc from cluster center. In the case of AXJ2019+1127, observed R_{max} is 0.5 Mpc. Then, adding to the gas mass actually detected in R_{max} of 0.5M pc, I plot extrapolated gas mass when assuming R_{max} of 1.0 Mpc in the figure.

This figure indicates significant difference between the nearby and distant clusters. The gas mass of the two very distant clusters, AXJ2019+1127 and MS1054-0321, are only 10% or 25% of those of nearby clusters when comparing at the same temperature. Two other distant clusters at the red-shift of around 0.5, 3C295 and MS0451.6-0305 also contain only 20%-50% gas masses of nearby clusters when comparing at the same temperature. Thus, I found a hint of evolution in this relationship.

It is suggestive that their gas masses are much smaller than those of nearby cluster although the no significant difference is seen in the temperature and X-ray luminosity relationship. It should indicate different distribution of ICM between the two groups. Small amount of gas can emit large luminosity if its distribution is compact. The β_{fit} of AXJ2019+1127 and MS1054-0321 determined with *ROSAT*/HRI are ~ 0.9 and $0.66 - 1.0$, respectively (Hattori et al. 1997a; Donahue et al. 1997). The values are larger than the typical value of $\beta = 0.6 - 0.7$ for nearby rich clusters, which indicates the two clusters are compacter than nearby clusters.

2.3. TEMPERATURE AND IRON ABUNDANCE

It has been already reported that no evidence for a change in the iron abundance and temperature relationship as a function of red-shift is seen at $z < 0.6$ (Tsuru et al., 1996; Mushotzky and Loewenstein 1997). In this report, I add the new data of MS1054-0321 and AXJ2019+1127 to the relationship in the figure 3. The iron abundance of MS1054-0321 is consistent with nearby clusters. However, that of AXJ2019+1127 is extremely higher than the relationship.

2.4. TEMPERATURE AND IRON MASS RELATIONSHIP

Next, I show the temperature and iron mass relationship in the figure 4. The iron mass of AXJ2019+1127 is consistent with the relationship of the nearby clusters. Since its low gas mass and high iron abundance counterbalances each other, the iron mass comes on the relationship. In the case of MS1054-0321, the abundance is consistent with nearby clusters but the gas mass is very low. Then, the iron mass becomes very low.

The result of AXJ2019+1127 indicates that the metal injection process into its ICM through its life had already finished before the red-shift of 1.0. On the other hand, it is suggested that the metal injection in MS1054-0321 has not started yet.

3. Summary

(1) No evidence for a change in the $k_B T$-L_X relationship or $k_B T$- iron abundance at $z < 0.6$ is found. (2) No significant change in $k_B T$-L_X relationship at $z > 0.6$ is seen, either. However, the gas masses of very distant clusters at $z \sim 0.8 - 1.0$ and some clusters around $z \sim 0.6$ are significantly lower than those of nearby clusters, which indicates a hint of evolution. (3) The iron masses of the two clusters at $z \sim 0.8 - 1.0$ are significantly different, which suggests that the epoch of the metal injection into ICM is different from cluster to cluster.

References

Collins *et al.* 1997, *Astrophys. J. Letters*, **479**, L117
David *et al.* 1993, *Astrophys. J.*, **412**, 479
Donahue 1996, *Astrophys. J.*, **468**, 70
Donahue *et al.* 1997, *astro-ph*/970710v2
Fukazawa 1997, Doctor Thesis, Univ. of Tokyo
Gioia *et al.* 1990, *Astrophys. J. Letters*, **356**, L35
Hattori *et al.* 1997a, *Nature*, **388**, 146
Hattori *et al.* 1997b, *Astrophys. J.*, submitted
Henry and Henriksen 1986 *Astrophys. J.*, **301**, 689
Hughes and Birkinshaw 1995, *Astrophys. J. Letters*, **448**, L93
Hughes 1997, private communication
Mushotzky and Scharf 1997, *Astrophys. J.*, **482**, 13
Mushotzky and Loewenstein 1997, *Astrophys. J. Letters*, **481**, L63
Schindler *et al.* 1997, *Astron. Astrophys.*, **317**, 646
Tsuru *et al.* 1996, in UV and X-ray Spectroscopy of Astrophysical and Laboratory Plasmas, ed. K. Yamashita and T. Watanabe (Tokyo: Universal Academy Press), 375

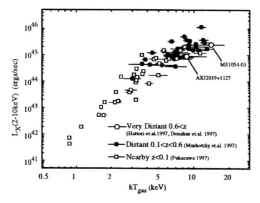

Figure 1. The temperature and X-ray luminosity relationship. Since all the data were obtained with *ASCA* observatory, the result is free from difficulty of cross-calibration (Fukazawa 1997; Mushotzky and Scharf 1997; Donahue et al. 1997; Hattori et al. 1997a).

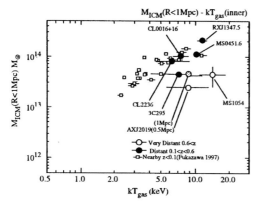

Figure 2. The temperature and gas mass relationship. All the temperatures were obtained with *ASCA*. The gas masses for the nearby clusters were calculated from the results of imaging analysis of *ASCA* data (Fukazawa 1997). The gas mass of 3C295 is derived with *Einstein* result (Henry and Henriksen 1986). The other masses are adopted from *ROSAT* results (Hughes and Birkinshaw 1995; Donahue 1996; Schindler et al. 1997; Donahue et al. 1997; Hattori et al. 1997a; Hattori et al. 1997b; Hughes 1997).

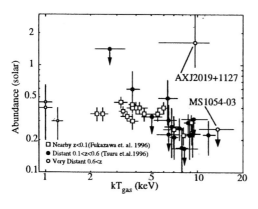

Figure 3. The temperature and iron abundance relationship. All the data were determined with *ASCA* (Fukazawa 1997; Tsuru et al. 1996; Donahue et al. 1997; Hattori et al. 1997a)

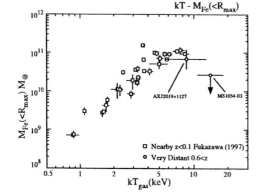

Figure 4. The temperature and iron mass relationship. The iron mass is defined as that with in R_{max} (Fukazawa 1997; Donahue et al. 1997; Hattori et al. 1997a).

Session 2: Future Space Programs

EARLY RESULTS FROM THE LOW-ENERGY CONCENTRATOR SPECTROMETER ON-BOARD BEPPOSAX

A.N. PARMAR, M. BAVDAZ, F. FAVATA, T. OOSTERBROEK,
A. ORR, A. OWENS, U. LAMMERS AND D. MARTIN
Astrophysics Division,
Space Science Department of ESA
ESTEC,
2200 AG Noordwijk,
The Netherlands

1. Introduction

SAX, an acronym for "Satellite Italiano per Astronomia a raggi X", now renamed "BeppoSAX" in honor of Giuseppe Occhialini, is the first X-ray mission sensitive in the very broad energy range between 0.1 and 300 keV (Boella et al. 1997a). The Narrow Field Instruments (NFI) have approximately 1° fields of view and consist of the imaging low- and medium-energy concentrator spectrometers (LECS, 0.1–10 keV, Parmar et al. 1997; and MECS, 1–10 keV, Boella et al. 1997b), and the non-imaging high pressure gas scintillation proportional counter (HPGSPC, 3–120 keV, Manzo et al. 1997) and Phoswich detector system (PDS, 15–300 keV, Frontera et al. 1997). All the NFI are coaligned and are normally operated simultaneously. In addition, the payload includes two wide field cameras (WFC, 2–30 keV; Jager et al. 1997) which observe in directions perpendicular to the NFI. These allow the detection of X-ray transient phenomena, as well as long-term variability studies.

2. The LECS

In order to achieve the extended low-energy response the LECS has an extremely thin (1.25 μm) entrance window and, in contrast to conventional GSPCs, does not have separate drift and scintillation regions, since this would unacceptably compromise the low-energy response (see Parmar et al. 1997). The LECS operates in the energy range 0.1–10 keV, but it is

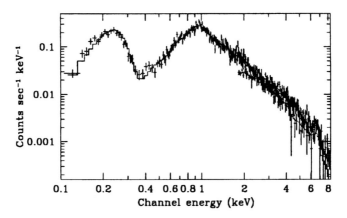

Figure 1. The observed LECS and MECS spectrum of VY Ari, together with the best-fit two temperature MEKAL spectrum.

≲0.5 keV where the instrument really comes into its own. Here, it provides better energy resolution than previous non-dispersive instruments such as the PSPC on ROSAT (Trümper 1983), or the SIS on ASCA (Tanaka et al. 1994), as well as providing time resolution of up to 16 μs.

2.1. CORONAL PLASMA STUDIES

In the line-dominated spectra produced by optically-thin plasma in collisional ionization equilibrium, such as in coronal sources, much of the emission is concentrated around and below ~1 keV (which includes the Fe L complex). An example is the LECS spectrum of the active binary VY Ari obtained with a 37 ks exposure on 1996 September 4–6 (Fig. 1; Favata et al. 1997). The LECS, with its good spectral resolution and low-energy response is well suited to the study of such emission. Current CCD detectors such as the ASCA SIS have little response ≲0.5 keV, and thus miss the intense soft continuum of coronal sources. While proportional counters are generally used at low energies because of their good efficiency, they have significantly worse spectral resolution than the LECS. Thus, the combination of good spectral resolution and low-energy response is allowing the LECS to make a unique contribution to several areas of coronal physics. In particular, results from ASCA show evidence for low metal abundances in many coronal X-ray sources. LECS spectra will allow the accurate determination of metal abundances using different diagnostics than with ASCA.

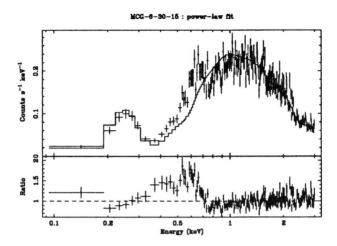

Figure 2. The LECS spectrum of MCG-6-30-15 fitted with an absorbed power-law model.

In particular, the ratio between the line-free continuum at low-energies and the strength of the Fe emission complex $\simeq 1$ keV is a sensitive diagnostic of the metal abundance.

2.2. WARM ABSORBERS IN AGN

Figure 2 shows the LECS spectrum of the bright Seyfert 1 galaxy MCG-6-30-15 obtained during a 49 ks exposure on 1996 July 29 to August 3 (Orr et al. 1997). The X-ray spectrum of this object is complex and can be well studied by the good spectral resolution and broad-band coverage of BeppoSAX. An absorption feature due to partially ionized O (the "warm absorber") is visible. Studies of this feature can provide information about the location and properties of the material surrounding the active nucleus as well its relation to the ionizing continuum. The continuous line in Fig. 2 shows an absorbed power-law model. Large residuals due to a blend of O VII and O VIII. absorption edges are clearly seen below 1 keV.

2.3. COMETARY X-RAY EMISSION

The LECS detected low-energy X-ray emission from Comet Hale-Bopp on 1996 September 10–11 (Owens et al. 1997). While this may be unsurprising, given earlier ROSAT observations of X-ray emission from Hyakutake and at least four other comets (Lisse et al. 1996), the mechanism for producing cometary X-rays remains unclear. Two leading models are charge exchange

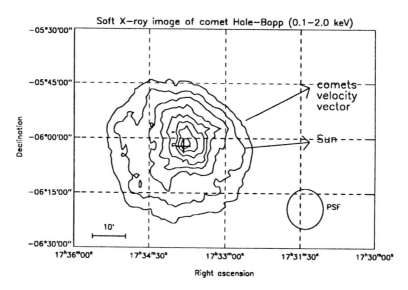

Figure 3. A contour plot of the X-ray emission observed from comet Hale-Bopp. The position of the nucleus is indicated by the cross.

excitations of highly ionized solar wind ions with neutral molecules in the comet's atmosphere and the scattering, fluorescence and bremsstrahlung of solar X-rays in attogram (10^{-18} g) dust particles.

The bulk of the emission detected by the LECS originates on the sunward side of the comet, in agreement with ROSAT images of comet Hyakutake. The extent of the emission is consistent with the LECS point spread function (PSF) of 9.5' FWHM at the mean energy of the detected emission, although the width normal to the Sun-nucleus axis appears wider than in the direction of motion. The 68% confidence limit to any source extent is <6.5'. The LECS spectrum is equally well fit by a thermal bremsstrahlung with temperature 0.29±0.06 keV or a power-law with photon index $3.1^{+0.6}_{-0.2}$.

References

Boella, G., et al. (1997a) *A&AS*, **122**, 299
Boella, G., et al. (1997b) *A&AS*, **122**, 327
Favata, F., et al. (1997) *A&A*, **324**, L41
Frontera F., et al. (1997) *A&AS*, **122**, 371
Jager, R., et al. (1997) *A&AS*, **125**, 557
Lisse, C.M., et al. (1996) *Science*, **274**, 205
Manzo, G., et al. (1997) *A&AS*, **122**, 357
Orr A., et al. (1997) *A&A*, **324**, L77
Owens, A., et al. (1997) *ApJ*, in press
Parmar, A.N., et al. (1997) *A&AS*, **122**, 309
Tanaka, Y., Inoue, H., and Holt, S.S. (1994) *PASJ* **46**, L37
Trümper, J. (1983) *Adv. Space Res.*, **2**, 241

THE ASTRO-E MISSION

Y. OGAWARA
Institute of Space and Astronautical Science, Kanagawa 229

1. Introduction

The past three decades have seen an explosion in high-energy astrophysics. We have found X-ray astronomy to be an indispensable tool in understanding our Universe. The discipline has become mature, and future X-ray observatories must be more highly specialized. High-resolution spectroscopic imaging in the band above 2 keV, systematically exploited by *ASCA*, has led to much new astrophysical knowledge. *ASCA* has also been playing a particularly important role in studying sources hidden behind dense material. *Astro-E*, the successor of *ASCA*, is scheduled for launch in the year 2000 by the Institute of Space and Astronautical Science (ISAS) with its newly developed M-V rocket (Ogawara & Inoue 1997). Our new *Astro-E* observatory features high energy resolution and high sensitivity over the broad energy range 0.5 keV to 600 keV. The general emphasis of the observatory is to provide large collecting areas at higher energies, with angular resolution good enough to avoid the confusion limit. This paper provides a brief description of the performance of the *Astro-E* instruments.

2. Scientific Instruments

Astro-E carries two focal-plane instruments for soft X-ray observations, and one collimated large-area counter array for hard X-rays (Table 1). A microcalorimeter array (X-ray Spectrometer – XRS) and four identical sets of X-ray CCDs (X-ray Imaging Spectrometers – XIS) cover the energy band from 0.5 keV to ~10keV with imaging capability. The Hard X-ray Detector (HXD) is a combination of scintillation detectors and silicon PIN detectors, which cover the hard X-ray band from 10 keV to 600 keV. These instruments are being prepared in an extensive collaboration between scientists from Japan and the United States.

Table 1. Comparison of *Astro-E* and *ASCA*

	Astro-E		*ASCA*
Focal Plane Detector	XRS	XIS	SIS/GIS
Number of Units	1	4	2/2
Focal Length	4.5 m	4.75 m	3.5 m
Outer Diameter	400 mm	400 mm	345 mm
Number of Layers	168	175	120
Surface	Ir or Pt	Au	Au
Effective Area (1.5 keV)	~ 460 cm^2	~ 460 cm^2	~ 310 cm^2
Effective Area (8.0 keV)	~ 280 cm^2	~ 250 cm^2	~ 120 cm^2
Half Power Diameter	1.5′		3.5′
Weight /Unit	20 kg		10 kg

Five telescope modules provide flux for the soft X-ray focal-plane instruments. One telescope module with a focal length of 450 cm feeds the XRS; the other modules have XIS units at their focal planes. The telescopes consist of nested conical thin-foil mirrors based on a similar design concept with the *ASCA* telescopes (Seremitos et al. 1995). They are made of replica foils and allow a half-power diameter of $\sim 1.5'$, which is better than *ASCA* by at least a factor of two. Additionally, the ratio of the focal length to the diameter is smaller, making the grazing angles smaller. This leads to better reflectivity for higher photon energies. The focal lengths of 4.5 m for the XRS and 4.75 m for the XIS give effective areas of ~ 500 cm^2 at 1.5 keV and ~ 300 cm^2 at 8 keV. The collecting area for the four XIS units is roughly twice as great as that of *ASCA*.

The microcalorimeter array employed in the XRS is the first such instrument to fly in a satellite and will have an ultimate energy resolution of 12 eV FWHM at 6.4 keV (McCammon et al. 1996). It allows us to perform precise diagnostics of cosmic hot plasmas at a level of physical knowledge never achieved in the past. In the microcalorimeter we take advantage of the low energies of phonons at very low temperature. The small heat capacities at low temperature lead to small fluctuations in the numbers of phonons, resulting in higher energy resolution compared to ordinary semiconductor devices (Moseley et al. 1984). The microcalorimeter must be operated at about 65 mK. It will be cooled by an adiabatic-demagnetization refrigerator embedded in a liquid helium/solid neon dewar system. The cryogenic system limits the life of the XRS to about two years (Voltz et al. 1996).

The *Astro-E* XRS array consists of 32 Si pixels containing ion-implanted thermistors. HgTe with high quantum efficiency through 10 keV is attached

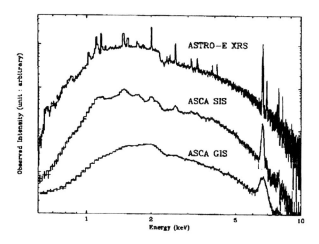

Figure 1. Spectroscopy with the *Astro-E* microcalorimeter array. The upper curve shows the 12-eV resolution expected from XRS; the lower two curves show the same spectrum at *ASCA* resolutions.

to the Si pixel as the X-ray absorber (Stahle et al. 1996). The characteristics of the XRS are summarized in Table 2 of Takahashi et al. (1998), and we illustrate in Figure 1 the remarkable improvement in spectral resolution to be achieved. Although the sky coverage is small ($2' \times 4'$ for 2×16 pixels), we also have imaging capability at a resolution of $\sim 0.5'$.

The imaging capability of the *Astro-E* CCDs (XIS) will play a key role in deep surveys. The SIS units on board *ASCA* were the first X-ray detectors in space using CCDs in a photon-counting mode. Many efforts have been made to improve the performance of the CCDs and their related electronics, based on *ASCA* experience. These CCDs, like *ASCA*'s, are front-illuminated frame-transfer devices with 1024×1024 pixels, developed by MIT. They have thicker depletion regions than those on *ASCA* ($\sim 60 \mu m$ vs ~ 30 μm). A weak radioisotope source at each corner of each chip provides known photon energies for gain calibration. In the XIS, we have the capability to monitor the dark current on a pixel-by-pixel basis, rather than globally as with ASCA. The lower operating temperature, 180 – 190K, is also expected to yield better performance in terms of charge transfer efficiency.

Astro-E also carries a Hard X-ray Detector (HXD) array (Takahashi et al. 1996; Kamae et al. 1996), covering the non-thermal energy range 10 keV to 600 keV. The well-type phoswich counters (GSO/BGO) provide nearly 4π sr shielding (see Table 4 of Takahashi et al., 1998). The detector will thus be characterized by a low background rate, expected to be $\sim 10^{-5}$ photons/s/cm^2/keV at 100 keV. It consists of 16 (4×4) modular units and has an overall photon collecting area of about 350 cm^2. Two layers of large (2 × 2 cm^2) 2mm-thick PIN (silicon) detectors are buried in the deep BGO well just above the detection part of the well-type phoswich counter. Softer X-rays are absorbed in the PIN diodes; harder ones penetrate the diodes and are absorbed by the GSO phoswich. The energy resolution of the HXD is ∼4 keV below 50 keV (by the silicon detector) and 8-9% at 662 keV. Although the HXD does not have imaging capability, the field of view of the detector for high energy photons is restricted to 4° × 4° by active collimators made of BGO, and the field of view for low energy photons (below ∼100 keV) is restricted to 0.5° × 0.5° by fine passive collimators made of phosphor-bronze sheets placed inside the BGO well. The sensitivity of the HXD for point sources should be substantially higher than any other past mission in the energy band between 10 keV to several 100 keV. We therefore expect to detect and study many new cosmic hard X-ray point sources.

3. Conclusions

With the three experiments described above, the *Astro-E* mission will be able to perform various kinds of studies for a wide variety of X-ray sources with the highest energy resolution and with the highest sensitivity over a wide energy range from 0.5 to 700 keV ever achieved. Because of the improved image quality and the large effective area, the sensitivity limit by the XIS will improve over *ASCA*'s and is expected to be ∼1 × 10^{-14} erg/cm^2/s in the 2-10 keV band. Wide-band spectroscopy with *Astro-E* should play an important role in studies of the nature of highly-absorbed sources and their contributions to the cosmic X-ray background. The construction of the flight model for *Astro-E* is underway at the time of writing. The first end-to-end test of the satellite is scheduled in July 1998, leading to a launch in early 2000.

References

Kamae, T., et al., 1996, SPIE 2806, 315
McCammon D., et al., 1996, NIM, A370, 266
Moseley, S.H., et al., J. Appl. Phys, 56, 1984, 1257
Ogawara, Y., and Inoue, H., 1997, in Proc. 7th ISCOPS Conference (American Astronautical Society, to be published)
Serlemitsos, P., et al., PASJ, 47, 1995, 105
Stahle, C.M., et al., 1996, NIM, A370, 173
Takahasi,T., et al., 1996, A&AS, 120, 645
Takahasi, T. Inoue, H., & Ogawara, Y., 1998, Astr. Nachrichten, to be published
Voltz, S.M., et al., 1996, Cryogenics, 36, 763

AXAF IN CONTEXT: A REVOLUTION

MARTIN ELVIS

Harvard-Smithsonian Center for Astrophysics
60 Garden St., Cambridge MA, 02138, USA

1. AXAF and the Promise of X-ray Astrophysics

AXAF has been a long time, 20 years, in coming. This has made AXAF so overfamiliar that it is hard to see how revolutionary it still is. Some even describe AXAF as the end of the line in X-ray technology; a never to be repeated venture to high resolution. This view is wrong. To see why we need to see where X-ray astrophysics is going.

It should not be controversial to say that X-ray Astrophysics has barely begun. Today's X-ray satellites have scored many successes; yet astronomers still extract very little of the information carried to us by X-ray photons; the subarcsecond spatial, and $R=1000$ spectral resolution *routinely* available at longer wavelengths, is not even begun in X-ray astronomy. AXAF is our first step into that wo ld.

Eventually X-ray telescopes will be built that have all three qualities needed to fully inhabit X-ray astrophysics: many sq. meters of effective area; sub-arcsecond angular resolution; and $R=1000$-10,000 spectral resolution. Then the riches of the atomic transitions in the X-ray band spectrum can be exploited for *all* classes of X-ray source (Elvis & Fabbiano 1996). Getting there is the problem.

The next NASA Great Observatory, the *Advanced X-ray Astrophysics Facility* takes the first step, combining two qualities: sub-arcsecond imaging (0.5″ HPD) with high spectral resolution ($R=1000$), albeit with a modest increase in area (~ 0.1 m^2). So AXAF will show us where X-ray astrophysics can go. In this context we can recognize AXAF for what it is - a revolution. This short paper tries to show just how revolutionary.

2. The Power of AXAF

Why is AXAF 'Advanced'? The AXAF optics put AXAF in a separate league from all other X-ray missions- past, present or planned: 75% of power within 1 arcsec *diameter* up to 10 keV. (This above spec. performance is thanks to Hughes-Danbury, Kodak, and the AXAF Mirror Scientist, Leon van Speybroeck). The AXAF *beam area* (the important quantity) is $\frac{1}{100}$ that of the ROSAT HRI, $\frac{1}{1000}$ of XMM and $\frac{1}{10,000}$ of ASCA. As a result AXAF has unprecedented sensitivity and resolution, angular and spectral.

AXAF has essentially zero background ($\sim$1 count/megasec/sq.arcsec) and hence high sensitivity. For point sources observations up to 2 weeks long will be photon limited. In 5 minutes AXAF will reach 10 times fainter than the ROSAT All Sky Survey, and will return a $\frac{1}{2}''$ position. At this flux there are 4 million sources available. The deepest, megasecond, AXAF surveys will reach $f_x \sim 5 \times 10^{-17}$ c.g.s., about $\frac{1}{20}$ of the ROSAT limit. This is uncharted territory. Our only guide is the ROSAT fluctuations analysis (Hasinger et al., 1993). This shows that we can confidently expect several thousand sources per square degree, i.e. about 1/sq.arcmin, or 250 per ACIS-I field. But what are these objects?

Whenever one can image $\frac{1}{100}$ times the detail of *any* previous telescope extraordinary things will be found. The step from ROSAT to AXAF is equal to that from ground-based astronomy to *Hubble Space Telescope*. *Hubble* images often give me the feeling of looking up the answer in the back of the book. These images tell us that there is structure on every scale in astrophysics. Many of the same *Hubble* objects are also bright X-ray sources. Surely they won't lose all structure when we look with X-rays?

To get a more terrestrial perspective on what better angular resolution can do, consider these images from spy satellites published in the New York Times (February 10 1997, page 1.) The ratio of resolution in the three images correspond (from top to bottom) to the ROSAT PSPC, the ROSAT HRI, and AXAF. While the ROSAT HRI shows a peculiar and puzzling

Figure 1. San Francisco Bay, SFO airport 747 at 30, 10, 1-meter resolution [N.Y.Times]

structure inside the large feature found in the PSPC, time-resolved imaging at AXAF resolution of the cross-shaped object in the lower panel makes everything clear. *Hubble* shows that astrophysics works the same way.

Choosing how to image a particular object with AXAF is complex. AXAF carries three imaging instruments with multiple modes: 2 types of CCD (P.I. G. Garmire, Penn State) optimized for (1) low energy (E<0.5 keV) response and high throughput (ACIS-S); (2) good energy resolution ($\Delta E \sim 80$ eV and large (16'×16') field-of-view (ACIS-I); and an HRC microchannel plate (P.I. S. Murray, SAO) which is best for fine detail (<0.5″), wide field ($\sim 25'$ dia.), and high time resolution (msec). The ACIS (CCD) instrument gives ~ 3-6 times, and the HRC $\sim$twice, the PSPC or SIS count rate.

The AXAF transmission grating spectrometers give the first high resolution data that other astronomers would call 'spectra', rather than broadband photometry, with useful area. The AXAF gratings have 100 times the spectral resolution of the ASCA SIS, ($R=E/\Delta E \sim 1000$) at 1 keV, and cover a seven times broader energy range, 0.07-10 keV. The line blending problems that limit current spectra are largely gone at this resolution, opening up fainter lines and so many physical diagnostics. The low energy (LETGS, P.I. B. Brinkmann, Utrecht) and high energy (HETGS, P.I. C. Canizares, MIT) AXAF gratings have 20-200 times greater area than their predecessors on *Einstein*. At high energies (E>0.5 keV) the HETGS gives count rates similar to the ROSAT PSPC. The LETGS includes a region (0.07<E<0.5 keV) only previously explored by the EUVE SWS spectrometer, and has an area some 10 times larger.

AXAF is the first X-ray telescope with good simultaneous spatial and spectral resolution. Each ACIS CCD chip will have 250,000 independent beam areas. (ASCA has perhaps 16.) Not that AXAF has the effective area to fill so many bins, but in a complex source it will be possible to isolate structures, even the sinuous shock fronts in supernova remnants and clusters of galaxies, and derive their distinctive spectra.

The transmission gratings have spatial resolution too. They are slitless spectrographs, familiar from optical prism surveys. The image of the source is diffracted, so e.g. a supernova remnant makes an image in each of the lines of its spectrum. Complex fields can provide enormous returns of data. A stellar cluster can yield dozens of spectra. This is no simple analysis task - the spectra overlap in space. Fortunately the CCD energy resolution provides a third axis, and in this data cube the spectra will almost all thread delicately past one another.

With AXAF's order of magnitude advances in angular resolution, in spectral resolution and in both at once we can expect surprises. Complex spectra will show up in unusual places; many spectra will show features that are simply unknown, since laboratory work has covered only a few of the transitions we will encounter with AXAF; and many sources now thought of as simple will show complex images, even whole new types of source. For example, *Hubble* has shown that the bright stars in the Orion Trapezium are surrounded by evaporating proto-planetary disks around nearby, newly forming stars (Bally et al., 1997). Since the bright stars are also bright X-ray sources, it is a simple prediction that these evaporating proto-planetary disks will be shining in fluorescent X-rays.

3. AXAF is a Culmination, AND a Beginning

Both technologically and scientifically AXAF is like *Hubble*: both achieved 10 times improved resolution by using heavy, rigid mirrors and were limited in area by the mirror weight. But both demonstrate that high resolution is possible, not end of the line. Scientifically, both let us see how complex, yet comprehensible, the universe is. There will be no going back to less resolution, once we have seen *Hubble* and AXAF images.

How do we get to more area with high resolution? Which axis should we push on first? A first step is HTXS. With $\sim$3 sq.meters of collecting area it pushes the area dimension hard, maintaining good spectral resolution. However angular resolution will be limited. What is next?

The stumbling block is X-ray optics: we need 10 sq.meters of effective area, yet must maintain arcsecond resolution (Elvis & Fabbiano 1996) and be light in weight. Work on this challenging goal is beginning in Europe, under the 'XEUS' banner (M. Turner, these proceedings). Discussions in the NASA community are just beginning. Certainly if we do not begin to develop the technology for such a mission X-ray astronomy will wait another 20 years before fulfilling the promise of the AXAF revolution.

4. Your Turn

After launch AXAF will quickly become a user driven observatory. PV and Cal. observations will be public at once. From month 5 onward 70% of the time will be for Guest investigators, increasing later to 85%. About 12 Msec of observing time is up for bids in the first NASA announcement of opportunity, with a deadline of 2 February 1998. A Proposers Guide and an AXAF simulator ('MARX') guide are available from the AXAF Science Center (http://asc.harvard.edu). AXAF is here at last. Enjoy it.

This work was supported in part by NASA contract NAS8-39073 (ASC).

Bally J., et al. 1997, http://www.cita.utoronto.ca/ johnston/orion.html
Elvis M., & Fabbiano G., 1996, in *'Next Generation X-ray Observatories'*, [U. Leicester], eds. M.J.L.Turner. & M.G.Watson XRA97/02, p. 33
Hasinger G. et al., 1993, A& A 288, 466

ABRIXAS

G. HASINGER

Astrophysikalisches Institut, D-14482 Potsdam

J. TRÜMPER

Max-Planck-Inst. f. extrat. Physik, D-85740 Garching

AND

R. STAUBERT

Institut für Astronomie und Astrophysik, Astronomie, D-72076 Tübingen

1. Introduction

All Sky Surveys are very useful for several reasons: Firstly, their unbiased view provides the potential for new discoveries. Secondly, they yield large unbiased samples of a variety of objects which can be used for statistical investigations. And thirdly, the unlimited field of view of telescopic sky surveys allows to study extended objects and large scale structures.

The first telescopic all sky survey in X-rays which was performed by ROSAT in the soft X-ray band (0.1-2.4 keV) has demonstrated all these advantages. The main objective of ABRIXAS (A BRoad Band Imaging X-ray All Sky Survey) will be to perform a telescopic sky survey at higher energies, between ~ 0.5 keV and ~ 12 keV (see Trümper, Hasinger & Staubert, 1998). Thus, compared with ROSAT, the energy range is shifted upwards by about a factor of five. By using a CCD detector ABRIXAS shall have a substantially better spectral resolution compared with the ROSAT PSPC, and a somewhat improved survey point-spread function (HEW < 1 arcmin).

2. Scientific Objectives

ABRIXAS will allow to detect several ten thousand X-ray sources among which should be at least 10.000 new sources, which are too absorbed to be detected in the ROSAT All Sky Survey. ABRIXAS will therefore have a role as pathfinder e.g. for the XMM and AXAF and ASTRO-E missions.

Figure 1. Artist's concept of ABRIXAS

ABRIXAS with its comparatively high spectral and angular resolution will also perform detailed spectroscopy of large-scale diffuse plasma sources like nearby supernova remnants, nearby clusters of galaxies or the diffuse emission region of the galactic ridge, which are too big to fit into the field of view of pointed X-ray telescopes.

Having identified the thermal galactic contributions through their line emission, ABRIXAS will obtain high-quality spectra and the angular distribution of the diffuse extragalactic background. It will be able to search for the redshifted iron line feature expected from the superposition of many active galactic nuclei as well as the fluctuations introduced by the large scale structure in the universe.

Finally, the regular pattern of the seven X-ray telescopes scanning the sky will produce information on the time variability of bright X-ray sources over time scales ranging from 10 seconds to 3 years.

Population synthesis models of the X-ray background (XRB), which are based on the unified AGN model, predict the XRB to be largely due to AGN with a wide distribution of intrinsic absorption column densities (Comastri et al., 1995). ABRIXAS will detect many of the nearby obscured AGN. The three year continuous X-ray survey of ABRIXAS will provide an unprecedented quality for the study of the X-ray background. Every piece of the sky will be covered many times by the surveys in seven different telescopes, thus allowing to correct for systematic errors. Therefore it will be possible to measure the energy spectrum of the X-ray background in a broad energy band with unprecedented quality, hopefully settling some of the systematic errors still hampering our current understanding. With the high statistical quality we will be able to search for the spectral feature around 2 keV, which is expected from the superposition of redshifted AGN iron lines (Matt and Fabian 1994).

The high spectral resolution will allow to discriminate the galactic thermal emission, which is thought to exist even at high galactic latitudes, from the isotropic extragalactic background. After removal of the galactic components we will be able to search for the XRB dipole due to the Compton-Getting effect. A multipole analysis of the residual background might be able to further constrain the power spectrum of cosmological density fluctuations. Finally, as soon as the future microwave background explorers (MAP and Planck Surveyor) will be operational, a crosscorrelation between the hard X-ray background and the high resolution microwave maps promises important cosmological results, e.g. on the integrated Sunyaev-Zeldovich effect and the X-ray/microwave source populations.

3. ABRIXAS Satellite and Payload

ABRIXAS is a scientific satellite project carried out by three institutes; the Astrophysical Institute Potsdam (AIP), the Institute of Astronomy and Astrophysics Tübingen (IAAT), and the Max Planck Institute for Extraterrestrial Physics (MPE). ABRIXAS will survey the whole sky in the way ROSAT did, by scanning it in great circles. The survey will take 3 years and provides full sky coverage in the energy band ~0.4 to 12 keV. The 3 years sensitivity will be comparable to that of the ROSAT all sky survey in the overlapping energy band, viz. at 1 keV. The angular resolution will be better than one arc min (see Döhring et al., 1998).

The optical system of ABRIXAS consists of seven Wolter type I mirror systems having 27 concentric shells each (see fig. 1). Their focal length is 160 cm. The optical axes of the seven telescopes are tilted with respect to each other by $7.25°$ fig. 1a. The corresponding fields of views form a hexagonal pattern in the sky, c.f. fig. 1b (see Friedrich et al., 1996).

The 7 telescopes share one CCD chip of 6x6 cm2 as the imaging detector. The chip used for ABRIXAS as well as the whole detector including electronics is identical with the EPIC-Maxi detector developed by MPE for XMM. However, because of the low earth orbit of ABRIXAS the CCD chip will be operated at a somewhat higher temperature. Therefore, the lowest ABRIXAS energy will be 0.5 keV (possibly 0.3 keV) instead of 0.1 keV for XMM. Unique features of the pn-CCD are its large sensitivity at high energies (95 % at 10 keV), and its ~70 millisecond time resolution in the full frame mode achieved by a parallel readout. This is important for

TABLE 1. ABRIXAS Summary

Mirror Systems	7 Wolter, 27 electroformed Nickel shells per system
Detector	$6 \times 6\ cm^2$ XMM pn-CCD, Pixel size 150μ, shared FOV
Operating Temperature	$-80\ ^oC$, ambient cooling
Spacecraft	3-axis angular momentum stabilized
Attitude Control	1 Momentum wheel, 3 magnetic torquers
Attitude Sensors	2 sun sensors, 3-axis magnetometer, 1 laser gyro
Attitude Determination	2 star trackers, GPS
Size	$2.5 \times 1.8 \times 1.2 m^3$
Mass	~ 460 kg
Power	~ 200 W
Orbit Height	~ 580 km circular, Inclination 51°
Launch	COSMOS rocket, Spring 1999 from Kapustyn Yar
Lifetime (goal)	3 years
Number of Surveys	6, Average exp. time 4.000 s

a scanning mission like ABRIXAS to avoid source blurring along the scan direction (70 ms correspond to a scan path of 17 arc sec).

ABRIXAS is a small (low cost) satellite with a mass of 460 kg to be launched by a Cosmos rocket purchased from Polyot. The satellite is developed and built by OHB Bremen with ZARM as the subcontractor for the attitude measurement and control system, c.f. Table 2. The X-ray mirrors are produced by Carl Zeiss and tested in MPE's Panter facility. The CCD camera is developed and built at the MPI/MPE Semiconductor Laboratory with parts of the readout electronics being provided by IAAT. The ground station will be operated by DLR/GSOC, while the overall project management will be done by DLR as the successor of DARA. The launch of ABRIXAS will take place in spring 1999 from Kapustyn Yar, the Russian equivalent of Huntsville, which is located east of Volgograd.

Acknowledgements

We are grateful for the enthusiastic support by the ABRIXAS teams in the funding agency, in industry and in the scientific institutes.

References

Comastri A., Setti G., Zamorani G., Hasinger G. 1995, A&A 296, 1
Döhring T., Friedrich P., Aschenbach B., et al., 1998, AN 319, in press
Friedrich P., Hasinger G., Richter G., et al., 1996, MPE report 263, 681
Matt G., Fabian A.C., 1994, MNRAS 267, 187
Trümper, J.E., Hasinger, G., Staubert, R., 1998, AN 319, in press

THE INTEGRAL MISSION

W. HERMSEN[1] AND C. WINKLER[2]
[1] *Space Research Organization Netherlands*
Sorbonnelaan 2, NL-3584 CA Utrecht, The Netherlands
[2] *ESA/ESTEC, Space Science Department, Astrophysics Div.*
NL-2200 AG Noordwijk, The Netherlands

On behalf of the INTEGRAL Science Working Team

1. Introduction

The International Gamma-Ray Astrophysics Laboratory (INTEGRAL) is dedicated to the fine spectroscopy (ΔE: 2 keV FWHM @ 1 MeV) and fine imaging (angular resolution: 12' FWHM) of celestial gamma-ray sources in the energy range 15 keV to 10 MeV. INTEGRAL was selected in 1993 as the next ESA medium-size scientific mission (M2) to be launched in 2001. ESA has the overall spacecraft and mission resposibilities, Russia will provide a PROTON launcher and launch facilities, and NASA will provide ground station support through the Deep Space Network. The scientific payload complement and the INTEGRAL Science Data Centre (ISDC) will be provided by large collaborations led by Principal Investigators (PI).

The INTEGRAL observatory will provide to the science community at large an unprecedented combination of imaging and spectroscopy over a wide range of X-ray and gamma-ray energies including optical monitoring.

2. Scientific Objections

INTEGRAL is a 15 keV - 10 MeV gamma-ray mission with concurrent source monitoring at X-rays (3 - 35 keV) and in the optical range (550 - 850 nm). All instruments - co-aligned with large FOV's - cover simultaneously a very broad energy range when observing high energy sources.

The scientific goals of INTEGRAL will be attained by fine spectroscopy (ΔE: 2 keV FWHM @ 1 MeV) with fine imaging (angular resolution: 12' FWHM) and accurate positioning of celestial sources of gamma-ray emission. Fine spectroscopy over the entire energy range will permit spectral features to be uniquely identified and line profiles to be determined for physical studies of the source region. The fine imaging capability of INTEGRAL within a large field of view will permit the accurate location and hence identification of the gamma-ray emitting objects with counterparts at other wavelengths, enable extended regions to be distinguished from point sources and provide considerable serendipitous science. In summary, the scientific topics to be addressed include: • Compact Objects (White Dwarfs, Neutron Stars, Black Hole Candidates) • Stellar Nucleosynthesis (Hydrostatic Processes, Supernovae, Novae) • High Energy Transients • Galactic Structure (Cloud Complex Regions, Mapping of continuum and line emission, ISM, CR distribution) • The Galactic Centre • Particle Processes and Acceleration • Transrelativistic Pair Plasmas • Extragalactic Astronomy (Nearby Galaxies, Clusters, AGN, Cosmic Diffuse Background) • Identification of High Energy Sources • Unidentified Gamma-Ray Objects as a Class • PLUS: Unexpected Discoveries.

3. Scientific Payload

The INTEGRAL payload consists of two main gamma-ray instruments: Spectrometer SPI and Imager IBIS, and of two monitor instruments, the X-ray Monitor JEM-X and the Optical Monitoring Camera OMC.

The design of the INTEGRAL instruments is largely driven by the scientific requirement to achieve - to the maximum extent possible - complementarity in fine spectroscopy and accurate imaging. Therefore, the main gamma-ray instruments, SPI and IBIS, are differently optimised in order to complement each other and to achieve overall excellent performance. This optimisation also takes recent observations at high energies into account, which show that - in general - line emissions do occur on a wide range of angular and spectral extent: That is, broad lines seem preferably to be emitted from point-like sources and narrower lines from extended sources. The INTEGRAL continuum and line sensitivities are increased by an order of magnitude or more compared to succesful earlier instruments like SIGMA (continuum) and OSSE and COMPTEL (lines). The two monitor instruments (JEM-X and OMC) will provide complementary observations at X-ray and optical energy bands. A short overview of the INTEGRAL payload follows:

• The *Spectrometer SPI* will perform spectral analysis of gamma-ray point sources and extended regions with an energy resolution of 2 keV (FWHM) at 1 MeV. This will be accomplished using an array of 19 hexagonal high purity Germanium detectors cooled by active cooling to an operating temperature of 85 K. The total detection area is 500 cm^2. A hexagonal coded aperture mask is located 1.7 m above the detection plane in order to image large regions of the sky (fully coded field of view = 16°) with an angular resolution of 2°.

In order to reduce background radiation, the detector assembly is shielded by an active BGO veto system which extends around the bottom and side of the detector almost completely up to the coded mask.

• The *Imager IBIS* provides powerful diagnostic capabilities of fine imaging (12 arcmin FWHM), source identification and spectral sensitivity to both continuum and broad lines over a broad (15 keV - 10 MeV) energy range. The energy resolution is < 7 keV @ 0.1 MeV and 60 keV @ 1 MeV. A tungsten coded aperture mask (located at 3.2 m above the detection plane) is optimised for high angular resolution imaging. Sources ($> 10\sigma$) can be located to $< 60''$. As diffraction is negligible at gamma-ray wavelengths, the angular resolution obtainable with a coded mask telescope is limited by the spatial resolution of the detector array. The IBIS design takes advantage of this by utilising a detector with a large number of spatially resolved pixels, implemented as physically distinct elements.

The detector uses two planes, a front layer (2600 cm^2) of CdTe pixels, each (4x4x2) mm (wxdxh), and a second one (3100 cm^2) of CsI pixels, each (9x9x30) mm. The aperture is restricted by a passive tungsten shield. The detector array is shielded in all other directions by a BGO scintillator veto system.

• The Joint European *X-Ray Monitor JEM-X* supplements the main INTEGRAL instruments and plays a crucial role in the detection and identification of the gamma-ray sources and in the analysis and scientific interpretation of INTEGRAL gamma-ray data. JEM-X will make observations simultaneously with the main gamma-ray instruments and provides images with 3' angular resolution in the 3 - 35 keV prime energy band.

The baseline photon detection system consists of two identical high pressure imaging microstrip gas chambers (Xenon at 5 bar) each viewing the sky through a coded aperture mask (4.8° fully coded FOV), located at a distance of 3.2 m above the detection plane. The total detection area is 1000 cm^2.

• The *Optical Monitoring Camera OMC* consists of a passively cooled CCD in the focal plane of a 50 mm lens. The CCD (1024 x 2048 pixels) uses one section (1024 x 1024 pixels) for imaging, the other one for frame transfer before readout. The FOV is 5° x 5° with a pixel size of 17.6''. The OMC will observe the optical emission from the prime targets of the INTEGRAL main gamma-ray instruments with the support of the X-Ray Monitor JEM-X. OMC offers the first opportunity to make long observations in the optical band simultaneously with those at X-rays and gamma-rays. Variability patterns ranging from 10's of seconds, hours, up to months and years will be monitored. The limiting magnitude will be 19.2^{m_v} (3σ, 10^3 s), which corresponds to $\sim$40 photons cm^{-2}s^{-1}keV^{-1} (@ 2.2 eV) in the V-band. Multi-wavelength observations are particularly important in high-energy astrophysics where variability is typically rapid. The wide band observing opportunity offered by INTEGRAL is of unique importance in providing for the first time simultaneous observations over seven orders of magnitude in photon energy for some of the most energetic objects.

4. INTEGRAL Science Data Centre

The ISDC, located in Versoix, Switzerland, will receive the complete raw science telemetry plus the relevant ancillary spacecraft data from the Mission Operation Centre in Darmstadt. Science data will be processed, taking into account the instrument characteristics, and raw data will be converted into physical units. Using incoming science and housekeeping information, the ISDC will routinely monitor the instrument science performance and conduct a quick-look science analysis. Most of the Targets of Opportunity (TOO) showing up during the lifetime of INTEGRAL will be detected at the ISDC during the routine scrutiny of the data. Final data products obtained by standard analysis tools will be distributed to the observer and archived for later use by the science community. Facilities will be provided to support the science community in the analysis of INTEGRAL data.

5. Observing Programme

INTEGRAL will be an observatory-type mission with a nominal lifetime of 2 years, an extension up to 5 years is technically possible. The parameters for the baseline orbit (PROTON launcher) are: period 48 hours, inclination 51.6°, perigee height 46 000 km, apogee height 75 000 km. In fact, 100% of time in orbit can be used for observations. Most of the observing time (65% during year 1, 70% (year 2), 75% (year 3+)) will be awarded to the scientific community at large as the General Programme. Proposals, following a standard AO process, will be selected on their scientific merit only by a single Time Allocation Committee. The remaining fraction of the observing time (i.e. 35% (year 1), 30% (year 2), 25% (year 3+)) will be reserved, as guaranteed time, for, mainly: (i) the institutes (PI collaborations) which have developed and delivered the instruments and the ISDC (guaranteed PI time), and (ii) for Russia and NASA for their contributions to the programme (PROTON launcher and Deep Space Network ground stations).

For more detailed information on INTEGRAL please consult the INTEGRAL pages on the WWW:

http://astro.estec.esa.nl/SA-general/Projects/Integral/integral.html

Session 3: Diagnostics of High Gravity Objects with X- and Gamma Rays

3-1. White Dwarfs and Neutron Stars

SUPER-EDDINGTON SOURCES IN GALAXIES

G. FABBIANO
Harvard-Smithsonian Center for Astrophysics
60 Garden St., Cambridge MA, 02138, USA

1. Introduction

Imaging X-ray observations have resolved X-ray sources in many nearby 'normal' galaxies. These sources are associated with different galaxian component, including: bulge and globular clusters(LMXB, e.g. in M31); the disk/arm component (LMXB, HMXB, SNR, SN, ISM shells); and low-luminosity AGN (see reviews in Fabbiano 1989; 1995; 1996). A number of these sources have X-ray luminosities in excess of the Eddington luminosity for 1 $M_\odot$ accreting object ($\sim 1.3 \times 10^{38}$ergs s^{-1}), and have been called Super-Eddington sources. Nearly 100 Super-Eddington sources have been reported so far, based on *Einstein* and *ROSAT* images (see Fabbiano 1989; 1995; 1996; Makishima 1994). The number of these sources increases with the maximum distance of the galaxy sample considered (e.g. Read et al 97). This is in part due to confusion, but it is also true that considering the larger sample of galaxies one obtains in a larger volume of space the amount of these relatively rare sources would increase as well.

Super-Eddington sources may occur in nuclear regions as well as elsewhere in a galaxy. They can reach luminosities of 10-100$\times L_{Edd}$. If they are single accreting objects, this points to $\geq 10 M_\odot$ black holes.

2. Nuclear Super-Eddington Sources

Bright point-like X-ray sources are normally associated with AGN. What we are concerned here with are sources with X-ray luminosities $\sim 10^{40}$ergs s^{-1}. This type of sources is sometimes associated with low-luminosity Seyfert or LINER nuclei (e.g. NGC4594, Fabbiano and Juda 1997), and their likely explanation is sub-Eddington accretion on a very massive nuclear black hole ($M \sim 10^{7-8} M_\odot$).

However, there are at least two instances in which the nature of the nuclear source is not totally understood, and a Super-Eddington accretion binary is a possibility. One such instance is the bright ($L_X \sim 10^{39}$ergs s^{-1}) source at the nucleus of M33 (Long et al 1981). This source was seen to vary with *Einstein* (Peres et al 1989), and later ASCA spectra showed a remarkable similarity with the spectra of galactic black hole candidates rather than those of AGN (Takano et al 1994). The variable X-ray source at the nucleus of NGC 3628 (Dahlem et al 1995) is the other example, although in this case the ASCA spectra may favor an AGN interpretation (Yaqoob et al 1995).

3. Non-nuclear Super-Eddington Sources

The majority of Super-Eddington sources in spiral galaxies is not in the nuclei. Before we speculate on their nature, we need to understand if these sources are individual point-like sources, unresolved extended emission regions, or unresolved source complexes. We have examples of all these categories, sometimes occurring in the same galaxy. In the Scd galaxy IC 342, 3 Super-Eddington sources ($L_X \sim 10^{39}$ergs s^{-1}) were discovered with the *Einstein* IPC (Fabbiano and Trinchieri 1987). One of these sources - at the nucleus - was later resolved into three sub-Eddington components plus diffuse emission with a higher resolution *ROSAT* HRI observation (Bregman et al 1993). However, another source (IC 342 A) was later found to vary with ASCA, confirming its nature as individual source (Okada et al 1994). The temporal/spectral characteristics of this source are very similar to those of LMXB, suggesting that we are indeed in the presence of an accreting black hole in a binary system (Makishima 1994). Variability also points to single objects (and thus black hole candidates) in source B of NGC 1313 (Petre et al 1994), and source X-3 in the Antennae galaxy (Fabbiano et al 1997). In the Antennae galaxy (the merging pair NGC 4038/4039), 12 sources, all more luminous than $4 \times 10^{39} ergs\ s^{-1}$ were detected with the *ROSAT* HRI, including the variable source X-3. Although 3 of these sources look point-like, the majority is likely to be part of complex emission regions, and to be perhaps due to bubbles of hot gas in star-forming regions (Fabbiano et al 1997).

There is a class of Super-Eddington sources, which are convincingly identified with SN or young SNR (see Fabbiano 1996). They include SN1980k in NGC 6946 (Canizares et al 1982), SN1978k in NGC 1313 (Schlegel et al 1996), SN1986j in NGC 891 (Bregman & Pildis 1992), and SN1993j in M81 (Zimmerman et al 1994). The large X-ray luminosity of these sources can be explained with either the interaction of a reverse shock with the ejecta, or the interaction of the SN shock with surrounding clouds (Schlegel et al

1996).

4. Future Prospects

Super-Eddington sources are clearly exceptional and interesting objects. To progress in our understanding of their nature we need to be able to study their temporal and spectral behaviour and to associate them accurately with galaxian features.

With the imminent AXAF launch (27 Aug. 1998), we are in a very good position to study these sources. We will be able to obtain spectra and spatial information with good statistics from sources in nearby galaxies (at a distance of 5 Mpc, 10^{39} ergs s^{-1} correspond to 2000 cts/10ks, and the 1/2 arcsec AXAF beam corresponds to a linear size of 12 pc). We will be able to identify accurately any such source with galaxian features out to 50 Mpc, where the AXAF beams corresponds to 120 pc.

XMM (due to be launched in Dec. 1999) will have four times the AXAF count rate, and thus will give us very good spectra and timing data on sources in nearby galaxies, but its beam is 15 arcsec, 30 times that of AXAF. This larger beam means that confusion becomes a serious problem: at a distance of 50 Mpc, this beam corresponds to a significant fraction of a galaxy (3.6 kpc). HTXS, which is currently in planning in the USA, with a projected launch around year 2007, is expected to have five times the XMM throughput, and a similar beam. Thus, while very detailed studies of sources in nearby galaxies will be possible, confusion will impede studying sources in galaxies outside of a few Mpc radius.

To really move forward in this field, we need to combine high throughput with high angular resolution. A Large X-ray Telescope, with a ~ 10 sqmeter collecting area and arcsec resolution (see Elvis and Fabbiano 1997) would open up the study of Super-Eddington sources with CCD-type spectra (1000 cts) out to 50 Mpc. This prospect is probably closer than I believed before attending this meeting. ESA is taking an active role in pushing forward such a mission (XEUS, see Turner, this volume).

This work was supported by NASA Contract NAS 8-39073 (AXAF Science Center).

References

Bregman, J.N., Cox, C.V., Tomisaka, K. 1993, Ap. J., 415, L79
Bregman, J.N., and Pildis, R.A. 1992, Ap. J., 380, L107
Canizares, C.R., Kriss, G.A., Feigelson, E.D. 1982, Ap. J., 253, L17
Dahlem, M., Heckman, T.M., Fabbiano, G. 1995, Ap. J., 442, L49
Elvis, M., and Fabbiano, G. 1997, in 'Next Generation X-ray Observatory', Workshop Proc., Univ. of Leicester, M.D.L. Turner ed., p. 33

Fabbiano, G. 1989, ARA&A, 27, 87
Fabbiano, G. 1995, in 'X-ray Binaries', W.H.G. Lewin et al eds., (Cambridge: University Press), p.390
Fabbiano, G. 1996, in 'Roentgenstrahlung from the Universe', MPE Report No.263, H.U. Zimmerman, J.E. Truemper and H. Yorke eds., p.347
Fabbiano, G., and Juda, J.Z. 1997, Ap. J., 476, 666
Fabbiano, G., Schweizer, F., Mackie, G. 1997, Ap. J., 478, 542
Fabbiano, G. and Trinchieri, G. 1987, Ap. J., 315, 46
Long, K.S., D'Odorico, S., Charles, P.A., Dopita, M.A. 1981, Ap.J., 246, L61
Okada, K, Mihara, T., and Makishima, K. 1994, in 'Horizon of X-ray Astronomy', F. Makino and T. Ohashi eds., (Tokyo: Universal Academy Press), p.515
Makishima, K. 1994, in 'Horizon of X-ray Astronomy', F. Makino and T. Ohashi eds., (Tokyo: Universal Academy Press), p. 171
Peres, G., Reale, F., Collura, A., Fabbiano, G. 1989, Ap. J., 336, 140
Petre, R., Okada, K., Mihara, T., Makishima, K., and Colbert, E.J.M. 1994, PASJ, 46, L115
Read, A.M., Ponman, T.J., Strickland, D.K. 1997, MNRAS, 286, 626
Schlegel, E.M., Petre, R., Colbert, E.J.M. 1996, Ap. J. 456, 187
Takano, M., Mitsuda, K., Fukazawa, Y., Nagase, F. 1994, Ap. J., 436, L47
Yaqoob, T., Serlemitsos, P.J., Ptak, A., Mushotzky, R., Kunieda, H. and Terashima, Y. 1995, Ap. J., 455, 508
Zimmerman, H.U. et al 1994, Nature, 367, 621

MASS MEASUREMENT OF ACCRETING MAGNETIC WHITE DWARFS WITH HARD X-RAY SPECTROSCOPY

M. ISHIDA AND R. FUJIMOTO
Institute of Space and Astronautical Science
3-1-1 Yoshinodai, Sagamihara, Kanagawa 229, JAPAN

1. Introduction

Accreting magnetic white dwarfs are usually found as component stars in Magnetic Cataclysmic Variables (MCVs), in which a white dwarf with $B = 10^{5-8}$ G accepts mass from a late type (secondary) star via Roche Lobe overflow. Matter from the secondary is funneled by the magnetic field and concentrates on the magnetic pole(s) of the white dwarf. Since the accretion flow becomes highly supersonic, a standing shock wave is formed close to the white dwarf. The temperature of the plasma at the shock front reflects the gravitational potential and can be denoted as a function of the mass (M) and the radius (R) of the white dwarf as:

$$kT_S = \frac{3}{8} \frac{GM}{R} \mu m_H = 22 \left(\frac{M}{0.6 M_\odot}\right) \left(\frac{R}{8.7 \times 10^8 \text{cm}}\right)^{-1} \quad [\text{keV}]. \quad (1)$$

Note here that the height of the shock is expected to be within 10 % of the white dwarf radius, and hence neglected here.

As the equation of state of the degenerate electron is well studied, M-R relation of the white dwarf is established (Hamada & Salpeter 1961, Nauenberg 1972). We can thus determine the mass and the radius of the white dwarf if we successfully measure T_S from the X-ray spectroscopy.

However, since the post-shock plasma is cooled by the optically thin thermal plasma emission as it descends the accretion column, the continuum spectrum cannot be represented by a single temperature thermal bremsstrahlung (Aizu 1973). The continuum spectral shape becomes further complicated because of the photoelectric absorption in the pre-shock accretion column. As the photons from the plasma travel along different paths through the accretion column, the absorption cannot be represented by a single column density (Ishida, Fujimoto 1995). In addition, it is also revealed that the reflection from the white dwarf surface makes significant contribution to the observed spectrum. As easily imagined, it is very difficult to evaluate all these effects consistently and derive T_S (although significant progress was made recently; Cropper et al. 1997, Done, Magdziarz 1997).

In this paper, we present a new method of measuring T_S and hence the mass of the white dwarf by means of the $K\alpha$ emissions lines from abundant heavy elements. The essential ideas presented in § 2 are already summarized in Fujimoto, Ishida (1997) in detail. In § 3, comparison of the masses so far determined with our method to those from optical observations are summarized. Finally, concluding remarks are given in $ 4.

2. Method and Results

Since the emission from the post-shock plasma is optically thin, the observed intensity of the $K\alpha$ emission line from the element Z at the ionization state z is expressed as

$$I_{Z,z}(T_S, T_B, A_Z, S, D) = \frac{S}{4\pi D^2} \int_{T_B}^{T_S} A_Z \varepsilon_{Z,z}(T) \, n(z(T))^2 \frac{dz}{dT} \, dT, \qquad (2)$$

where T_B is the temperature at the base of the optically thin accretion column, A_Z, S and D are the abundance of the element Z in the unit of Solar, cross section of the accretion column, and the distance to the source, respectively. $\varepsilon_{Z,z}(T)$ is the volume emissivity (photons cm^3 s^{-1}) of $K\alpha$ line, which is tabulated by Mewe et al. (1985). Density and temperature profiles of the accretion column is calculated by Aizu (1973) by assuming pure thermal bremsstrahlung cooling. Although eq.(2) depends upon several uncertain factors such as A_Z, S and D, we can eliminate all of them by using the intensity ratio of the hydrogenic to the He-like $K\alpha$ line of the same atom. Since this ratio $R_Z = I_{Z,H}/I_{Z,He}$ depends only upon T_S and T_B, we can constrain them if we can measure R_Z for two elements at least.

In Fig. 1, we show the allowed area of T_S and T_B obtained by the $K\alpha$ emission line ratios resolved by the *ASCA* SIS for the three MCVs — EX Hya, AO Psc and V1223 Sgr. The Masses of the white dwarf determined from eq.(1), referring to the resulting range of T_S, are labeled on top of each panel (Fujimoto 1996).

3. Comparison with Optical Measurements

We have successfully measured the mass of the white dwarf with the hard X-ray line spectroscopy. Our method just uses the depth of the gravitational potential at the surface of the white dwarf, and hence potentially applicable to all the MCVs. Masses of binary components are usually measured by optical line velocity measurements. This is, however, reliable only for eclipsing systems, because of uncertainty of the inclination angle. Our method is, of course, free from this uncertainty.

To establish reliability of our method, it is necessary to compare the results with those from optical line velocity measurements in eclipsing MCVs. A remarkable agreement is already obtained for EX Hya from which velocity amplitudes of the white dwarf and the secondary are both measured to be 69 ± 9km s^{-1} (Hellier et al. 1996) and 356 ± 4km s^{-1} (Smith et al. 1993). With these numbers, the mass of the white dwarf in EX Hya is obtained to be $0.49 \pm 0.03 M_\odot$, which should be compared with $0.48^{+0.10}_{-0.06} M_\odot$ shown in Fig. 1.

There are two more eclipsing MCVs whose X-ray intensity is strong enough for the line spectroscopy. One of them is XY Ari which is discovered by *Ginga*

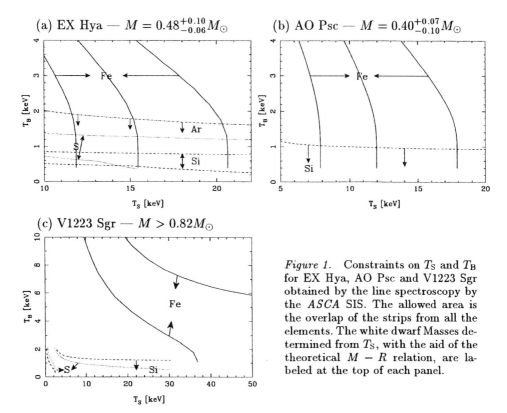

Figure 1. Constraints on T_S and T_B for EX Hya, AO Psc and V1223 Sgr obtained by the line spectroscopy by the ASCA SIS. The allowed area is the overlap of the strips from all the elements. The white dwarf Masses determined from T_S, with the aid of the theoretical $M - R$ relation, are labeled at the top of each panel.

(Koyama et al. 1991, Kamata, Koyama 1993). Fig. 2 show (a) the X-ray spectrum obtained by the SIS with roughly 90 ks observation, and (b) the allowed range for T_S and T_B. As XY Ari is behind the molecular cloud Lynds 1457, the low energy X-ray intensity is attenuated. As a result, we can obtain only the line intensity ratio of iron. Assuming that the plasma is cooled in the same manner as in the sources shown in Fig. 1, we have tentatively obtained the mass of the white dwarf in XY Ari to be $1.1^{+0.3}_{-0.5} M_\odot$ as shown in Fig. 2(b). Note that the upper boundary is Chandrasekhar limit. On the other hand, there is no optical measurement yet of the mass of the white dwarf in XY Ari. Optical observation is strongly encouraged.

The other candidate MCV is 1H1752−081 (V2301 Oph), which is strong enough but not observed yet by ASCA. The mass of the white dwarf is estimated to be 0.5–$0.8 M_\odot$ by the optical observation (Silber et al. 1995).

4. Concluding Remarks

We have successfully measured the masses of the white dwarf in the three MCVs — EX Hya, AO Psc, V1223 Sgr, and possibly in XY Ari. All these sources belong to so-called intermediate polars, in which the magnetic field of the white dwarf is relatively weak. In polars, which include a stronger field white dwarf, optical cyclotron emission dominates over the thermal bremsstrahlung close to the shock front. In this case, T_S and hence the mass measured from the hard X-ray line

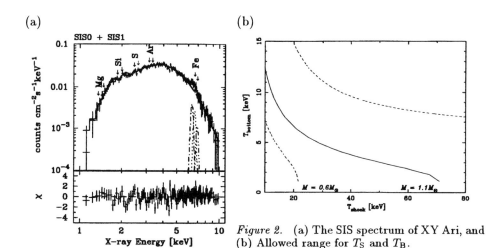

Figure 2. (a) The SIS spectrum of XY Ari, and (b) Allowed range for T_S and T_B.

spectroscopy become lower limits to them. To evaluate the contribution of the cyclotron emission relative to the thermal bremsstrahlung, it is necessary to measure the density of the post-shock hot plasma, which will be possible by the *ASTRO-E* micro-calorimeter.

The other MCVs are generally too faint for *ASCA* to perform a high quality X-ray spectroscopy. Mass determination of such faint objects with our method will be taken over to the next generation X-ray astronomy satellites — *AXAF*, *XMM* and *ASTRO-E*.

References

Aizu, K. (1973) *Prog. Theoret. Phys.*, **49**, 1184
Cropper, M., Ramsay, G. and Wu, K. (1997) *MNRAS*, in press
Done, C. and Magdziarz, ?? (1997) *MNRAS*, submitted
Fujimoto, R. and Ishida, M. (1997) *ApJ*, **474**, 774
Fujimoto, R. (1996) Ph.D. Thesis, University of Tokyo
Hamada, T. and Salpeter, E. E. (1961) *ApJ*, **134**, 683
Hellier, C. (1996) in *IAU Colloq. 158, Cataclysmic Variables and Related Objects*, eds. A. Evans and J. H. Wood, Kluwer, Dordrecht, p. 143
Ishida, M. and Fujimoto, R. (1995) in *Cataclysmic Variables*, eds. A. Bianchini, M. Della-Valle, M. Orio, Kluwer, Dordrecht, p. 93
Kamata, Y. and Koyama, K. (1993), *ApJ*, **405**, 307
Koyama, K., Takano, S., Tawara, Y., Matsumoto, K., Noguchi, K., Fukui, Y., Iwata, T., Ohashi, N., Tatematsu, K., Takahashi, N., Umemoto, T., Hodapp, K. W., Rayner, J. and Makishima, K. (1991) *ApJ*, **377**, 240
Mewe, R., Gronenschild, E. H. B. M. and van den Oord, G. H. J. (1985) *A&A Suppl.*, **62**, 197
Nauenberg, M. (1972) *ApJ*, **175**, 417
Silber, A. D., Remillard, R. A., Horne, K. and Bradt, H. V., (1994) *ApJ*, **424**, 955
Smith, R. C., Collier-Cameron, A. and Tucknott, D. S. (1993) in *Cataclysmic Variables and Related Objects*, eds. O. Regev and G. Shaviv (Ann. Israel Phys. Soc., Vol.10). p. 70

PHOTOIONIZED PLASMAS IN X-RAY BINARY PULSARS: ASCA OBSERVATIONS

FUMIAKI NAGASE
*The Institute of Space and Astronautical Science,
3-1-1 Yoshinodai, Sagamihara, Kanagawa 229, Japan*

1. Introduction

Massive X-ray binary pulsars have often evolved early-type companion stars which emanate strong stellar winds. X-rays emitted from the accreting neutron star irradiate and ionize the surrounding stellar wind, thus forming a photoionized sphere surrounding the neutron star. The photoionization structure of matter surrounding the neutron star was calculated by Hatchett and McCray (1977) and McCray et al. (1984), for Cen X-3 and Vela X-1 respectively.

Emission lines from He-like and hydrogenic ions are expected to be emitted from such plasmas due to radiative recombination followed by cascades. The spectroscopy of emission lines in X-ray binary systems is useful for investigating the ionization structure of the photoionized region surrounding the neutron star.

Before the *ASCA* observations, however, only an iron $K\alpha$ emission line at 6.4 keV was observed for several X-ray pulsars (e.g., Nagase 1989). Using data obtained with the SIS sensors (CCD cameras) onboard *ASCA*, emission lines from various elements, such as Mg, Si, S, Ar, Ca, and Fe are detectable and even $K\alpha$ lines from ions of different ionization states become resolvable owing to the high sensitivity and energy resolution of the CCD cameras.

2. Emission Lines from Photoionized Plasma

A prominent fluorescent $K\alpha$ line was detected at 6.40 keV in the spectrum of GX 1+4, yielding the estimate for the ionization degree of iron to be lower than Fe IV (Kotani et al 1997). From the SIS spectrum of GX 301−2 it was found that this source exhibits fluorescent Kα lines of Si, S, Ar, and Ca as well the Kα line of iron (Saraswat et al. 1996). This result yields an estimate for the ionization parameter of the effective emission site to be $\log \xi \approx 1.5$, thus suggesting a cold matter for reprocessing.

In contrast to these results, an interesting feature of the emission lines was observed in the spectrum of Vela X-1, which was observed during the eclipse phase (Nagase et al. 1994). The spectrum unfolded with the best-fit model is shown in Figure 1.

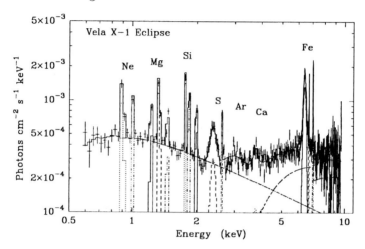

Figure 1. The *ASCA* unfolded SIS spectrum of Vela X-1 obtained during eclipse phase. The model consists of two component continua and fifteen lines.

From this spectrum we can resolve 15 lines superposed on the flat continuum which can be modeled by two (with high and low absorption column densities) power laws. Kα lines of Ne, Mg, Si, S, Ar, Ca, and Fe are detected and mostly dominated by the recombination lines of He-like ions for these elements, except for the iron for which the 6.4-keV fluorescent Kα line by cold matter is dominant. This result suggests that radiative recombination followed by cascades is the dominant process in the X-ray irradiated stellar wind, since such a stellar wind forms highly ionized zones of He-like and hydrogenic ions with a relatively low electron temperature of about 100 eV (Kallman & McCray 1982). From the present result, the photoionization parameter of the effective emission site is estimated to be $\log \xi \approx 2.3$, which is consistent with the calculation by McCray et al. (1984).

From the SIS spectrum of Cen X-3 obtained with *ASCA*, three iron lines are clearly resolved at 6.4 keV, 6.7 keV, and 6.96 keV, identifying the fluorescent Kα line of cold matter, and the recombination Kα lines of He-like and hydrogenic ions of iron, respectively. Interestingly, the relative intensities of the three lines change along the orbital phase. In addition to the iron lines, lines of Ne, Mg, Si, and S are also observed at lower energies and these are identified to the recombination Kα lines of hydrogenic ions. From the *ASCA* spectrum, the ionization parameter of the effective emission site is estimated as $\log \xi \approx 3.4$ for Cen X-3 (see Ebisawa et al. 1996 for details). Hence, the degree of photoionization of Cen X-3 is much higher than that of Vela X-1. The ξ-value derived is consistent with the calculation by Hatchett and McCray (1977).

The different emission line features typically seen in the three HMXB pulsars, GX301−2, Vela X-1, and Cen X-3 can be understood by the difference of the binary size and the X-ray luminosity of the source (see e.g., Nagase 1997).

3. Recombination Continuum

In the above consideration of Vela X-1 spectrum, we did not mention the emission due to the transition of free electrons directly to the ground level of the He-like or hydrogenic ions. The probabilities of these free-bound transitions are not negligible compared with the cascade recombination probabilities. Since the electron temperature of the photoionized stellar wind is as low as ~ 100 eV, such free-bound transitions will produce broad emission-line-like features at the recombination edges.

The broad features at the S and Ar bands (2.4 and 3.0 keV) in the Vela X-1 spectrum (see Fig. 1) are the evidence of the free-bound emission of He/H-like ions of Si and S. A broad emission line feature observed at 1.42 keV in the 4U1626−67 spectrum is consistent with the free-bound emission of hydrogenic neon (Angelini et al. 1995). Better and clearer evidence of the recombination continuum emission that mimics a line feature was obtained from the Cyg X 3 spectrum (Liedahl and Paerels 1996; Kawashima and Kitamoto 1996). From these broad line features the plasmas surrounding these system are confirmed to be in highly photoionized states due to X-ray irradiation.

4. 4U1538−52 and OAO1657−415

4U1538−52 and OAO1657−415 are eclipsing X-ray binary pulsars with binary parameters and X-ray lunminosities similar to those of Vela X-1. *ASCA* observed the two pulsars during their eclipses with the hope that they would show recombination-lines dominated spectra similar to that of Vela X-1.

As seen in Figure 2, the observed spectra are both dramatically different from each other and from the Vela X-1 spectrum (Fig. 1) against expectation. Excess of soft X-ray continuum is seen in the 4U1538−52 eclipse spectrum. This is due to contamination by interstellar dust scattering. No conspicuous line was detected at Mg, Si or S. Only an intense 6.4-keV fluorescent line of iron was detected and He-like and hydrogenic Ne lines are marginally seen (see Nagase et al. 1997 for details).

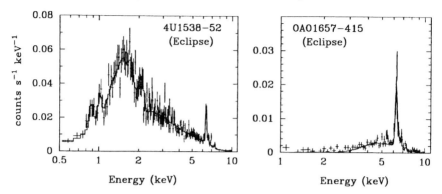

Figure 2. The *ASCA* SIS spectra of 4U1538−52 and OAO1657−415 obtained during their eclipse phases. The solid lines represents the best fit model to the data.

The spectrum of OAO1657−415 observed during eclipse shows a further drastically different feature. It exhibits only an intense iron line at 6.4 keV. Even the scattering continuum is weak and highly absorbed at energies below 3 keV. No evidence of recombination lines of Mg, Si, and S is seen.

These differences in eclipse spectra may be due to the delicate balance of their circumstellar environments, such as the wind density, the X-ray luminosity and chemical abundance of elements. Present results suggest that the *ASCA* spectroscopy is extremely suitable to investigate the ionization structure of matter surrounding the neutron star in wind-fed pulsars.

References

Angelini, L., et al. (1995) *Astrophys. J.* **449**, L41
Ebisawa, K., et al. (1996) *Publ. Astron. Soc. Japan* **48** 425
Hatchett, S. & McCray, R. (1977) *Astrophys. J.* **211**, 552
Kawashima, K., & Kitamoto, S. (1996) *Publ. Astron. Soc. Japan* **48**, L113
Kotani, T., et al. (1997) *Astrophys. J.* submitted
Kallman, T.R., & McCray, R. (1982) *Astrophys. J. Suppl.* **50**, 263
Liedahl, D.A., & Paerels, F. (1996) *Astrophys. J.* **468**, L33
McCray, R., Kallman, T.R., Castor, TJ.I., & Olsen, G.L. (1984) *Astrophys. J.* **282**, 245
Nagase, F., (1989) *Publ. Astron. Soc. Japan* **36**, 1
Nagase, F., (1997) in *X-Ray Imaging and Spectroscopy of Cosmic Hot Plasmas*, Proc. of the Internat. Symp. on X-ray Astronomy (March 11-14, 1996, Waseda Uni. Tokyo), eds., F. Makino and K. Mitsuda (Universal Academy press, Inc.), p. 419
Nagase, F., et al. (1994) *Astrophys. J.* **436**, L1
Nagase, F., et al. (1997) in preparation
Saraswat, P., et al. (1996) *Astrophys. J.* **463**, 726

BINARY STUCTURE OF ACCRETING NEUTRON STARS

D.A. LEAHY
University of Calgary
Calgary, Alberta, Canada T2N 1N4

1. Introduction

The study of X-ray binaries has made great progress with the advent in the past few years of a number of very capable X-ray astronomy missions. These are reviewed, for example, by Bradt et al 1992, and a set of recent relevant papers in Makino and Mitsuda, 1997. For example, ASCA has allowed a significant increase in sensitivity and spectral resolution in 0.5-10 keV X-rays (Tanaka et al 1994). Many recent Compton/GRO results on X-ray binaries are reviewed in the proceedings of the Second Compton Symposium (Fichtel et al 1994). Another source of recent results from analysis of data from several satellite missions is the proceedings of the Evolution of X-ray Binaries (Holt & Day, 1994). In this short paper, the emphasis is on guiding the reader to some relevant literature.

2. Theory and Observations of Accreting Neutron Stars

The source of energy is gravitational potential energy. Various thermalization processes occur which result in temperatures for the X-ray emitting region of several keV (e.g. Joss & Rappaport 1984, Parmar, 1994). Material can flow slowly from the companion through the Lagrangian point to form a large disk around the compact object. Or, if the companion is well within the Roche lobe, a stellar wind can cause material to flow at high speed past the Lagrangian point and form a small disk. The small disk can be transient and reverse its spin in a quasiperiodic manner (Fryxell & Taam, 1988, Blondin et al, 1991). The disk interacts with the magnetosphere of the neutron star, then the infalling matter is thermalized in an accretion mound at the surface or by a strong shock above the neutron star surface. The opacity is dominated by the cyclotron opacity (e.g. Meszaros and Nagel, 1985).

Prince et al (1994) gives a recent review of BATSE results on X-ray pulsars, in particular his Table 2 lists orbital parameters for 15 X-ray pulsars determined by pulse timing. The low mass systems show a regular spin down over the 3 years of monitoring. The intermediate mass system Her X-1 shows a random walk behaviour in period. The high mass systems show a significant random walk component but also a secular trend: spin-up for Cen X-3 and spin-down for OAO 1657-415 and Vela X-1. The details of the spin behaviours are not understood, but are related to the accretion torque on the neutron star and the various components of angular momentum of the neutron star and their coupling forces.

Valuable results on the state of cicumstellar matter in Vela X-1 are given by Nagase 1994 (see also these proceedings for Vela X-1, GX301-2, Cen X-3 and other systems). New results also come from analysis of archival data. E.g., GINGA observations of Her X-1 give interesting results (Leahy, 1997, Scott, 1994).

In summary, numerous physical processes can be studied, ranging from the dynamics of neutron stars to conditions in circumstellar and accreting plasmas (temperatures, ionization states, elemental abundances). Current and future observations of this nature are giving further insight into the structure and physical state of accretion flows in X-ray binaries.

References

Blondin, J., Kallman, T., Fryxell, B. & Taam, R. 1990, ApJ, 356, 591

Bradt, H., Ohashi, T. & Pounds, K. 1992, ARAA, 30, 391

Fichtel, C., Gehrels, N. & Norris, J. 1994, Proceedings of the Second Compton Symposium (AIP Conference Proceedings 304)

Fryxell, B. & Taam, R. 1988, ApJ, 335, 862

Holt, S. & Day, C. 1994, Proceedings of the Evolution of X-ray Binaries (AIP Conference Proceedings 306)

Joss, P. & Rappaport, S. 1984, ARAA, 22, 537

Leahy, D. 1997, M.N.R.A.S. 287, 622

"X-ray Imaging and Spectroscopy of Cosmic Hot Plasmas" ed. Makino, F., Mitsuda, K. (Universal Academy Press, Tokyo, 1997)

Meszaros, P. & Nagel, W. 1985, ApJ, 299, 138

Nagase, F. 1994, in Proceedings of the Evolution of X-ray Binaries (AIP Conference Proccedings 306), p567

Parmar, A. 1994 in Proceedings of the Evolution of X-ray Binaries (AIP Conference Proceedings 306), p415

Prince, T. et al 1994 in Proceedings of the Evolution of X-ray Binaries (AIP Conference Proceedings 306), p235

Scott, J.M. 1994 PhD Thesis (Washington State University)

Tanaka, Y., Inoue, H. & Holt, S. 1994, PASJ, 46, L37

MILLISECOND TIME VARIATIONS OF X-RAY BINARIES

J. H. SWANK
NASA/GSFC
Greenbelt,MD 20771, U.S.A.

Abstract.

Millisecond time-scales are natural for some neutron star and black hole processes, although possibly difficult to observe. The Rossi X-Ray Timing Explorer (RXTE) has found that for the neutron stars in low-mass X-ray binaries (LMXB) there are flux oscillations at high frequencies, with large amplitudes. Z sources and bursters tend to exhibit oscillations in the range 300-1200 Hz. Persistent emission may exhibit one or both of two features. In bursts from different bursters, a nearly coherent pulsation is seen, which may be the rotation period of the neutron star. For some the frequency equals the difference between the two higher frequencies, suggesting a beat frequency model, but in others it is twice the difference. The sources span two orders of magnitude in accretion rate, yet the properties are similar. The similar maximum frequencies suggests that it corresponds to the Kepler orbit frequency at the minimum stable orbit or the neutron star surface, either of which would determine the neutron star masses, radii and equation of state. Theories of accretion onto black holes predict a quasi-periodic oscillation (QPO) related to the inner accretion disk. The two microquasar black hole candidates (BHCs) have exhibited candidates for this or related frequencies.

Dynamical Time-scales of Compact Stars

The Newtonian Keplerian frequency in an orbit at radius R about a compact mass M is a measure within a factor of a few of the time scales of various other phenomena, such as free fall, or oscillations, that can occur closest to the compact object. For a $1.4 M_\odot$ neutron star with a 10 km radius, the frequency is 2500 Hz, corresponding to a period of 0.4 ms. For a $16 M_\odot$ black hole and $R = \frac{6GM}{c^2}$, the frequency would be 127 Hz with a 7 ms period. The effective area of the Proportional Counter Array (PCA) (0.7 m^2), the time resolution (1μs), the data modes, and the telemetry (30 – 512 kps are possible), were all required to study this regime.

It is clear from the results of RXTE that dynamical phenomena close to neutron stars and stellar black holes do generate X-ray signals which vary on the dynamical time-scales and that, furthermore, the signals are not greatly smeared by scattering before escaping. Thus we are able to see oscillations of various sorts which are in the expected frequency regimes.

The signals that have been found for neutron stars in the frequency range 50 Hz to 1300 Hz, starting with the discovery in 4U 1728-34 [1], are very significant (in many cases more than 10 sigma). The highest quasiperiodic signal that has been reported is about 1220 Hz. Many of the QPO features are narrow; they have $Q = \frac{\Delta \nu}{\nu} = 100 - 1000$.

During the first year of the RXTE mission, two BHCs were very bright, GRO J1655-40 and GRS 1915+105. In both of these, a feature which qualifies as related to the dynamical time-scale of the inner accretion disk has been seen, 300 Hz and 67 Hz, respectively [2]. While for the LMXB QPO the fractional root mean square (rms) power of the QPO were 1-20%, for these BHC features the fractional rms amplitudes were less than 1%.

High Frequencies from Neutron Stars

Three kinds of frequencies higher than previously known have been discovered in neutron star systems with RXTE observations. There are now at least 6 low-mass X-ray binary sources of Type I bursts in which almost coherent oscillations have been seen during the bursts. There are at least 14 low-mass X-ray binaries in which oscillations at frequencies ranging from 300 Hz to 1220 Hz have been seen in the persistent emission [3].

The pattern of oscillation behavior is similar in many bursts. When the data is folded on the period, the pulsed light curve is sinusoidal. A spot slightly hotter than the rest of the neutron star and rotating with it in and out of view would generate such an oscillation. The fractional root mean square amplitude of the oscillation can exceed 50% at the beginning of the burst and decreases as the burst flux rises, as would be consistent with the model of a burning front spreading on the neutron star. We think the periods (1.7-2.7 ms) probably indicate the pulsars in these LMXB[4].

The kilohertz oscillations in the persistent emission from LMXB present a simple spectrum in comparison to lower frequency power spectral densities (PSDs) of these sources. Just two frequencies dominate over other signals. Changes in the source luminosity usually cause frequency change and the two features move together with the difference approximately constant. An obvious candidate for the higher frequency is the Kepler frequency of matter

TABLE 1. LMXB with Kilohertz Oscillations

Source	$L_x(10^{38})$	Hz_L	Hz_H	Δ	Hz_B
4U 0614+09	0.05	326,400-800	730-1145	326	
4U 1728-34	0.06	650-790	500-1150	363	363
Aql X-1	0.08	750			549
4U 1608-52	0.1	690,800-900	900-1125	60-230	
4U 1702-43	0.1	900	1180	280	330
4U 1636-53	0.2	800-900	1170-1216	255	581
4U 1735-44	0.4		1149		
KS 1731-26	0.2?	898	1159-1207	260	524
X 1744-29?	0.2?				589
4U 1820-30	0.6	546-796	1066	270	
GX 17+2	1	880-682	988	306	
Cyg X-2	1	490-530	730-1020	343	
Sco X-1	1	634	926	292	
	2	794-820	1062-1133	247	
GX 5-1	4	325-448	567-895	325	

orbiting the neutron star at the inner edge of the accretion disk. Various mechanisms could define the inner edge of the disk. It could be (See [5] for extended discussion.) interaction with the magnetosphere, a sonic radius, or the marginally stable orbit, if the neutron star's surface is inside it. A lower frequency would be seen at an alias with the neutron star rotation. If there is a clump circulating at the inner radius of the disk, accretion from it along field lines to the neutron stars magnetic pole would be most enhanced at the nearest conjunction of the pole and the clump.

The sources with reported high frequency signals in the persistent emission are summarized in Table 1 (luminosity in ergs s^{-1}, frequencies in the persistent emission(L and H), the difference(Δ), and in the bursts(B)). It can be seen that while for 4U 1728-34 the difference between the 2 high frequencies equals that of the bursts, for 4 other sources, the difference is approximately half the frequency seen during the bursts. However the burst period does not appear interpretable as the harmonic of the spin because of the high pulsed amplitude and the sinusoidal form.

The maximum frequencies appear to be all very close to each other.

This suggested that the determination of the frequency must be dominated by neutron star characteristics and roughly independent of the accretion rate and the optical depths of the formation region. We argued that most likely the equation of state of nuclear matter does lead to neutron stars with radii inside the inner-most stable orbit for a neutron star, and the disk cannot approach closer to the neutron star than this[7]. The observed frequencies of about 1200 Hz would imply neutron star masses of $1.8-2M_\odot$. The existence of an orbit which is marginally stable would be a verification of an important consequence of strong gravity[8].

BHC oscillations

The frequency of fast oscillations in the inner disk around a black hole depends on the black hole mass, its angular momentum, and the model. Instabilities of the disk which lead to circulating clumps would imply the masses of 7 and 33 $M_\odot$ for GRO J1655-40 and GRS 1915+105, respectively, if they are slowly rotating. Epicyclic frequencies would match those observed for smaller slow black holes, but for the more massive black holes if they have near maximum rotation rate[6]. The mass and the angular momentum of the hole imply the inner radius of the optically thick accretion disk and the inner radius is more consistent with the mass of GRO J1655-4 deduced from optical measurements if the hole is rotating fast[9].

References

1. Strohmayer, T. E., Zhang, W., Swank, J. H., Smale, A. P., Titarchuk, L., & Day, C. (1996) Millisecond X-Ray Variability from an Accreting Neutron Star System, *Ap. J. Letters*, **469**, L9.
2. Remillard, R. E., Morgan, E. H., McClintock, J. E., Baiyln, C. D., Orosz, J. A., & Greiner, J. (1997) Multifrequency Observations of the Galactic Microquasars GRS 1915+105 and GRO J1655-40, *Proceedings of 18th Texas Symposium on Relativistic Astrophysics*, Eds. A. Olinto, J. Frieman, and D. Schramm, World Scientific Press.
3. van der Klis, M. (1997) Kilohertz Quasi-Periodic Oscillations in Low-Mass X-Ray Binaries, *Proc. NATO ASI "The Many Faces of Neutron Stars"*, in press.
4. Strohmayer, T. E., Zhang, W., & Swank, J. H. (1997) 363 Hz Oscillations during the Rising Phase of Bursts from 4U 1728-34: Evidence for Rotational Modulation, *Ap. J. Letters*, **487**, L77.
5. Miller, M. C., Lamb, F. K., &Psaltis, D. (1998) Sonic-Point Model for High-Frequency QPOs in Neutron Star Low-Mass X-Ray Binaries, *Ap. J. Suppl.*, in press.
6. Nowak, M. A., Wagoner, R. V., Begelman, M. C., & Lehr, D. E. (1997) The 67 Hz Feature in the Black Hole Candidate GRS 1915+105 as a Possible Diskoseismic Mode, *Ap. J. Letters*, **477**, L91.
7. Zhang, W., Strohmayer, T., & Swank, J. H. (1997) Neutron Star Masses and Radii as inferred from Kilohertz Quasi-Periodic Oscillations, *Ap. J. Letters*, **482**, L167.
8. Kaaret, P., Ford, E. C., & Chen, K. Strong-Field General Relativity and Quasi-periodic Oscillations in X-Ray Binaries, *Ap. J. Letters*, **480**, L27.
9. Zhang, S. N., Cui, W., & Chen, W. (1997) Black Hole Spin in X-Ray Binaries: Observational Consequences, *Ap. J. Letters*, **482**, L155.

GRO J1744–28 AND THE RAPID BURSTER; BIZARRE OBJECTS!

WALTER H.G. LEWIN
MIT
Cambridge, MA

The bursts from GRO J1744–28 are due to accretion instabilities as is the case for type II bursts in the Rapid Burster. Both sources are transient Low-Mass X-ray Binaries, and they both exhibit unusual quasi-periodic-oscillations in their persistent X-ray flux following several (not all) of the type II bursts. There are important differences too. GRO J1744–28 is an X-ray pulsar; the Rapid Burster is not. In addition, the pattern of bursts and the burst peak luminosities are very different for the two sources. Time intervals between the rapidly repetitive bursts in the Rapid Burster can be as short as 10 sec, in 1744–28 they are as short as 200 sec. The peak luminosities of the bursts from GRO J1744–28 can exceed the Eddington luminosity (for assumed isotropic emission) by one to two orders of magnitude. The QPO centroid frequencies (see above) differ by an order of magnitude ($\sim$0.04 Hz for the Rapid Burster, and 0.3 Hz for GRO J1744–28). The difference in behavior probably lies in the difference in the magnetic dipole field strength of the accreting neutron stars (for GRO J1744–28 it is almost certainly much higher than for the Rapid Burster). It remains puzzling, why GRO J1744–28 and the Rapid Burster are the only known sources which exhibit rapidly repetitive type II bursts.

Unlike GRO J1744–28, the Rapid Burster also emits type I bursts. No kHz near-coherent oscillations have been found in a handful of type I bursts from the Rapid Burster, observed with *RXTE*. However, kHz oscillations have been observed in several type II bursts from the Rapid Burster. For more details see Guerriero, Lewin, and Kommers, IAUC 6689, (1997); also Guerriero et al., manuscript in preparation.

The following papers compare in some detail GRO J1744–28 with the Rapid Burster: Lewin et al., ApJ 462, L39 (1996) and Kommers et al., ApJ 482, L53 (1997).

X-RAY AND GAMMA-RAY OBSERVATIONS OF ISOLATED NEUTRON STARS

D.J. THOMPSON
Laboratory for High Energy Astrophysics
NASA Goddard Space Flight Center
Greenbelt, Maryland USA

1. Introduction

Pulsars provide useful diagnostics of isolated neutron stars, because the timing information allows many physical parameters to be derived. This brief review describes some of the X-ray and gamma-ray properties of pulsars.

2. Light Curves

As shown in Figure 1, pulsar light curves at high energies are varied in shape. The preponderance of double pulses with a bridge of emission suggests emission from a cone or surface above a single pole of the neutron star. Very young pulsars (Crab, PSR B0540−69, PSR B1509−58) and at least one old, ms pulsar (PSR B1821−24) show only magnetospheric emission associated with accelerated charged particles, as indicated by the similarity of the pulse shape across the entire X-ray and gamma-ray band. Intermediate-age pulsars (Vela, Geminga, PSR B0656+14, PSR B1055−52) also reveal a thermal component in soft X-rays, probably from near the surface of the star. This component differs in both shape and phase from the higher-energy, nonthermal radiation.

3. Broadband Energy Spectra

Figure 2 shows the energy spectra of several pulsars across the entire electromagnetic spectrum. In all cases the observable output is dominated by the X-ray and gamma-ray emission, indicating the importance of these high-energy radiations for pulsars. For the youngest and oldest pulsars, represented here by the Crab (age about 1000 y) and PSR B1821−24 (age greater than 10^7 y), the emission peaks in the hard X-ray band. The intermediate-age pulsars (Vela has a timing age of about 1×10^4 y; Geminga's age is about 3×10^5 y) are seen most prominently in the high-energy gamma-ray range. All spectra have some high-energy cutoff.

4. Trends and Models

The energy spectra of Figure 2, integrated over energy, can be used along with the distance estimates from radio and optical observations to construct a broadband luminosity L for these high-energy pulsars. A trend is seen for L to be linearly proportional to $\dot{E}^{1/2}$, the spin-down luminosity. This quantity is also proportional to the open-field-line voltage or the polar cap current (Harding, 1981). The broadband luminosity trend contrasts with the trend for X-rays alone, which shows a linear dependence on $\dot{E}$ (Becker and Trümper, 1997). As shown by Goldoni and Musso (1996), other possible phenomenological trends are less convincing.

Both polar-cap and outer-gap models have evolved with the discoveries of new X-ray and gamma-ray pulsars. Although both types can explain many observed features, neither appears to offer a complete enough picture to warrant broad generalizations. For recent summaries, see Daugherty and Harding (1996) and Romani (1996). Future observations, from operating telescopes such as ASCA and RXTE, and from future telescopes such as AXAF, XMM and GLAST, should resolve many of the open questions.

References

Becker, W., and Aschenbach, B. 1995, in *The Lives of Neutron Stars*, (eds A. Alpar, U. Kiziloğlu & J. van Paradijs), Kluwer Academic Publishers, p. 47.
Becker, W., & Trümper, J. 1997, *A&Ap*, in press
Daugherty, J. K. & Harding, A. K. 1996, *ApJ*, **458**, 278.
Goldoni, P, & Musso, C. 1996, *A&ApS*, **120**, 103.
Halpern, J.P. & Wang, F. Y.-H. 1997, *ApJ*, **477**, 905.
Harding, A.K. 1981, *ApJ*, **245**, 267.
Kanbach, G. 1997, Proceedings of NATO Advanced Study Institute, *The Many Faces of Neutron Stars*, Lipari, October, 1996, in press
Kuiper, L. et al. 1997a, in Proceedings of NATO Advanced Study Institute, *The Many Faces of Neutron Stars*, Lipari, October, 1996, in press
Kuzmin, A.D. & Losovsky, B. Y. 1997, IAU Circ. 6559
Much, R. et al. 1997, *Proc. 4th Compton Symposium*, (ed. C. Dermer and M. Strickman), AIP, in press
Nel, H.I. et al. 1996, *ApJ*, **465**, 898
Romani, R.W. 1996, *ApJ*, **470**, 469
Saito, Y. et al. 1997, *ApJ*, in press
Shitov, Yu. P. & Pugachev, V.D. 1997, preprint
Thompson, D.J. 1996, in *Pulsars: Problems and Progress*, IAU Colloquium 160 (ed. M. Bailes, S. Johnston, M. Walker)

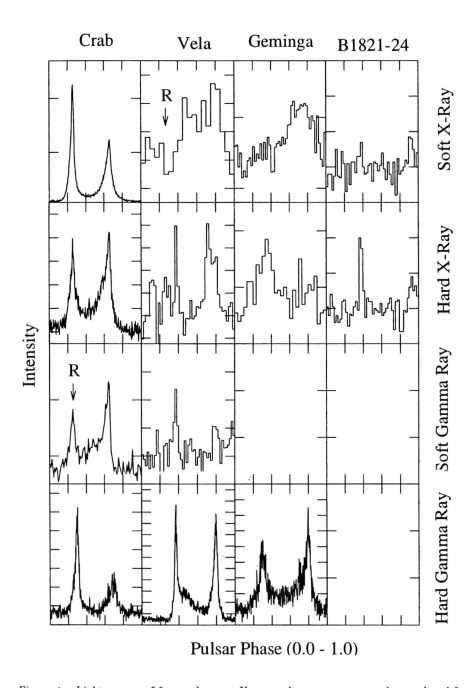

Figure 1. Light curves of four pulsars at X-ray and gamma-ray energies, updated from Thompson (1996). New references: Crab: Becker and Aschenbach (1995), Much et al. (1997), Kanbach (1997); Vela: Harding and Strickman, priv. com., Kuiper et al. (1997); PSR B1821−24: Saito et al. (1997)

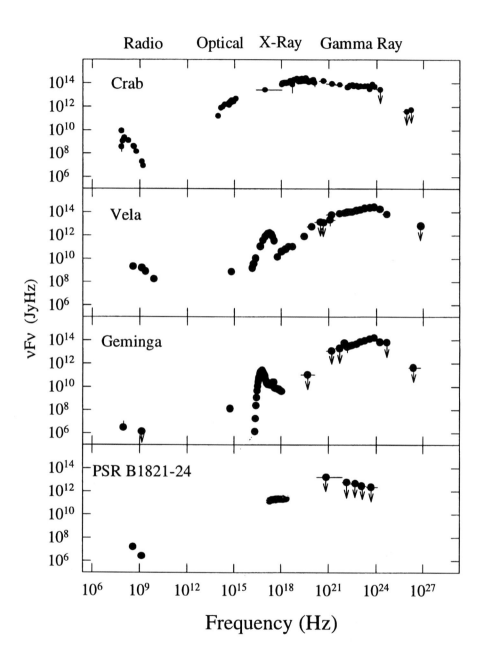

Figure 2. Energy spectra of four pulsars, updated from Thompson (1996). New references: Geminga: Kuzmin and Losovsky (1997), Shitov and Pugachev (1997), Halpern and Wang (1997), Vela: Harding and Strickman, priv. com., Kuiper et al. (1997); PSR B1821−24: Saito et al. (1997), Nel et al. (1996)

SEARCH FOR NONTHERMAL X-RAYS FROM SUPERNOVA REMNANT SHELLS

R. PETRE, J. KEOHANE, U. HWANG, G. ALLEN, E. GOTTHELF
NASA/GSFC
Greenbelt, MD 20771 USA

1. Introduction

The suggestion that the shocks of supernova remnants (SNR's) are cosmic ray acceleration sites dates back more than 40 years. While observations of nonthermal radio emission from SNR shells indicate the ubiquity of GeV cosmic ray production, there is still theoretical debate about whether SNR shocks accelerate particles up to the well-known "knee" in the primary cosmic ray spectrum at $\sim$3,000 TeV. Recent X-ray observations of SN1006 and other SNR's may have provided the missing observational link between SNR shocks and high energy cosmic ray acceleration. We discuss these observations and their interpretation, and summarize our ongoing efforts to find evidence from X-ray observations of cosmic ray acceleration in the shells of other SNR's.

2. ASCA and the Mystery of SN1006

The X-ray spectrum of SN1006 was for a long time a puzzle. Contrary to the prevailing wisdom that the X-ray emission in shell like SNR's is produced by the shock heating of ISM material and metal-rich ejecta (and thus should be dominated by lines), the integrated spectrum of SN1006 above 1 keV is featureless (Becker et al. 1980), and best fit by a power law. While a model was put forward suggesting the X-rays arise as synchrotron radiation from electrons accelerated to high energy in the shock (Reynolds & Chevalier 1981), the generally accepted explanation was that they arise from a thermal plasma in an extreme state of ionization nonequilibrium (Hamilton et al. 1986). The ASCA observations solved this mystery, and in so doing produced the first strong evidence for the production of high energy cosmic rays in SNR shocks. Spatially resolved ASCA spectra showed that throughout most of the remnant, the spectrum is dominated by emission lines, with inferred metal abundances characteristic of Type Ia SN ejecta. In contrast, the spectra of the bright NE and SW rims are dominated by a power law component with $\alpha \sim 2$. Thus we can understand SN1006 as being an ordinary thermal remnant, whose spectrum above 1 keV is dominated by a second, nonthermal component that is restricted to the bright limbs. The most reasonable model for this is synchrotron emission from electrons accelerated within the shock. At these highly relativistic energies, shock acceleration processes make no distinction between positively and negatively charged particles; thus we can infer the presence of high energy protons and nucleons as well, and conclude that cosmic ray acceleration occurs in the shell of SN1006. Reynolds (1997) has provided a shock acceleration model that reproduces both the broad band (radio through X-ray) spectrum of SN1006 and the X-ray morphology.

3. Nonthermal X-rays from the Shells of Other SNR

At least three other SNR's are now known to have nonthermal emission associated with their shock fronts: Cas A, G347.5-0.5, and IC 443. The existence of nonthermal tails in this many remnants (coupled with the discovery of TeV γ-rays from SN1006 – Tanimori et al. 1997), firmly establishes SNR's as a source of high energy cosmic rays.

Cas A: The striking morphological similarity between the ASCA 4-8 keV continuum map of Cas A and the radio image, and its contrast with the maps in all other X-ray bands, suggested that the hard continuum might arise from nonthermal processes (Holt et al. 1995). Recent RXTE and SAX spectra show a nonthermal spectrum that extends beyond 100 keV (Allen et al. 1997; Favata et al. 1997). This is almost certainly synchrotron emission from shock accelerated electrons.

G347.5-0.5: Koyama et al. (1997) found that the shell-like supernova remnant G347.5-0.5 (= RX J1713.7-3946), discovered in the ROSAT All-Sky Survey, has a featureless power law spectrum with $\alpha \sim 2.5$. Its similarity to SN1006 led Koyama et al. to propose that G347.5-0.5 is the second example of a shell-like SNR dominated by synchtron emission from shock acclerated electrons.

IC 443: ASCA imaging reveals that the hard component is localized to the region where the shock is interacting most strongly with a foreground molecular cloud (Keohane et al. 1997). The shock velocity in IC 443 is too low to support the production of TeV particles in the way they are produced in SN1006. Keohane et al. propose instead that the synchrotron electrons are produced by enhanced shock acceleration arising directly from the cloud/shock collision (Jones & Kang 1993).

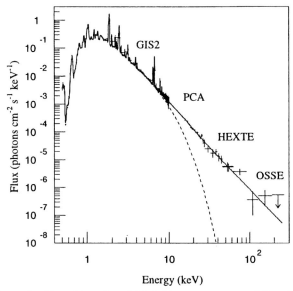

Figure 1. Broad band unfolded X-ray spectrum of Cas A, showing hard power law tail beyond 100 keV (from Allen et al. 1997)

We have embarked on a systematic search through the ASCA data for nonthermal tails in young Galactic SNR's. Our preliminary results for some remnants (W49B, 3C 397, Kepler, Tycho) have been compiled in Table 1, along with the remnants for which results have been published. While one might expect some correlation between the existence of a nonthermal shell component and age (shock velocity), radio brightness (magnetic field strength, electron density), or a X-ray/radio morphology match, no clear pattern has yet emerged.

4. The LMC Remnants

The young, X-ray luminous remnants in the Large Magellanic Cloud allow for the study of a full population of SNR's without the biases introduced within the Galaxy by distance and column density. We have systematically searched the spectra of the luminous LMC remnants observed by ASCA for evidence of hard tails. In general, remnants with small diameters have hard components and those with large diameters (e.g., N49B) do not. This suggests that the mechanism producing the tails is restricted to the younger remnants. While the presence of an Fe K line at 6.7 keV demonstrates that the hard X-ray emission in some remnants (N103B, N132D, N63A) is at least partly thermal, in other remnants ( )no Fe K line is present, suggesting that the tail could be nonthermal. The spectral index of these tails is consistent with that found in SN1006 and Cas A. If the hard X-ray emission in the LMC remnants arises from the same mechanism operating in the young Galactic remnants, then it provides the first evidence of high energy cosmic ray production in SNR's outside the Milky Way, and allows an opportunity to track the evolution of the cosmic ray production process.

Table 1. Preliminary results for Galactic SNR.

Object	Age (ky)	Hard Tail?	Radio/X-ray Match?	Type	High L_R?	Result
Cas A	0.3	Y	Y	II	Y	Hard X cont. has $\alpha \sim 2$ to 100 keV
						Hard continuum morphology matches radio
Kepler	0.4	–	Y	II?	–	NT component improves fit
						No strong constraint on α
						or nonthermal flux fraction from ASCA
Tycho	0.4	Y	–	Ia	Y	nonthermal $L_x < 0.3\ L_{1006}$ from ASCA
SN1006	1.0	Y	Y	Ia	N	Obvious nonthermal shell; $\alpha \sim 2$ to ~ 20 keV
G347.5	~ 1.0	Y	?	II?	N	Nonthermal – "SN1006 twin" $\alpha \sim \alpha_{1006}$
3C 397	~ 1.0	Y	Y	II?	Y	Hard X-ray morphology matches radio
W49B	1-3	Y	Y	?	Y	Hard X-ray has $\alpha \sim \alpha_{1006}$
						Hard X-ray morphology matches radio
IC 443	1-4	Y	–	II	Y	Hard X isolated; $\alpha \sim \alpha_{1006}$
						near MC interaction site

5. Summary of Current Results

The search for nonthermal components and hence cosmic ray acceleration in the shells of SNR has barely begun. Nevertheless, we can already formulate some conclusions. (1) The hard X-ray flux in at least four SNR's is dominated by synchrotron emission from their shells. (2) Several other SNR's have properties that suggest the presence of nonthermal shell emission. Among these are Tycho, Kepler, W49B and 3C 397. (3) A systematic search through the archival ASCA spectra of LMC remnants produces suggests that the strength of the hard component, whether thermal or nonthermal, diminishes with SNR age.

6. References

Allen, G.E., et al. 1997 *Ap. J. Lett.* **487**, L97.
Becker, R.H., et al., 1980 *Ap. J. Lett.* **235**, L5.
Hamilton, A.J.S., Sarazin, C.L., & Szymkowiak, A.E., 1986, *Ap. J.* **300**, 698.
Holt, S.S., et al., 1994, *Publ. Astron. Soc. Japan* **46**, L151.
Jones, T.W., & Kang, H., 1993, *Ap. J.* **402**, 560.
Keohane, J.W., et al., 1997, *Ap. J.* **484**, 350.
Koyama, K., Petre, R., Gotthelf, E.V., Hwang, U., Matsuura, M., Ozaki, M., & Holt, S.S., 1995, *Nature* **378**, 255.
Koyama, K., et al., 1997, *Publ. Astron. Soc. Japan* **49**, L7.
Reynolds, S.P., 1996, *Ap. J. Lett.* **459**, L13.
Reynolds, S.P., & Chevalier, R.A., 1981, *Ap. J.* **245**, 912.
Tanimori, T., et al., 1997, *IAU Circular* 6706.

DETECTION OF TEV GAMMA RAYS FROM SN1006

TORU TANIMORI
Department of Physics, Tokyo Institute of Technology
Ohokayama, Meguro, Tokyo 152, Japan

1. Introduction

In spite of the recent progress of high energy gamma-ray astronomy, there still remains quite unclear and important problem about the origin of cosmic rays. Supernova remnants (SNRs) are the favoured site for cosmic rays up to 10^{16} eV, as they satisfy the requirements such as an energy input rate. But direct supporting evidence is sparse. Recently intense non-thermal X-ray emission from the rims of the Type Ia SNR SN1006 (G327.6+14.6) has been observed by ASCA (Koyama et al. 1995)and ROSAT (Willingale et al. 1996), which is considered, by attributing the emission to synchrotron radiation, to be strong evidence of shock acceleration of high energy electrons up to ~100 TeV. If so, TeV gamma rays would also be expected from inverse Compton scattering (IC) of low energy photons (mostly attributable to the 2.7 K cosmic background photons) by these electrons. By assuming the magnetic field strength (B) in the emission region of the SNR, several theorists (Pohl 1996; Mastichiadis 1996; Mastichiadis & de Jager 1996; Yoshida & Yanagita 1997) calculated the expected spectra of TeV gamma rays using the observed radio/X-ray spectra. Observation of TeV gamma rays would thus provide not only the further direct evidence of the existence of very high energy electrons but also the another important information such as the strength of the magnetic field and diffusion coefficient of the shock acceleration. With this motivation, SN1006 was observed by the CANGAROO imaging air Čerenkov telescope in 1996 March and June, also 1997 March and April.

2. Analysis and Result

SN1006 was observed with the 3.8m diameter Čerenkov imaging telescope of the CANGAROO Collaboration (Patterson & Kifune 1992; Hara et al. 1993) near Woomera, South Australia (136°47' E and31°06' S). The 3.8

m, alt-azimuth mounted, telescope had a ~3 TeV threshold for detecting gamma rays near 70° elevation in 1996 observation. After 1996 observation the 3.8m mirror was recoated in 1996 October, and its reflectivity improved from 45% to more than 80%, which decreased the threshold energy about a half (~1.5 TeV). SN1006 was observed in 1996 and 1997.

The imaging analysis was carried out using standard parameterization of the elongated shape of the Čerenkov light image (Weekes et al. 1989). The α (image-orientation angle) peak appearing around the origin ($\alpha \leq 15°$) in the on-source data are attributed to γ-rays from the target position, and the number of background events under the α peak was estimated from the flat region of the α distribution (30° − 90°) in the on-source data. This technique has been, to date, applied for the detections of TeV gamma rays from PSR1706-44 (Kifune et al. 1995) and the nebula surrounding the Crab (Tanimori et al. 1998). From the results on the previously observed objects, the point spread function (PSF) of the CANGAROO telescope is estimated to have a standard deviation of 0°.18 when fitted with a Gaussian function.

The hard X-ray morphology of the NE rim observed by ASCA suggests that the TeV gamma rays may emanate from an extended area over a few times the PSF of the CANGAROO telescope, (several tenths of a degree in extent). Then we approximate the diffuse emission by a superposition of a small number of point sources separated by about 0°.3. In order to search the emission region of TeV gamma rays in SN1006, significances of peaked events of $\alpha \leq 15°$ were calculated at all grid-points in 0°.09 steps in the FOV which is a half pitch of a standard deviation of the PSF. The contour map of significances in 1996 and 1997 data are shown in Figs. 1a and 1b, in which the contours of the hard X-ray flux of ASCA also are overlaid as solid bold-lines. The region showing significant TeV gamma-ray emission clearly extends along the ridge of the NE rim over the PSF of the telescope, and matches the X-ray image fairly well. The source point in the NE rim giving the most significant α peak ($\alpha \leq 15°$) was found close to the maximum flux point in the 2–10 keV band of the ASCA data. The statistical significances of these peaks are estimated to be 5.3σ in 1996 data and 7.7σ in 1997 data at the maximum hard X-ray flux point of ASCA. The much improvement of the detection significance in 1997 data was due to the twice increase of the reflectivity of the mirror. Thus the TeV gamma-ray emission from the NE rim of SN1006 has been confirmed (Tanimori et al. 1997).

The threshold energies for the 1996 data and 1997 data were determined from Monte Carlo simulations to be 3 ± 0.9 TeV and 1.7 ± 0.5 TeV, respectively. Although Figs. 1a and 1b suggest the emission of TeV gamma rays extends along the ridge of the NE rim, we do not yet have a reliable method to estimate the number of gamma-ray events from an extended source. Approximating the emission as coming from a single point source

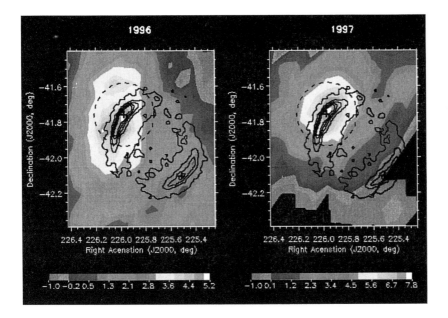

Figure 1. (a) The contour map of statistical significance in the sky around SN1006, where the intensity contours of hard X-rays by ASCA (courtesy M. Ozaki) is overlaid as solid line, and the dashed circle is the PSF of the CANGAROO telescope. (b) The same contour map obtained from the 1997 data.

at the maximum flux point in the NE rim, the integral gamma-ray fluxes for the 1996 and 1997 observations were calculated to be $(2.4\pm0.44)\times 10^{-12}$ cm^{-2} s^{-1} and $(4.6\pm0.6)\times 10^{-12}$ cm^{-2} s^{-1}, respectively. A larger flux would be obtained for an extended emission wider than the PSF of our detection.

No significant excess is evident in Figs. 1a and 1b near the position of the maximum X-ray flux from the SW rim. We set a upper limit on the TeV gamma-ray emission from the SW rim, estimated to be 1.5×10^{-12} cm^{-2} s^{-1} ($\geq 3 \pm 0.9$ TeV, 95% CL) from the α distribution in 1996 data at the position of the maximum ASCA flux in the SW rim.

3. Discussion

The detection of TeV gamma rays from SN1006 presents a convincing demonstration of the the shock acceleration mechanism for very high energy particles up to ~100 TeV in a SNR. The TeV gamma ray emission region is observed to extend over ~30 arcmin along the ridge of the NE rim, providing the first detection of very high energy gamma rays from an extended source. From non-thermal X-ray observation, the detected TeV gamma rays are readily presumed to be generated by IC scattering of very high energy electrons and 2.7K cosmic background photons. All of the calculated fluxes of TeV gamma rays based on those assumption are consistent with the obtained TeV gamma-ray fluxes by assuming the magnetic field strength (B) in the emission region of the SNR with around $6\mu G$. This strength of B is well consistent with the previous estimation of 6 - $10\mu G$ (Reynolds 1991).

The other candidate of the production mechanism of the TeV gamma-ray emission is a decay of neutral pions induced by high energy protons accelerated in the SNR. However, we can neglect the flux from the π^0 decay due to following arguments. Since SN1006 (G327.6+14.6) is located above the galactic plane, the matter density at the shock is low (~ 0.05 cm^{-3}) so that the expected flux will be about a factor of ten less than the observed flux (Mastichiadis 1996). The upper limit for GeV gamma-ray emission from the EGRET archive data is also consistent with the IC model. Thus, the detected gamma-rays are likely to be explained by IC radiation from electrons, and our result testifies to the existence of the very high energy electrons of more than several times 10 TeV in SN1006. The highest energy of non-thermal electrons can be estimated from bending point in synchrotron spectrum and the resultant magnetic fields. Although the bending energy of SN1006 is not precisely determined yet in recent observations, 1 keV photon from synchrotron radiation corresponds to electron energy of $\sim$ 60 TeV for $B \sim 6\mu G$. On the hand, there are no distinctions between electrons and protons for the basic theoretical treatment of shock acceleration via first-order Fermi process. Our observation is therefore the first convincing proof of the shock acceleration of high energy cosmic rays in a SNR.

References

Hara, T. et al. (1993), *Nuc. Inst. Meth. Phys. Res. A*, **332**, 300 - 309
Kifune, T. et al. (1995), *Astrophys. J.*, **438**, L91 - L94
Koyama, K. et al. (1995), *Nature*, **378**, 255 - 258
Mastichiadis, A. (1996), *Astr. Astrophys.*, **305**, L53 - L56
Mastichiadis, A. & de Jager, O.C. (1996), *Astr. Astrophys.*, **311**, L5 - L8
Patterson, J. R., & Kifune, T. (1992), *Australian and New Zealand Physicist*, **29**, 58 - 62
Pohl, M. (1996), *Astr. Astrophys.*, **307**, 57 - 395
Reynolds, S.P. & Ellison, D.C. (1991), *Proc. 22nd Int. Cosmic ray Conf.*, 2 404-407
Tanimori, T. et al. (1997), *IAU Circ.* No. 6706
Tanimori, T. et al. (1998), *Astrophys. J.*, in press
Yoshida, T.& Yanagita, S. (1997), *Proc. 2nd INTEGRAL Workshop*, **ESA SP382** 85-88
Weekes, T.C. et al. (1989), *Astrophys. J.*, **342**, 379 - 395
Willingale, R. et al. (1996) *Mon. Not. R. astr. Soc.*, **278**, 749 - 762

VERY HIGH ENERGY GAMMA RAYS FROM PLERIONS : CANGAROO RESULTS

T.KIFUNE
for CANGAROO Collaboration, Institute for Cosmic Ray Research, University of Tokyo, Tanashi, Tokyo 188, Japan

Abstract. The current status of very high energy gamma ray astronomy (in ~ 1 TeV region) is described by using as example results of CANGAROO (Collaboration of Australia and Nippon for a GAmma Ray Observatory in the Outback). Gamma rays at TeV energies, emitted through inverse Compton effect of electrons or π^0 decay from proton interaction, provide direct evidence on "hot" non-thermal processes of the Universe, as well as environmental features, such as the strength of magnetic field in the emission region, for the non-thermal processes.

1. Introduction

The number of Very High Energy Gamma Ray (VHEGR) sources is rapidly increasing from only one (Crab) in 1990 to the current value of more than five. VHEGR observation experienced a breakthrough by the Imaging Air Čerenkov Telescope (IACT; *e.g.* review by Cawley 1996), almost simultaneously with the launching of Compton Gamma Ray Observatory (CGRO). The GeV sources discovered by EGRET of CGRO are also likely VHEGR emitters and have served as a guide for VHEGR observation.

Energetic electrons (and positrons) are found abundant in point(-like) sources, plerions and Active Galactic Nuclei (AGN) to illuminate the sky bright at GeV and TeV energies (for review, Weekes et al. 1997 and references therein), while the diffuse emission from the Galactic disk is considered due to protons. As electrons radiate through inverse Compton and synchrotron processes into TeV band and longer wavelengths, respectively, the importance of multi-wavelengths study is built-in for VHEGR astronomy. The VHEGRs from proton progenitor of producing π^0 decay gamma rays are estimated to be of detectable fluxes from supernova remnant (SNR). Protons are abundant there and must be injected into the "accelerator" in the supernova shell (Drury et al. 1994; Naito and Takahara 1994).

2. Galactic sources and CANGAROO results

The CANGAROO with IACT in South Australia observes gamma rays at TeV energies from the southern sky, and has discovered VHEGRs from Galactic objects of plerion and SNR, while other IACTs such as the 10 m one of Whipple Observatory are for the northern sky, having revealed VHEGR activity of AGN such as Mrk 421 and 501.

Six gamma ray pulsars discovered by EGRET emit their largest portion of energy output into $\geq$ 100 MeV as pulsed radiation which is modulated with the spin period of neutron star. The pulsed signal dissapears and only unpulsed component remains as observed at $\geq$300 GeV energies (for review e.g. Kifune 1996 and references therein); the region of emission changes from the pulsar magnetosphere to outer region, where pulsar wind collides with the circumstellar medium. CANGAROO has detected VHEGRs from several pulsar nebulae formed in such a region, as well as from a SNR, SN 1006.

The Crab nebula is the unique case that the whole spectrum of synchrotron and inverse Compton radiation is measured, giving estimate of the magnetic field to be $\sim$ 270μG close to the equi-partition value (Weekes et al. 1997). The strong magnetic field and huge spin-down luminosity results in intense synchrotron radiation which are then converted to have higher energies by Compton scattering, making the Crab nebula only one unpulsed GeV plerion detected so far by EGRET. The energy spectrum observed can be compared in detail with model calculations (e.g. Atoyan and Aharonian 1995). The spectrum at higher energies is more sensitive to emission models, because of the effects of the maximum energy of accelerated electrons and/or Klein-Nishina fall-off of the electron-photon scattering cross section in inverse Compton process. The CANGAROO observation of the Crab, thanks to good sensitivity at $\geq$ 10TeV enabled by the observation at large zenith angles (53° − 60°) from the observation site in the southern hemisphere, has reported a constant power index of energy spectrum up to at least 50 TeV and possibly to $\sim$100 TeV (Tanimori et al. 1997).

Following the results on PSR B1706-44 (Kifune et al. 1995), VHEGRs from Vela pulsar direction was found. The emission region is displaced by about 0.13° from the pulsar, apparently in accordance with the birth place of the Vela pulsar (Yoshikoshi et al. 1997). The ratio $\eta = L_s/L_{ic}$ of luminosity in X-ray synchrotron radiation to inverse Compton is equal to the ratio $W_{mag}/W_{2.7K}$ of energy density of magnetic field to that of ambient taget photons for inverse Compton scattering. For Vela and PSR B1706-44,

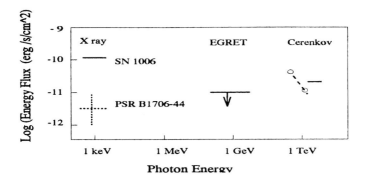

Figure 1. Comparison of fluence to X-ray and GeV band.

magnetic field B in the nebula is estimated as weak as that in the interstellar medium, $\sim 3\mu G$. The synchrotron life time of electrons calculated from the magnetic field allows the electrons generated at the birth time of Vela pulsar to survive until the present time (Harding et al. 1997). If B is $\sim 3\mu G$ in the pulsar nebulae of PSR B1706-44 and Vela, the energy density of progenitor electrons is higher, by an order of magnitude, than $B/8\pi^2$. The two energy densities are not in equi-partition.

However, the VHEGR data can imply a higher B with a constraint on life time and spatial confinement of electrons. By deviding the nebula into two region, then $\eta = \frac{1}{1+A} \cdot (\eta_2 + A \cdot \eta_1)$, where $A = \tau_e/\tau_2$ is the ratio of escape time τ_e of progenitor electrons (out of the central region 1 of stronger B) to the total life time τ_2 of electrons (in the peripheral region 2). When $A \ll 1$, η approaches to η_2. In the case of PSR B1706-44, $B_1 \sim 20\mu G$ can be compatible with both of the CANGAROO and X-ray data, with the escape time $\tau_e \sim 10 \cdot (E/20\text{TeV})^{-\delta}$ yrs and $\delta = 0 \sim 0.5$ (Aharonian et al. 1997).

3. Discussions

The sensitivity of IACT is at a level somewhat below 10^{-11} erg s^{-1} cm^{-2}, which corresponds to the luminosity 10^{33} erg s^{-1} at a distance of 1 kpc. In Fig. 1, observed fluence is compared between X-ray, GeV and TeV gamma rays for SN 1006 (solid line) and PSR B1706-44 (dashed line). The upper limit in GeV region is consistent with a spectrum having two peaks in X-ray and TeV band, each peaked one corresponding respectively to synchrotron and inverse Compton radiation. The spectrum implies abundant production and acceleration of electrons and positrons with a condition of B not much greater than $\sim 10\mu G$. The comparison of the two bands of GeV and TeV can be consistent with a spectrum of power index flatter than -2, which proton progenitor by shock acceleration can hardly explain. In order to know an allowed possible fraction of contribution from π^0 decay gamma rays due to protons, the energy spectrum of better accuracies is required through GeV to TeV bands.

The SNRs associated with unidentified EGRET sources attract interest of VHEGR observations, because the shock acceleration theory predicts detectable TeV fluxes when we attribute the EGRET flux to π^0 decay gamma rays from protons that are accelerated in the SNR shell and then collide with the matter of enhanced density in the nearby molecular cloud. The efforts are so far not successful to detect VHEGRs (Lessart et al. 1997; Hess et al. 1997). However, the complex features associated with those SNRs leave room for a variety of mechanisms other than π^0 decay process to explain the GeV emission and no detection at TeV. On the other hand, SN 1006 is of comparatively simple, "pure" shell type SNR, and the CANGAROO result has shown that particles with Lorentz factor $\gamma \geq 10^8$ undoubtedly exist as suggested from non-thermal X-rays (Koyama et al. 1995).

A step is forwarded towards the "origin of cosmic rays" through the study of VHEGRs. The progress will also provide us with the means to understand wealth of phenomena like the complex structure of SNRs. The Vela, PSR B1706-44 nebulae and SN 1006 appear brighter at TeV energies than in GeV region of EGRET. There is some indication of detection (Aharonian and Heinzelmann 1997 on micro quasar GRS 1915+105; Turver et al. 1997 on Cen X-3) that suggests new types of time-variable sources, encouraging further watching on the VHEGR sky.

References

Aharonian, F.A., Atoyan, A.M., and Kifune, T. (1997) *MNRAS*, **Vol. 291**, 162
Aharonian, F.A. and Heinzelmann, G. (1997) *Nucl. Phys. B*, to be published (astro-ph/9702059)
Atoyan, A.M. and Aharonian, F.A. (1996) *MNRAS*, **Vol. 278**, 525
Cawley, M.F. (1996) *Nuovo Cimento*, **Vol. 19C**, pp. 959–962
Drury, L.O'C., Aharonian, F.A., and Völk, H.J. (1994) *Astron. Ap.* **Vol. 287** 959
Harding, A.K., De Jager, O.C., and Gotthelf, E. (1997) *Proc. 25th ICRC (Durban)* **Vol. 3** 325
Hess, M. et al. (1997) *Proc. 25th ICRC (Durban)* **Vol. 3** 229
Kifune, T. et al. (1995) *Ap J Lett.* **Vol 438**, L91
Kifune, T. (1996) in *Pulsars: Problems and Progress, IAU Coll. 160*, pp. 339–346
Koyama, K. et al. (1995) *Nature* **Vol 378**, 255
Lessard, R.W. et al. (1997) *Proc. 25th ICRC (Durban)* **Vol. 3** 233
Naito, T. and Takahara, F. (1994) *J. Phys. G.: Nucl. Part. Phys.* **Vol. 20** 477
Tanimori, T. et al. (1997) to be published in *Ap J Lett.*
Turver, K.E. et al. (1997) to appear in *Proc. Kruger Workshop "Towards major atmospheric Čerenkov Telescope V*
Weekes, T.C., Aharonian, F.A., Fegan, D.J., and Kifune, T. (1997), to appear in *Proc. of the 4th Compton Symposium*
Yoshikoshi, T. et al. (1997) *Ap J Lett.* **Vol 487**, L95

Session 3: Diagnostics of High Gravity Objects with X- and Gamma Rays

3-2. Black Hole Binaries

GALACTIC RADIO-JET SOURCES: MULTIWAVELENGTH OBSERVATIONS

PH. DUROUCHOUX
CE. Saclay, DSM, DAPNIA, Service d'Astrophysique
91191 Gif sur Yvette, Cedex France

AND

D. HANNIKAINEN
Observatory, PO Box 14
University of Helsinki, Finland

1. ABSTRACT

So far, seven X-ray sources have been identified to be radio-jet sources. We present a review of the observations of 1E 1740-2942, GRS 1758-258, GRS 1915+105, GRO J1655-40, SS 433, Cir X-1, Cyg X-3.

2. Introduction

About 200 X-ray binaries have been discovered and, a radio counterpart has been detected for 15% of them. Of these 30 radio binary sources, four exhibit either periodically or continuously a radio-jet structure. In the three other cases (among the seven identified jet sources), binarity has not (yet?) been found, but due to their location (Galactic center or Galactic disk region) the large visual absorption could explain the lack of detection of a companion.

3. SS 433

SS 433, a 13.1-day orbital period X-ray binary, is a weak X-ray source at the center of a very large complex of radio (W50) emission (Konigl 1983; Baum & Elston 1985). This radio emission is persistent and occasionally undergoes huge radio bursts with the flux increasing by factors of 10. The most important characteristic of SS 433 is the continuous ejection of matter at a velocity of 0.26c in the form of radio jets (Hjellming & Johnson,

1988). In X-rays, Watson et al. (1983) found two bright diffuse lobes with the EINSTEIN satellite, symmetrically displaced east and west of SS 433, and aligned along the axis of W50. Recent observations performed with the X-ray ASCA satellite of the eastern lobe of W50 (see Fig. 1 in Yamauchi, Kawai and Aoki, 1994) reveal the nonthermal nature of the lobes, as evidenced by the absence of emission lines. More recently, Safi-Harb and Ogelman (1997) reanalyzed 1991 ROSAT observations of the SS 433 eastern lobe, and found a knotty structure (see Fig. 2 in Safi-Harb and Ogelman, 1997), with additional soft X-ray emission from this lobe at 1 degree east of SS 433 and coincident with the radio "ear", which is interpreted as the region associated with the terminal shock of the SS 433 jet.

4. Cyg X-3

Cyg X-3 is also a persistent radio source, with huge outbursts (factor 50) on a time scale of days. During such a flaring period, Hjellming et al. (1997) found Cyg X-3 to be an expanding radio source with an expansion velocity of 0.9c.

5. Cir X-1

The radio emission of Cir X-1 is less known. The source was observed several times in the 70's, exhibiting repeating radio flares every 16.6 days (the orbital period of the system). By the end of 1990 and the beginning of 1991, Steward et al. (1993) observed Cir X-1 and the nearby supernova remnant G321.9-0.3, and found clear evidence for jet-like structure within the nebula surrounding Cir X1. As mentioned by the authors, the jets originate at a compact source at the position of the binary, then extend outwards about 30 arcsec before curving back several arcmin towards the nearby supernova remnant G321.9-0.3. The optical counterpart of Cir X-1 is a faint red star (Moneti 1992), and the high ratio of the X-ray to optical luminosity indicates that Cir X-1 is a low mass X-ray binary (Oosterbroek et al. 1995). Recently, Shirey et al. (1996) studied the timing and spectral evolution of Cir X-1 versus orbital phase with the ROSSI X-ray Timing Explorer and found quasi-periodic oscillations.

6. GRO J1655-40

GRO J1655-40 was detected as an X-ray transient in July 1994 (Zhang et al. 1994). Soon after this observation, Tingay et al. (1995) reported the detection of two radio components moving away from each other at an angular speed of 65 ± 5 mas d^{-1}, corresponding to superluminal motion at the estimated distance of 3.2 kpc (Hjellming et al. 1995). A multiwavelength analysis showed delays between the X-ray and radio outbursts interpreted by Tingay et al. (1995) to be the ejection of material at relativistic speeds occuring during a stable phase of accretion onto a black hole, followed by an unstable phase with high accretion rate. The existence of a black hole as the compact star of a binary system with a 2.62 day orbital period was verified after measurement of its mass function(Bailyn et al. 1995).

7. 1E 1740.7-2942 and GRS 1758-258

Both 1E1740.7-2942, the Great Annihilator, and GRS 1758-258 are bright Galactic Center hard X-ray sources as well as clearly variable radio sources with arcmin scale radio jets. No optical/IR counterparts have been discovered probably because of the high visible absorption towards the Galactic Center. This does not rule out the possibility that these sources are binary systems.

8. GRS 1915+105

The X-ray transient GRS 1915+105 was discovered by the WATCH instrument onboard the GRANAT satellite in 1992 (Castro-Tirado et al. 1994), and soon after, SIGMA measured an accurate position allowing radio observations. VLA observations led to the discovery of relativistic ejections of plasma with an ejection velocity of 0.92c (Mirabel and Rodriguez, 1994). Analyzing carefully the timing of the ejections, we found a repetitivity around 28 days. Ryle radio observations (15 GHz) as well as Nancay (3.2 & 1.4 GHz) and GBI (2 & 8 GHz) exhibit large outbursts associated with these ejections, fitting with this repetitivity relatively well and leading to the idea that these ejections could be associated with the saturation of an accretion disk, filled by a hidden companion orbiting the compact object with a 28-day orbital period (Hannikainen and Durouchoux 1998).

9. Conclusions

It appears that, for binary systems, the jet sources have different types of compact objects (NS and BH), companions (early and late type stars) and orbital periods (from hours to tens of days). From Table 1 we also note two ranges of ejection velocities: the lower velocities could be associated with neutron stars and the higher with black holes. If large radio outbursts seem to be correlated with plasmoid ejections, there is no obvious correlation between radio and X/gamma-rays, and more correlated observations are strongly needed to better understand the physics which drives the behavior of the radio-jet sources.

10. References

Baum & Elston, 1985, in The Crab Nebula and Related Supernova Remnants, eds M.C Kafatos and R. Henry (Cambridge U. Press), P 251
Bailyn et al. 1995, IAU Circ. No. 6173
Castro-Tirado et al. 1994, ApJS, 92, 469
Fender et al. 1995, MNRAS, 274, 633
Fender, Brocksopp & Pooley, 1997, IAU Circ 6544
Han and Hjellming, ApJ, submitted
Hannikainen and Durouchoux, 1998, to be published
Hjellming & Wade. ApJ, 1971, 168, L21
Hjellming & Johnson, 1981, ApJ, 246, L141
Hjellming & Johnson, 1988, ApJ, 328, 600
Hjellming et al. 1995, Nature, 375, 464
Hjellming et al. 1997, this symposium
Konigl, 1983, MNRAS, 205, 471
Margon, 1984, Ann. Rev. Astron. Astrophys., 22, 507
Marti, Mirabel, Rodriguez, Parades, 1997, Vistas in Astronomy, vol 41
Mirabel & Rodriguez, 1994, Nature, 371, 46
Oosterbroek, van der Klis, Kuulkers, van Paradijs, Lewin 1995, A&A, 297, 141
Shirey et al. 1996, ApJ, 469, L21
Steward, Caswell, Haynes, Nelson, 1993, MNRAS, 261, 593
Watson et al., 1986, MNRAS, 222, 261
Zhang et al. 1994, IAU Circ. No. 6052

PROPERTIES OF THE RADIO-JET X-RAY SOURCES (1)

SOURCE	DISTANCE (Kpc)	COMPANION	COMPACT OBJECT	PERIODICITIES	REMARKS
SS 433	5.1	OB (?)	NS (?)	13.1 d (o), 164 d (p)	ejection ≈ 0.26 c
CYG X3	8.5-12	WR	NS (?)	4.8 h (o)	ejection 0.3-0.5 c
1E 1740.7-2942	8.5 (GC)	binary (?)	BH (?)		
GRS 1758-258	8.5 (GC)	binary (?)	BH (?)		
CIR X1	> 5.5	LMXRB (?)	NS	16.6 d (o)	
GRS 1915+105	12.5	Be (?)	BH (?)	28.2 d (o)	superluminal 0.92 c
GRO J1655-40	3.2	F or G	BH	2.6 d (o)	superluminal 0.92 c

ACCRETION DISKS IN BLACK HOLE CANDIDATES OBSERVED WITH ASCA

T. DOTANI
Institute of Space and Astronautical Science
3-1-1 Yoshino-dai, Sagamihara, Kanagawa 229, Japan

Abstract. Structure of the accretion disk is compared between the soft and hard states of Cyg X-1 using the ASCA data. Large uncertainty of the disk parameters in hard state prevent us from drawing clear conclusion, but the data are consistent with a factor of 3 larger (optically thick) inner disk boundary in the hard state than in the soft state.

1. Introduction

Accretion disk is believed to play an important role to power galactic and extra-galactic high energy objects. Gravitational potential energy is converted to radiation and kinematic energy through the accretion disk, but the conversion mechanisms are not fully understood yet.

Stellar mass black holes in Galaxy may be suited to investigate the accretion disk because of its proximity. We analyzed the ASCA data of Cyg X-1, which is a prototypical and the best studied black hole candidate (BHC). We compared the hard/soft state data to investigate how the disk structure changes with the state transition. Data from another BHC, LMC X-3, are also investigated.

2. Analysis and Results

2.1. SOFT STATE

2.1.1. *Cyg X-1*

Cyg X-1 is mostly found in hard state, but rare transition to soft state was observed in 1996 May, which continued for about 3 months. ASCA observation of Cyg X-1 in the soft state was made in May 30-31 for net

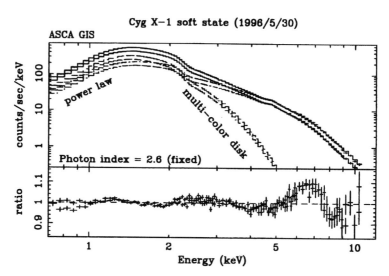

Figure 1. Energy spectrum of Cyg X-1 in the soft state obtained with ASCA GIS. Best-fit continuum model, i.e. multi-color disk model plus a power law, is also shown. Residual structure around 7 keV may be due to a broad iron line.

exposure of 33 ksec. Spectral analysis of the data is described in Dotani et al. (1997) and Kubota et al. (1997). We just show here the energy spectrum with the model function (figure 1). The apparent temperature of the inner disk boundary was 0.43 ± 0.01 keV and the apparent inner disk radius to be 71^{+13}_{-2} km for the assumed distance of 2.5 kpc. This apparent radius corresponds to $3R_s$ (R_s: Schwarzschild radius) of 12 $M_\odot$ object, which coincides with the best mass estimate of Cyg X-1. Thus, optically thick accretion disk is considered to be extended down to $3R_s$ in the soft state (Balucinska-Church et al. 1997).

A power law component become prominent above 4 keV; structures are noticed around 6–7 keV as shown in the residual plot of figure 1. The structures may be interpreted as a broad line, which is expected from highly ionized accretion disk (Cui et al. 1997).

2.1.2. *LMC X-3*

We observed LMC X-3, which always stay in soft state, in 1993 September and 1995 April. The energy spectra are well fitted with a power law plus a multi-color disk model. The inner disk temperature shows significant change between the two observations (0.83 ± 0.03 keV in 1993 and 0.98 ± 0.02 keV in 1995), but the apparent disk radius stayed constant. This is just expected if an optically thick accretion disk is extended down to $3R_s$.

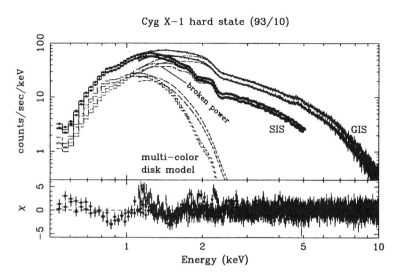

Figure 2. Energy spectrum of Cyg X-1 in the hard state obtained with ASCA. Best-fit model function, multi-color disk model plus a broken power law, is also shown.

2.2. HARD STATE

ASCA observed Cyg X-1 in hard state many times, because it is one of the instrumental calibration sources of the satellite. We analyzed here the data in 1993 October and November, because SIS data were available for these data. The same data were already analyzed by Ebisawa et al. (1996).

The energy spectrum and the best fit model functions are shown in figure 2 for the data in 1993 October. We used a model consisting of a multi-color disk model and a broken power law modified by the low-energy absorption and the dust scattering. We used parameters of the dust scattering halo obtained by ROSAT (Predehl and Schmitt 1995). As shown in the figure, the disk and the power law components have similar contribution in lower energy bands, and the uncertainty of the spectral shape of the power law component prevents us from determining the precise parameters of the disk emission. Allowed range of the disk parameters, in which systematic errors and time variations are also included, is shown in figure 3.

3. Discussion

Based on the wide band observations of Cyg X-1, Zhan et al. (1997) concluded that bolometric luminosity is almost same between hard and soft states. This means that mass accretion rate does not change during the

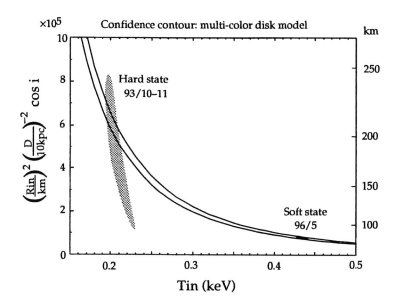

Figure 3. Allowed regions of disk parameters of Cyg X-1 determined with ASCA data. Abscissa is the color temperature at the inner disk boundary and the ordinate is a quantity proportional to the square of the inner disk radius. The regions include not only the statistical errors but also systematic errors due to the time variations and uncertainty in the power law component. See text for the solid lines.

hard/soft transition. Solid lines in figure 3 indicate the local disk temperature as a function of disk radius when the mass accretion rate is constant. As seen from the figure, hard/soft state data are consistent with only the change of inner disk radius, but the large uncertainty in the hard state data does not exclude other possibilities (Balucinska-Church et al. 1997).

Acknowledgment   Present work is based on the collaboration with following people: H. Inoue, F. Nagase, K. Mitsuda, Y. Ueda, K. Asai, Y. Aruga (ISAS), K. Makishima, A. Kubota (Tokyo Univ), H. Negoro (Riken), K. Ebisawa (NASA/GSFC), W. Cui (MIT), M. Balucinska-Church, M. Church (Birmingham Univ).

References

Dotani, T. et al. 1997, ApJL, 485, L87
Cui, W. et al. 1997, this volume.
Ebisawa, K. et al. 1996, ApJ, 467, 419
Kubota, A. et al. 1997, this volume.
Balucinska-Church, M. et al. 1997, this volume.
Predehl, P. and Schmitt, J. H. M. M., 1995, A&A, 293, 889
Zhan, S. N. et al. 1997, ApJL, 477, L95

Session 3: Diagnostics of High Gravity Objects with X- and Gamma Rays

3-3. AGNs

EVIDENCE FOR STRONG GRAVITY IN THE AGN PLASMAS

K. IWASAWA
Institute of Astronomy
Madingley Road, Cambridge CB3 0HA, UK

Abstract. X-ray spectroscopy of the broad iron line has revealed some relativistic effects caused by strong gravity about a black hole in active galactic nuclei (AGN). Recent results from ASCA observations of AGNs are reviewed.

1. Introduction

The impoved spectral resolution with the CCD X-ray spectrometer onboard ASCA has enabled the iron line profile in active galactic nuclei (AGN) to be resolved. The breadth and its skewed profiles found in Seyfert galaxies are in good agreement with those distorted by relativistic effects due to strong gravity about a black hole. They can therefore be used as a direct probe to the innermost part (e.g., $\leq 20 r_g$) of the accretion disk in AGNs.

The detection of spectral features, a high energy hump above 10 keV and an iron K line at 6.4 keV, with Ginga in Seyfert 1 galaxies (Pounds et al 1990; Matsuoka et al 1990) is consistent with reflection from a cold disk irradiated by an X-ray source above it (e.g., George & Fabian 1991). Rapid X-ray variability commonly seen in AGNs suggests that the X-rays are produced in regions very close to a central massive black hole. X-ray irradiation of the disk occuring at such immediate neighbourhood of a black hole would lead considerable relativistic effects on line emission produced in the disk, as the speed of gas motion is approaching to the speed of light near the event horizon.

The line shape expected from such a relativistic disk have been computed for a stational (Schwartzschild) and a spinning black holes (Fabian et al 1989; Kojima 1991; Laor 1991; see also Chen, Halpern & Filippenko 1989 for emission line profiles from the outer disk). Major effects are Doppler boosting and gravitational redshift. Importance of those effects depends

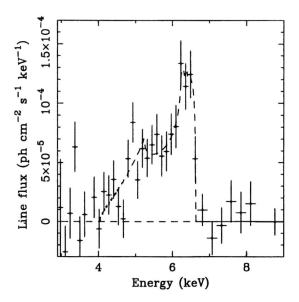

Figure 1. The iron K line profile of the Seyfert galaxy MCG–6-30-15 observed with the ASCA SIS (Tanaka et al 1995). The dashed-line shows the best-fit relativistic disk-line model.

on inclination angle of the disk. Since the iron line is the most prominent feature in the reflection spectrum, X-ray spectroscopy of the line can probe to strong gravity around a black hole in AGNs. Here ASCA studies of the broad iron line in AGNs are reviewed.

2. The ASCA results

The 4-day long observation of the bright Seyfert 1 galaxy MCG–6-30-15 provided high quality data for measuring the iron line profile (Fig. 1, Tanaka et al 1995). The line shape is clearly asymmetric and skewed to lower energy below 5 keV. The stronger blue (higher energy) peak is due to Doppler beaming originating in orbital motion of the disk. The sharp drop of the line emission around 6.6 keV constrains the inclination of the disk to be $\sim 30°$. A fit with the diskline model suggests that the emission line is produced within 20 gravitational radii (r_g). Gravitational redshift plays the most important role to shift the emission to lower energies at such a small inclination angle. The relativistic diskline emission appears to be a currently most plausible interpretation for the broad iron lines (Fabian et al 1995).

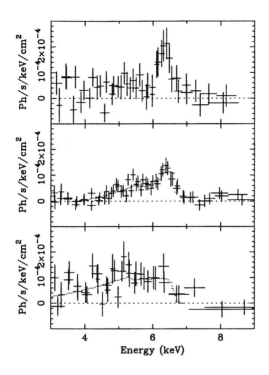

Figure 2. The iron K line profiles observed during the long-look at MCG–6-30-15. Top: bright flare; bottom: deep minimum; and middle: the others.

Variability of the iron line has also been found in several objects (MCG–6-30-15, Iwasawa et al 1996a; NGC7314, Yaqoob et al 1996; NGC3516, Nandra et al 1997a). A detailed study of the ASCA long-look data on MCG–6-30-15 revealed complicated behaviours of the line in response to the continuum changes (e.g., Fig. 2). Particularly, the line profile observed when the X-ray source is extremely faint indicates even more redshift than the others. In order to explain the enormous redshift, the line emission should be produced very close to a black hole (a few r_g) which is well within the last stable orbit of the disk in a stational black hole ($\sim 6r_g$). One possible way to account for this is having a black hole spinning. In a maximally spinning ($a/M = 0.998$) Kerr black hole, for instance, the accretion disk stretches down to $1.24r_g$. Dabrowski et al (1997) fitted the very broad profile of MCG–6-30-15 and demonstrated that the black hole is indeed rapidly spinning ($a/M > 0.9$). Reynolds & Begelman (1997) however proposed an alternative model with a Schwartzschild black hole, considering emission from gas within the stable orbit. In either case, the primary X-ray source must be located close to a central black hole, where strong gravity is operating.

Broad, skewed iron line profiles have been found in many other Seyfert galaxies (e.g., Mushotzky et la 1995; Yaqoob et al 1995; Iwasawa et al 1996; Weaver et al 1997; Nandra et al 1997b; Reynolds 1997). Quasars, however, appear to have different iron line properties from their lower luminosity counterparts. Nandra et al (1997c) investigated X-ray luminosity dependence of the line profile and intensity, using an ASCA data sample. The decreasing trend of equivalent width (EW) of the line towards higher luminosity AGNs (X-ray Baldwin effect, which was originally suggested by Iwasawa & Taniguchi 1993 based on Ginga data) is found, although there are a few counter-examples (e.g., 3C109, Allen et al 1997). The line emission is peaked at energies above 6.4 keV, suggesting an ionized disk in quasars. They interpret this as a consequence of a high accretion rate.

References
Allen S.W., et al, 1997, MNRAS, 286, 765
Chen K, Halpern J.P., Filippenko A.V., 1989, ApJ, 339, 742
George I.M., Fabian A.C., 1991, MNRAS, 249, 352
Fabian A.C., Rees M.J., Stella L., White N.E., 1989, MNRAS, 238, 729
Fabian A.C. et al, 1995, MNRAS, 277, L11
Iwasawa K & Taniguchi Y., 1993, ApJ, 413, L15
Iwasawa K., et al, 1996a, MNRAS, 279, 837
Iwasawa K., et al 1996b, MNRAS, 282, 1038
Kojima Y., 1991, MNRAS, 250, 629
Laor A., 1991, ApJ, 376, 90
Matsuoka M., et al, 1990, ApJ, 360, L35
Mushotzky R.F., et al, 1995, MNRAS,
Nandra K., et al, 1997a, MNRAS, 284, L7
Nandra K., et al, 1997b, ApJ, 477, 602
Nandra K., et al, 1997c, ApJ, 488, L91
Pounds K.A., et al, 1990, Nat, 344, 132
Reynolds C.S., 1997, MNRAS, 286, 513
Reynolds C.S., Begelman M.C., 1997, ApJ, 488, 109
Tanaka Y., et al, 1995, Nat, 375, 659
Yaqoob T., e al, 1995, ApJ, 453, L81
Yaqoob T., et al, 1996, ApJ, 470, L27
Weaver K.A., et al., 1997, ApJ, 474, 675

X-RAY ASPECTS OF THE IRAS GALAXIES

E.J.A. MEURS

Dunsink Observatory
Castleknock, Dublin 15, Ireland

Abstract. Several IRAS galaxies have been detected at X-rays, with a variety of satellite observatories. About half of these are classified optically as Seyfert galaxies. Among those not (convincingly) classified as AGN, many have X-ray luminosities for which stellar evolution products offer convenient explanations. Some non-active IRAS galaxies display anomalously high levels of X-ray emission for which several conceivable origins are investigated: optical misclassification, X-ray misidentification, hidden AGN, incidental activity, starburst, environmental sources. X-ray spectral studies and temporal variations constitute important tools for further investigation, for instance to assess the strength of a starburst or to establish signatures of an active core.

1. Introduction

IRAS galaxies figure regularly in the timelines of the X-ray satellite observatories. Often this refers to active galaxies of some sort and it is not obvious *a priori* that IRAS galaxies generally would be suitable X-ray targets. In this talk the question is considered why non-active IRAS galaxies could be interesting at X-rays at all. One strong point is that they offer homogeneously selected samples for study. The high levels of starformation experienced by many of these objects constitutes an area of specific interest, which has to be contrasted with the case for hidden active nuclei, given the sometimes strong X-ray emission from these objects. In a wider context, galaxies as considered here have been suggested as significant contributors to the X-ray background.

2. Homogeneous samples for study

The galaxies included in the IRAS Point Source Catalog tend to exhibit increased levels of starformation, as their IR colours demonstrate (Helou 1986). Most of the IRAS galaxies are spiral galaxies and a number of them is classified as active (Seyferts). The IR most luminous cases are referred to as Ultraluminous and a link with QSOs has been suggested (Sanders et al. 1988).

The selection of objects by the IRAS satellite is rather homogeneous and thus provides attractive samples for study, distributed all over the sky. At X-rays, several galaxies from the IRAS Bright Galaxy Survey were studied by David et al. (1992), using data from the Einstein satellite. A larger compilation of Einstein data was published by Green et al. (1992). Based on a multivariate selection of likely extragalactic IRAS sources (Meurs et al. 1988; Boller et al. 1992a), a correlation with ROSAT All Sky Survey data yielded 242 proposed detections out of 14708 target positions (Boller et al. 1992b); optical data on these cases were taken from the NASA Extragalactic Database. Moran et al. (1996) published a dedicated optical follow-up of the majority of objects in this sample.

The Boller et al. results included a dozen of surprisingly X-ray luminous and apparently non-active spiral galaxies, reaching L_X values well into the range normally exhibited by proper active galaxies. Initial guesses by Boller et al. suggested active nuclei (possibly hidden), starbursts, or a combination of these two. Subsequent optical spectroscopy has produced contradictory results for the classification of these objects as regards their level of activity (*cf.* Meurs et al. 1992; Moran et al. 1994). Evidence for non-active galaxies emitting X-rays at such high levels has been put forward by Fruscione and Griffiths (1991), Griffiths et al. (1992, 1995) and Boyle et al. (1995), while a different view was expressed by Moran et al. (1994, 1996) and Halpern et al. (1995). Taking the extensive Moran et al. (1996) classification results at face value, trends for decreasing L_X and increasing Δ_{ox} (optical/X-ray positional difference) are observed when going from Sy1 via Sy2 and Starbursts to normal galaxies (Meurs, to be submitted). This suggests that, from active nuclei towards stellar evolution products, the bodies of the galaxies are increasingly contributing to the observed X-ray emission. This is consistent with ROSAT PSPC results that along the same sequence of objects the X-ray source size increases (Meurs et al., to be submitted).

Archival Rosat PSPC data for 7 X-ray luminous IRAS galaxies have been analysed, concentrating on (1) the exact position of the X-ray source, (2) the X-ray morphology, (3) companion X-ray sources (Meurs and Norci 1996). It turns out that not in all cases the X-ray source is consistent with the optical nucleus of the IRAS galaxy. The morphology of the X-ray

TABLE 1. Results of archival ROSAT PSPC study

Object	Central	Extended	Companions
01590-3158	+	□	?
10126+7339	+	□	+
10257-4338	-	+	-
10303+7401	-	-	+
11395+1033	p	-	?
13224-3809	?	-	?
16155+6831	p	+	+

sources indicates that some exhibit extended X-ray sources, while several have companion sources close to them. The table shows the results of this (pilot) investigation. Confirmations are indicated by +, rejections by -, weak support by □, uncertain cases by ? in the table. If no improvement of position with astrometry (based on corresponding optical and X-ray point sources) was possible but the X-ray position from the satellite pointing appears consistent with the galaxy nucleus, then this is indicated by p.

The implication of these results is that the evidence is not generally in favour of unrecognized active nuclei, but again that the disk (or perhaps halo) of the galaxy may be responsible for a significant fraction of the X-ray emission. The source extensions and the occurrence of nearby sources may be meaningful in view of the fact that available catalogued optical data in some cases refer to membership of small groups of galaxies.

In view of some ambiguous results from optical spectroscopy, it is interesting to remember that active galaxies might change category due to change of activity class (*e.g.* NGC4151, Penston and Pérez 1984) or to temporary activity (Rees 1988). Following this suggestion in Meurs and Norci (1996), this very possibility was quoted for the IRAS galaxy NGC5905 (Bade et al. 1996) and also for another case, NGC3256 (Moran et al. 1996).

3. X-ray diagnostics

Regarding X-ray luminosities, the apparently non-active IRAS galaxies occupy a position between active and normal galaxies. A particularly interesting question in this connection is how high an X-ray luminosity can be reached by an evolving young stellar population alone (*i.e.* without invoking an active nucleus). This issue was already considered by David et al (1992) and more progress will be obtained with the help of high-energy population synthesis work (*e.g.* Lipunov et al. 1996; Guillout et al 1996; Norci and Meurs, in preparation).

For individual IRAS galaxies, the current X-ray satellites (ASCA, SAX) provide sufficiently high spectral resolution and wide enough spectral bandwidth that several useful diagnostics can be applied. These are especially interesting for assessing whether active nuclei, hidden at optical wavelengths, could be mainly responsible for the observed X-ray emission. Spectral modelling informs about abundances and physical conditions in starburst components. Hard X-ray tails may be relevant to contributions from nuclear activity, but for completely obscured nuclei scattered Fe 6keV lines are expected to be seen (*e.g.* Iwasawa et al. 1993; Ueno et al. 1996; Moran and Lehnert 1997). Cold and possibly also ionized absorbing columns can be measured from the X-ray spectra and refer to conditions near the source. Other X-ray characteristics, like luminosity, morphology and temporal modulation, remain of importance as with previous satellite data.

References

Bade et al. (1996) AA 309, L35
Boller et al. (1992a) AA 259, 101
Boller et al. (1992b) AA 261, 57
Boyle et al. (1995) MN 272, 462
David et al. (1992) ApJ 388, 82
Fruscione and Griffiths (1991) ApJ 380, L13
Green et al. (1992) MN 254, 30
Griffiths et al. (1992) MN 255, 545
Griffiths et al. (1995) MN 275, 77
Guillout et al. (1996) AA 316, 89
Halpern et al. (1995) ApJ 453, 611
Helou (1986) ApJ 311, L33
Iwasawa et al. (1993) ApJ 409, 155
Lipunov et al. (1996) ApJ 466, 234
Meurs et al. (1988) *Astronomy from large databases: scientific objectives and methodological approaches*, Murtagh & Heck (Eds), p. 49
Meurs et al. (1992) *Star forming galaxies and their interstellar medium*, Franco et al. (Eds), p. 168
Meurs and Norci (1996) *Röntgenstrahlung from the Universe*, Zimmermann et al. (Eds), MPE Report 263, p. 481
Moran et al. (1994) ApJ 433, L65
Moran et al. (1996) ApJS 106, 341
Moran and Lehnert (1997) ApJ 478, 172
Penston and Pérez (1984) MN 211, P33
Rees (1988) Nature 333, 523
Sanders et al. (1988) ApJ 328, L35
Ueno et al. (1996) PASJ 48, 389

THE PROPERTIES OF BLAZARS DETECTED BY EGRET

J.R. MATTOX
Boston University Astronomy Department
725 Commonwealth Avenue, Boston, MA 02215
http://bu-ast.bu.edu/~mattox/

1. The Detection of Blazars by EGRET

One of the most exciting recent developments in astrophysics is the detection of members of the blazar class of AGN at ~1 GeV by the EGRET instrument aboard the *Compton Observatory*. Although a weak detection of 3C 273 was obtained by the *COS B* mission, the very strong detection of 3C 279 by EGRET (Hartman *et al.* 1992) two months after launch was not anticipated.

Another ~50 blazars have subsequently been identified with EGRET sources. Detected blazars include BL Lacertae objects, "Highly Polarized Quasars" (HPQs), and "Optically Violent Variable" quasars (OVVs). Analysis of the EGRET data for the initial 3C 279 observation revealed strong variability on ~1 day timescales (Kniffen *et al.* 1993). Variability is also apparent for all of the other bright EGRET blazars (Mattox *et al.* 1997a). The brightest EGRET blazar, PKS 1622−297, shows variability on a time scale of several hours (Mattox *et al.* 1997b), as rapid as possibly resolved with the meager EGRET statistics. The EGRET blazar detections led to Čerenkov observations which have detected TeV γ-rays from nearby BL Lac objects Mrk 421 (Gaidos *et al.* 1996) and Mrk 501 (Catanese *et al.* 1997) which have both been observed to flare dramatically. Von Montigny *et al.* (1995) and Mukherjee et al. (1997) provide compendia of EGRET blazar results. Chiang *et al.* (1996) report evidence for positive evolution, and present a luminosity function.

A careful analysis of EGRET source identifications is appropriate because the location of γ-ray emission can't be accurately determined with EGRET. In the first EGRET catalog (Fichtel *et al.* 1994), the reliability of the blazar identification was not addressed. In the second EGRET catalog (Thompson *et al.* 1995 & 1996), blazar identifications were classified as "high-confidence" if the radio source was located within the 95% confidence contour of the EGRET likelihood position estimate (Mattox et al.1997a). A similar classification is made by Punsley (1997). Thompson et al. (1995 & 1996) classified some radio sources beyond the 95% confidence contour as "lower-confidence" identifications. Mattox et al. (1997a) offer a thorough analysis of the identification of EGRET sources with radio sources. They find that there is no doubt that EGRET is detecting blazars — they specify 16 EGRET blazar identifications for which the confidence of a correct identification exceeds of 99%. They also tabulate an additional ~50 plausible identifications. The third EGRET catalog (Hartman et al. 1997) will change the status of some plausible EGRET blazar identifications because additional EGRET exposure has been used to refine position estimates of previously detected sources, and new EGRET sources have been detected.

2. The Emission of Gamma Rays by Blazars

The apparent γ-ray luminosity of the EGRET blazars (as quantified by νF_ν) is often 1–2 orders of magnitude greater than that observed at other energies. Therefore, unless the γ-ray emission is more highly beamed than that at lower frequencies, most of the non-thermal energy emerges as $E \gtrsim 100$ MeV photons during the γ-ray high states. The rapid variability of the flux observed by EGRET implies that the emission takes place in a compact region. Because the opacity to $\gamma - \gamma$ pair production must be low for the γ-rays to escape, relativistic beaming is required if the observed x-rays originate in the same volume as the γ-rays (Dondi & Ghisellini 1995). Blazar 3C 279 was observed in 1996 to flare simultaneously at γ-ray and x-ray energies (Wehrle et al. 1998). This indicates that x-rays do originate in approximately the same volume as γ-rays, and that the γ-rays are produced in a relativistic jet.

Most models of blazar γ-ray emission feature inverse Compton scattering as the γ-ray emission mechanism, but there is not a consensus as to the origin of the low energy photons which are scattered. It has been suggested that they might originate within the jet as synchrotron emission (Maraschi Ghisellini & Celotti 1992). This is designated as the synchrotron self-Compton (SSC) process. Another possibility is that the scattered low-energy photons are ambient. This is designated as the external Compton scattering (ECS) process. Dermer, Schlickeiser, & Mastichiadis (1992) suggested that they come directly from an accretion disk around a blackhole at the base of the jet. It was subsequently proposed that the dominant source of the low energy photons for scattering could be due to re-processing of disk emission by broad emission line clouds (Sikora, Begelman, & Rees 1994; Blandford & Levinson 1995). Ghisellini & Madau (1996) suggest that the dominant source of scattered low-energy photons is broad-line-cloud re-processing of jet synchrotron emission.

Maraschi et al. (1994) reported that the change in the multiwavelength spectrum of 3C 279 between the 1991 flare and a low state observed 18 months later was consistent with the SSC model for γ-ray emission. However, Hartman et al. (1996) noted that the spectra of 3C 279 in both these high and low states could also be fit with an ECS model. The situation is now further complicated by the finding that the spectral change for the 1996 flare of 3C 279 is different than for the 1991 flare (Wehrle et al. 1998). At the peak of the 1996 flare, the EGRET flux is a factor of 3 larger than at the peak of the 1991 flare, but the optical flux is only 70% of that at the peak of the 1991 flare. Wehrle et al. (1998) note that the γ-ray emission increases by more than the square of the observed increase of the infrared-optical flux during the course of the flare. This is probably not consistent with a one-zone SSC model. It is also not consistent with any of the ECS models except that of Ghisellini & Madau (1996). There model can be tested with sensitive multi-epoch observations of the Lyα flux of 3C 279 in conjunction with simultaneous observations of the synchrotron flux of the jet (Wehrle et al. 1998).

3. The Properties of EGRET Blazars

Even though EGRET has obtained some exposure for all directions on the sky, only ~10% of blazars are detected. Insight into the γ-ray emission process of blazars may potentially be gained by an examination of the properties that distinguish the EGRET blazars from the general blazar population. All EGRET blazars which have been observed with VLBI sufficiently well show superluminal motion (Barthel et al. 1995). Several groups are making additional VLBI observations of EGRET blazars (Tingay et al. 1996; Bower et al. 1997; Marchenko et al. 1997).

Impey (1996) reported that the total radio flux of the EGRET blazars tends to be larger than the radio flux of other blazars in the parent population. This difference is also apparent for the milliarcsecond-scale radio emission of EGRET blazars (Moellenbrock et al. 1996, Mattox et al. 1997). This supports arguments that the γ-ray emission is taking place at the base of a relativistic jet. Although it is expected to be of insufficient strength to cause the observed difference, the increase in the security of EGRET blazar identifications with radio flux (as quantified by Mattox et al. 1997a) is a selection effect which causes some bias in the observed direction.

4. Future γ-ray Observations

The γ-ray emission of blazars is not yet fully understood. Because its spark-chamber gas is nearly depleted, EGRET is unlikely to provide definitive new blazar data. Data from the proposed GLAST satellite, which offers a sensitivity 1—2 orders of magnitude better than EGRET, will be very important for further blazar studies, e.g., the GLAST data will provide for a determination of whether the diffuse extragalactic high-energy γ-ray flux (Sreekumar et al. 1997) is exclusively emitted by blazars. In the mean time, much can probably be learned from Čerenkov γ-ray blazar observations in conjunction with simultaneous observations at lower energies.

References

Barthel, P. D., et al., 1995, ApJL, 444, L21
Blandford, R.D., & Levinson, A., 1995, ApJ, 441, 79
Bower, G.C., et al., 1997, ApJ, 484, 118
Catanese, M., et al., 1997, ApJ, 487, L143
Chiang, J., et al., 1996, ApJ, 452, 156; errata 465, 1011
Dermer, C. D., Schlickeiser, R., & Mastichiadis, A., 1992, A&A 256, L27
Dondi, L., Ghisellini, G, 1995, MNRAS, 273, 583
Fichtel, C.E., et al., 1994, ApJ S, 94, 551
Gaidos, J.A., et al., 1996, Nature, 383, 319
Ghisellini, G., & Madau, P., 1996, MNRAS, 280, 67
Hartman, R.C., et al., 1992, ApJ, 385, L1
Hartman, R.C., et al., 1996, ApJ, 461, 698
Hartman, R.C., et al., 1997, in preparation
Impey, C., 1996, AJ, 112, 2667
Kniffen, D.A., et al. 1993, ApJ, 411, 133
Maraschi, L., Ghisellini, G., & Celotti, A. 1992, ApJ 397, L5
Maraschi, L., et al., 1994, ApJ 435, L91
Marchenko, S.G., Marscher, A.P., et al., 1997, IAU Coll. 164, ASP, in press
Mattox, J.R., et al., 1997a, ApJ, 481, 95
Mattox, J.R., et al., 1997b, ApJ, 476, 692
Moellenbrock, G.A., 1996, AJ, 111, 2174
von Montigny, C., et al., 1995, ApJ, 440, 525
Mukherjee R., et al., 1997, ApJ, 490, 116
Punsley, B., 1997, AJ, 114, 544
Sikora, M., Begelman, M. C., & Rees, M. J. 1994, ApJ 421, 153
Sreekumar, P., et al. 1997, ApJ, in press
Thompson, D. J., et al., 1995, ApJ S, 101, 259
Thompson, D. J., et al., 1996, ApJ S, 107, 227
Tingay, S.J., et al., 1996, ApJ, 464, 170
Wehrle et al., 1998, ApJ, in press

HIGH ENERGY PHENOMENA IN AGN JETS

F. TAKAHARA
Department of Earth and Space Science, Osaka University
Toyonaka, Osaka 560, Japan

1. Introduction

BL-Lac objects and optically violent variable quasars (OVVs), called together blazars, are characterized by rapid time variability, strong optical polarization, superluminal expansion and strong gamma-ray emission. Such properties are understood in the framework of a relativistic jet emanated from the central powerhouse. Blazars are considered to be objects for which the direction of the jet is very close to the line of sight.

Four major issues are addressed on the physics of blazars. The first one is the bulk acceleration mechanism of the jets; radiative acceleration has been shown to be ineffective as a result of strong radiation drag and magnetic acceleration is still ad hoc and not a compelling one. The second one is the energetics of the jets; do particles or magnetic fields dominate the dynamics of jets? The third issue is the composition of the jets; is the jet composed of ususal proton-electron plasma or electron-positron plasma? The fourth issue is the particle acceleration mechanism in the jets; we need an efficient and ubiquitous mechanism to explain observations of blazars.

Important keys to these questions are provided by recent gamma-ray and multiwavelength observations. As is well known, to explain rapid time variabilities and gamma-ray transparency, we need relativistic beaming and a small size of the gamma-ray emitting region [1]. The inferred size of the gamma-ray emitting region is around 0.003pc to 0.03pc. This implies that gamma-ray emission and bulk acceleration of jets take place at a few hundred Schwardschild radii unless the ratio of the size to the distance from the central black hole is extremely small. Thus, the gamma-ray emission probes deep inner part of the relativistic jets, one or two orders of magnitudes deeper than the radio VLBI observations.

Broad band emission spectra from blazars are composed of distinctly two components. While radio through optical emission is believed to be produced by the synchrotron radiation of relativistic electrons, gamma-rays are most naturally produced by the inverse Compton scattering of soft photons by the same electron population [1, 2, 3, 4]. X-ray emission is due to either synchrotron radiation or inverse Compton depending on the sources. In the following, I discuss implications for the electron acceleration mechanism and the nature of the relativistic jets implied by these observations. Here, I only describe the main results; the details can be seen in [5, 6, 7].

2. ELECTRON ACCELERATION

As for the particle acceleration mechanism, diffusive shock acceleration has shown some success in reproducing the observed radiative properties of blazars, although its simple application to electron acceleration in AGN environments may not be so trivial. Comparing acceleration time scale, radiative cooling time and advection time, we expect the universal power law spectrum with an index of 2 below the break Lorentz factor γ_{br} and a power law spectrum with an index of 3 above γ_{br}. The resultant emission spectra of synchrotron radiation and inverse Compton scattering of external soft photons are a power law with an energy index of 1.0 at high energies, while 0.5 at lower energies. Those break features may be indentified with the MeV break observed for many of OVV quasars. Note that since the synchrotron self-Compton spectra are rather smooth, reflecting the broad energy distribution of target photons, a clear break will be seen only for Compton scattering of external photons.

The derived spectral shape implies that the dissipated energy into the particle acceleration is efficienty converted into the beamed radiation, because for the universal energy spectra the dissipated energy is equally shared among electrons with a unit logarithmic energy scale.

Note that while in most of other works, electron spectra are arbitrarily assumed to fit the observed spectra, the standpoint here is that the shock acceleration can well reproduce such spectra provided that the shock velocity is near the speed of light.

3. MODEL PREDICTIONS

For simplicity, adopting a uniform sphere of a radius R with a bulk Lorentz factor of Γ and assuming given energy densities of magnetic field u_{mag}, relativistic electrons u_{rel} and external soft photons u_{ext}, we can predict the energy densities of various radiation components [6, 7]. It is found that as long as the shock is relativisitic, energy density of the radiation becomes comparable to that of electrons, which means a large radiative efficiency.

An important inference can be obtained for the classification of radiative properties of relativisic jets based on the relative importance of $u_{\rm rel}$ and $u_{\rm mag}$. When there are negligible external photons, we have an SSC dominated source if $u_{\rm rel}$ dominates over $u_{\rm mag}$, while we have a synchrotron dominated source if $u_{\rm mag}$ dominates over $u_{\rm rel}$. Effect of external soft photons is also easily incorporated. When $u_{\rm ext}$ dominates over $u_{\rm mag}$, external Compton always dominates over synchrotron radiation. The relative importance SSC and external Compton luminosities is determined by the ratios of $u_{\rm rel}$ to $u_{\rm mag}$ and $u_{\rm mag}$ to $u_{\rm ext}$. Generally we have Compton dominated sources for a large ratio of $u_{\rm rel}$ to $u_{\rm mag}$. Thus the radiative properties are neatly classified according to the relative importance of energy densities of magnetic field, particles and external photons.

Some limiting cases deserve further examinations. For sources in which cooling is dominated by synchrotron radiation, the jet is dominated by the Poynting flux and is not radiatively decelerated. For sources cooling mainly through SSC, the dominant power is particle kinetic flux but the jet does not suffer from strong radiation drag, either. When external Compton radiation dominates, jets suffer from strong radiation drag and are decelerated.

Above arguments are further extended to the estimation of the jet power. The power of each component is determined by $L_i = \pi R^2 c u_i \Gamma^2$ except for the external soft photons. For the radiation components, L_i is smaller than the observed isotropic luminosity $L_{i,\rm obs}$ by a factor of $4\delta^2$ if we take $\delta \approx \Gamma$, where δ is the beaming factor. Because of efficient radiative cooling, as far as the shock is relativistic, the kinetic power is equal to the radiative power within a factor of a few.

4. OBSERVATIONAL COMPARISONS

Conversely, we can infer the properties of the gamma-ray emission region from multiwavelength data. Applying to typical objects such as 3C279 and Mkn421, we have some general results. First, the observed properties are consistent with predictions of diffusive particle acceleration by (mildly) relativistic shocks. Second, kinetic power is about 30 times greater than the Poynting power, so the jet is kinetic power dominated. When we compute the kinetic power, we only take account of relativistic electrons with $\gamma_{\rm min} = 1$. This puts a strong constraint on the magnetic mechanism of the bulk acceleration of jets. Third, the derived kinetic power amounts to about 10% of the Eddington luminosity for a typical black hole mass. If the jet is composed of electrons and protons, associated proton power would succeed the Eddington luminosity by a few orders of magnitudes, which may suggest that the jet is composed of electron positron pairs. One caveat is the value of $\gamma_{\rm min}$; if $\gamma_{\rm min}$ is as large as 100, the proton power would be comparable to the Eddington luminosity, which may be reasonable. Detailed observations of the region, where electrons with Lorentz factor of 1 to 100 are resposible, will discriminate these two cases. Also the kinetic power derived from radio observations should be compared.

Since the model predicts that γ_{max} and γ_{br} are inversely proportional to the energy density of soft photons, TeV emission is expected for low luminosity objects with a high cutoff frequency of the synchrotron radiation. These sources roughly correspond to X-ray selected BL Lac objects. OVV quasars reveal a high luminosity with a low cutoff frequency of synchrotron radiation.

5. CONCLUDING REMARKS

I discussed several theoretical aspects of relativistic jets in AGNs, motivated by gamma-ray observations of blazars. Physical conditions of the jets are deduced by analyzing multiwavelength spectra in terms of the synchrotron and inverse Compton model. The results show that the jets are particle dominated and that they are largely comprised of electron-positron pairs. The electron spectra are explained by diffusive shock acceleration with radiative cooling. Maximum energy of electrons turns out to be as high as TeV only for low luminosity objects, while it is an order of 10GeV for high luminosity objects. I show that radiative properties of the jets are neatly classified by the relative importance between the kinetic and Poynting power.

Acknowledgements

This work is supported in part by the Scientific Research Fund of the Ministry of Education, Science and Culture under Grant Nos. 09640323 and 09223215.

References

1. Maraschi L., Ghisellini G. and Celotti A. 1992, Ap.J.L., 397, L5.
2. Sikora M., Begelman M. C. and Rees M. J. 1993, Ap.J., 421, 153.
3. Dermer D. C. and Schlickeiser R. 1993, Ap.J., 416, 458.
4. Inoue S. and Takahara F. 1996, Ap.J., 463, 555.
5. Takahara F. 1994, in Towards a Major Atmospheric Cerenkov Detector III, ed. T. Kifune (Tokyo: Universal academy Press) 131.
6. Takahara F. 1996, J.Korean A.S., 29, S99.
7. Takahara F. 1997, to appear in "Relativistic Jets in AGNs", (Lecture Notes in Physics, Springer).

Session 3: Diagnostics of High Gravity Objects with X- and Gamma Rays

3-4. Gamma-Ray Bursts

GAMMA-RAY BURST OBSERVATIONS WITH BATSE

GERALD J. FISHMAN

Space Sciences Laboratory, Code ES-81
NASA-Marshall Space Flight Center
Huntsville, AL 35812 USA

1. Introduction

Gamma-ray bursts (GRBs) will be recorded as one of the outstanding new phenomena discovered in astronomy this century. About once per day, a burst of gamma rays appears from a random direction on the sky. Often, the burst outshines all other sources of gamma-rays in the sky, combined. This paper reviews some of the key observed phenomenon of bursts in the hard x-ray/gamma-ray region, as observed with the BATSE experiment [4] on the Compton Gamma Ray Observatory. The observed time profiles, spectral properties and durations of gamma-ray bursts cover a wide range. Recent breakthroughs in the observation of gamma-ray burst counterparts and afterglows in other wavelength regions have marked the beginning of a new era in gamma-ray burst research. Those observations are described in following papers in these proceedings.

BATSE has been in operation since its launch by the Space Shuttle Atlantis in April 1991 and it is planned to continue operation at least until the year 2005. A comprehensive gamma-ray burst catalog from the BATSE experiment is available from the CGRO Science Support Center at the NASA/Goddard Space Flight Center. The field of gamma-ray bursts has undergone a dramatic change since BATSE. This has resulted primarily from more sensitive observations of the gamma-ray burst intensity and sky distributions [8]. Prior to these observations, the source of gamma-ray bursts were considered by most workers in the field to be relatively nearby neutron stars in the Galactic plane [6]. They are now generally considered to come from sources at cosmological distances.

2. Temporal and Spectral Characteristics

The most striking feature of the time profiles of gamma-ray bursts is the diversity of their time structures and the wide range of their durations. Coupled with this diversity is the difficulty of placing gamma-ray bursts into well-defined types, based on their time profiles. Many bursts have multiple characteristics and many other bursts are too weak to classify. Some burst profiles are chaotic and spiky with large fluctuations on all time scales, while others show rather simple structures with few peaks. No periodic structures have been seen from gamma-ray bursts. There is, however, one general characteristic. At higher energies, the overall burst durations are shorter and sub-pulses within a burst tend to have shorter rise-times and fall-times (sharper spikes). Most bursts also show an asymmetry, with the leading edges being of shorter duration than the trailing edges. There are many bursts which have similar shaped sub-pulses within the burst. The durations of gamma-ray bursts range from about 30ms to over 1000s. However, the duration of a gamma-ray burst, like the burst morphology, is difficult to quantify since it is dependent upon the sensitivity and the time resolution of the experiment.

Almost all of the power from gamma-ray bursts is emitted above 50 keV. Typical spectra from Compton Observatory experiments covering a wide energy range are shown by Tavani (1996) [11]. Most spectra are well fit by a relatively simple analytical expression which has become known as the Band expression [1]. A search for line features (either absorption or emission features) with the detectors of BATSE-Compton Observatory has thus far been unable to confirm the earlier reports of spectral line features from gamma-ray bursts.

3. Intensity and Sky Distributions, Dilation and Redshift

The spatial distribution of the sources of gamma-ray bursts is derived from the observed intensity and sky distributions. The angular (sky) distribution provides two of the dimensions of the spatial distribution, while the intensity distribution is a convolution of the unknown luminosity function and the radial distribution. Even though the luminosity function is unknown, the intensity distribution can still provide constraints on the allowable spatial distributions of gamma-ray burst sources. The BATSE data show a significant deviation of the observed intensity distribution from that expected for a homogeneous, Euclidean distribution.

Since the launch of the Compton Observatory, burst locations have become available for a large sample of weak bursts. BATSE determines directions to burst sources by comparing the count rates of individual detectors, whose response varies approximately as the cosine of the angle from the detector normal. The systematic error of these locations is presently about 1.5°, as determined by comparison with burst locations known via interplanetary timing. The statistical error varies according to the burst intensity; it is around 13° near the BATSE threshold. The BATSE sky distribution, as shown in Figure 1 indicate that this distribution is consistent with isotropy and that this isotropy extends to the weakest bursts [3].

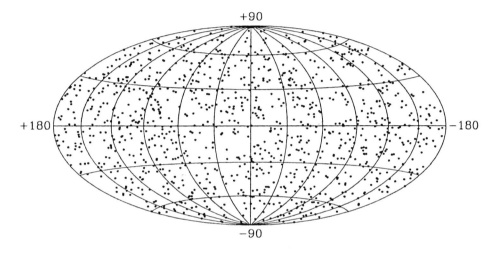

Figure 1. The distribution in galactic coordinates of the 1637 4B bursts.

In accordance with standard cosmology, the more distant bursts are fainter and they are receding faster. Thus they would show a larger time dilation than the nearer, more intense bursts. The entire burst would be "stretched" so that the fainter bursts (and presumably farther) would be, on the average, longer. Individual pulse structures within bursts and the time intervals between these pulse structures would be similarly stretched. Due to the complexity of the gamma-ray burst time structures and the wide range of their durations, any dilation effects can only be tested in a statistical sense. Initial work in these efforts and a positive result was announced by a group at NASA/Goddard Space Flght Center, using BATSE data [10]. An additional indicator of distance can be found in the continua spectra; fainter, more distant bursts would be redshifted, which, in essence, is a time dilation of the wavelength of emission in the observer's frame. Mallozzi et al. [7] show a systematic shift in the energy of the peak power of emission from a large sample of gamma-ray bursts that is consistent with a cosmological redshift.

4. A New Era in Gamma Ray Burst Research

Ever since the initial discovery of gamma-ray bursts, there has been a quest to discover a counterpart to a gamma-ray burst in any other wavelength region before, during, or after the gamma-ray event. These searches have taken many forms, including searches for statistical associations of known objects with bursts with poorly known locations as well as searches of archival plates and other data bases for transient or unusual objects within the error boxes of well-determined burst locations.

BATSE has a quick alert capability which was developed to provide burst locations within several hours, under favorable conditions. A near-real-time burst location system utilizing BATSE data, BACODINE (BAtse COordinates DIstribution NEtwork) [2], is in operation. BACODINE can provide GRB locations to external sites within about 5 s of their detection. When BACODINE is linked to a rapid-slewing optical telescope, there is the possibility of obtaining optical images of burst regions while the burst is in progress. This past year, a rapid burst response system has also been developed to provide more accurate burst locations than BACODINE within about fifteen minutes. The Rossi X-ray Timing Explorer (RXTE) can use this capability to scan a region in search of x-ray afterglow emission.

In 1997, the first x-ray and optical, and radio afterglows from gamma-ray bursts were observed, as described in subsequent papers in these proceedings. Optical observations were used to derive a lower limit for the redshift of the emitter and radio observations were used to derive an angular size. These breakthroughs were made possible by the accurate and precise locations of several gamma-ray bursts observed with BeppoSAX. A new era in gamma-ray burst research has begun.

Note: Some sections of this paper were derived from a review article on gamma-ray bursts [5]. A comprehensive set of references may be found in that paper.

References

1. Band, D., et al. (1993) *Astrophys. J.*, **413**, 281
2. Barthelmy, S., et al. (1994) in AIP Conf. Proc. **307**, Fishman, G., Brainerd, J., Hurley, K., eds. (New York, AIP Press), 643
3. Briggs, M.S., et al. (1996) *Astrophys. J.*, **459**, 40
4. Fishman, G.J., et al. (1989) in GRO Science Workshop, NASA/GSFC, 2-39
5. Fishman, G. & Meegan, C. (1995) *Ann. Rev. Astron. Astrophys.*, **33**, 415
6. Higdon, J., & Lingenfelter, R. (1990) *Ann. Rev. Astron. Astrophys.*, **28**, 401
7. Mallozzi, R., et al. (1996) *Astrophys. J.*, **454**, 597
8. Meegan, C., et al. (1992) *Nature*, **355**, 143
9. Meegan, C., et al. (1996) *Astrophys. J.*, **106**, 65
10. Norris, J.P., et al. (1994) *Astrophys. J.*, **424**, 540
11. Tavani, M. (1996) *Astrophys. J.*, **466**, 768

DISCOVERY OF X-RAY COUNTERPARTS TO GAMMA RAY BURSTS BY BEPPOSAX

L. PIRO
Istituto Astrofisica Spaziale, C.N.R.
Via Fosso del Cavaliere, 00131 Roma, Italy

ON BEHALF OF THE BEPPOSAX TEAM

1. Introduction

The nature of Gamma-Ray Burst (GRB) has been the object of many investigations but their origin has remained a mistery primarily for the difficulties in finding a counterpart. This difficulty derived from the intrinsically poor positioning capability of available GRB detectors.

In the first part of 1997 BeppoSAX has provided the real breakthrough in this domain. BeppoSAX payload is a combination of different instruments (Piro et al. 1995; Boella et al. 1997). While the NFI telescopes (Parmar et al. 1997; Boella et al. 1997b) probe a narrow field with high sensitivity, performing its observation plan as any regular observatory, the two Wide Field Cameras (40 x 40o at zero response) (Jager et al. 1997) are continuously monitoring in the 2–30 keV band, with good time and energy resolution, two other regions of the sky. The Gamma–Ray Burst Monitor (Frontera et al. 1997) electronics provide an onboard trigger on suspect GRBs and high temporal resolution recording of data under these trigger conditions. The combination of these two was very effective. On the GRBM trigger time the Science Operation Center looks for an excess in the WFCs ratemeters. If this is found an image is made at time corresponding to this excess. The burst appears as a point source right at that time while other excesses such as scattered solar flare or magnetospheric events is cancelled by the coded mask image reconstruction process.

The peculiarity of having all the instruments onboard BeppoSAX under the same scientific and operation control has allowed for the development of a fast follow-up pointing of the NFIs in the case of serendipitous occurrence of a GRB in the field of view of one of the two Wide Field Cameras.

In these cases BeppoSAX can provide 1) a first light curve starting from 10 s before the trigger in the band 40–700 keV. 2) An X-ray light curve (2–30 keV) from long time before the GRB, during it and after, only limited by the sensitivity to the reduced fluxes of the GRB. 3) A light curve with the Narrow Field Instruments of the afterglow source from about 6 hours until it has decayed to levels of the order of $3 \ 10^{-14}$ erg/cm^{2}s.

2. X-ray afterglows of GRB

The first detection of an X-ray counterpart to a GRB (hereafter *afterglow* happened on Feb. 28, 1997 (Costa et al. 1997). Fig. 1 shows the images recorded by the MECS in the 2-10 keV range 8 hours and 3 days after the GRB. The previously unknown X-ray source lying in the 3 arcmin WFC error circle decreased by a factor of 20. The source was observed to decay already during the first pointing (Fig.2) and continue to decrease after the second BeppoSAX observation, as recorded by ASCA (Yoshida et al. 1997) and then ROSAT (Frontera et al. 1997b). These (2-10 keV) data are well reproduced by a power law with a slope of 1.3. By extrapolating this curve to the time of the GRB we found a value surprisingly consistent with the flux of a train of pulses following the main event. A similar behaviour was then observed in the second afterglow discovered by BeppoSAX in April 97 in GB970402 (Piro et al. 1997).

In the case of GB970111 (Butler et al. 1997, Feroci et al. 1997) we found a marginal detection at a flux much below one would have expected extrapolating the evolution of the February and April events.

Light curve of GRB970508 (Piro et al. 1997b) is less regular. During the first pointing (after 6 hours) the source was varying but not with a

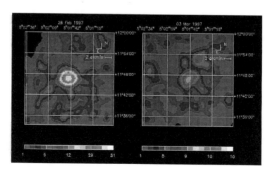

Figure 1. GB970228: MECS images of the afterglow observed on Feb.28, 8 hours after the burst (left image). On Mar.3 (right image) the source had decreased by a factor of 20

monotonic decay. In subsequent pointing, as shown in fig.3, it decayed with a light curve very far from a power law. This could suggest the existence of a gap between the GRB and its afterglow and possibly a behaviour similar to that of the optical counterpart that peaked two days after the GRB. But in this case data from Wide Field Camera after the initial burst show that a faint afterglow source is present at levels that can be reasonably connected with those of the subsequent NFI observations.

In the present very incomplete picture it seems more likely that the power law is an average trend, with a slope that can be substantially different in each GRB, and the actual light curve can include significant flare activity around this trend.

TABLE 1. Simultaneous detection of GRB's with the BeppoSAX GRBM and WFC and follow up with NFI

GRB	peak flux (40-700 keV)	peak flux (2-26 keV)	TOO[2] Δ T	TOO result F(2-10 keV)
960720[1]	1.0×10^{-6}	2.5×10^{-8}	1038	1.0×10^{-13} QSO 4C49.29
970111	4.5×10^{-6}	1.4×10^{-7}	16	$\leq 5 \times 10^{-14}$
970228	3.0×10^{-6}	1.4×10^{-7}	8	3×10^{-12} 1SAXJ0501.7+1146
970402	2.5×10^{-7}	1.6×10^{-8}	8	2×10^{-13} 1SAXJ1450.1-6920
970508	3.5×10^{-7}	6×10^{-8}	6	8×10^{-13} 1SAXJ0653.8+7916

Note: fluxes in cgs units; [1]: found during off-line analysis (Piro et al. 1997c) [2]: TOO start in hrs after the GRB trigger

References

Boella G. et al., A&A Supp. Ser., **122**, 299-307 (1997)
Boella, G., Chiappetti, L. et al., 1997b, A&AS, 122, 327
Butler, R.C., et al. 1997, IAU Circular n. 6539.
Costa, E. et al. 1997, Nature, 387, 783
Jager, R., et al. 1997, A&A Suppl. Ser., **125**, 557
F. Frontera, et al., A&A Supp. Ser., **122**, 357-369 (1997)
Frontera et al. 1997b, IAUC 6637
Parmar A.N. et al. 1997, A&AS, 122, 309
Piro L., Scarsi L. & Butler R.C. 1995 SPIE 2517, 169
Piro, L. et al. 1997, IAU Circ. 6617
Piro, L. et al. 1997b, A&AL, in press (astro-ph/9710355)
Piro et al. 1997c, A&A, in press (astro-ph/9707215)
Yoshida A et al. 1997 IAUC 6593

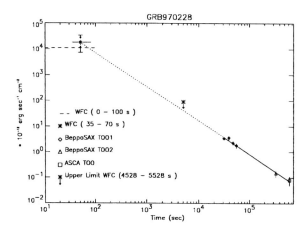

Figure 2. GB970228: X-ray evolution of the GRB from the initial event to its afterglow

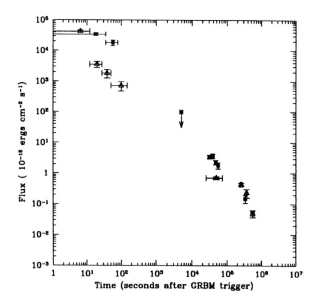

Figure 3. GRB970228 X-ray light curves from BeppoSAX Wide Field Cameras and Narrow Field Instruments of GRB970228 (squares) and GRB970508 (triangles). For the first part of the burst the average flux is displayed

A REVIEW OF GRB COUNTERPART SEARCHES

C. KOUVELIOTOU
Universities Space Research Association
ES-84, NASA/MSFC, Huntsville, Alabama 35812, USA

1. Introduction

Gamma-Ray Burst (GRB) research has been recently revitalized with exciting new results that have effectively started a new era in this thirty year old field, marked by the launch of the Italian–Dutch satellite BeppoSAX [1] on April 1996.

In the following, I will briefly (due to the limited space available) describe the searches and their results for each of the seven GRBs that have been rapidly followed-up in multiwavelength observations. As BeppoSAX, CGRO and RXTE are fully operational, we hope that the GRB counterpart sample will increase significantly within the next year, thus providing new insights to the properties of their parent population.

2. GRB Counterpart Searches

GRB970111: The event was detected with three satellites (BeppoSAX, BATSE, Ulysses). The original WFC error box (10′ in diameter) [5] was observed with the NFIs $\sim$21 hours after the burst trigger, and three X-ray sources were detected [3, 32]. A variable radio source was found within this error box [9] that coincided with one of the X-ray sources. A refined WFC error box [20] reduced the source location to a 3′ radius region, which although well within the original location, clearly excluded all X-ray and the radio source as the GRB counterparts. The radio non-detection may be used to set limits to GRB distances [12].

GRB970228: This is the first counterpart detection of a GRB after 30 years of searches [33]. The burst was detected and located with BeppoSAX [6], and it was observed with optical telescopes $\sim$21 hours after detection. Eight hours after the burst trigger, the NFIs detected an X-ray source associated with the WFC position, which decayed with an -1.1 power law index until the next NFI observation on 3 March 1997 [7]. Optical observations of the WFC error region made on 28 February and 8 March 1997 revealed a counterpart [16, 33] that was within the 0.75 arcmin^2 area defined by the WFC, NFI and the IPN annuli between Ulysses/SAX and Ulysses/WIND [19]. Subsequent optical observations on 13 March 1997 with ground-based telescopes revealed the existence of a 'fuzz' at the position of the transient [17]; the 'fuzz' was resolved with the Hubble Space Telescope (HST) Planetary Camera observations (on 26 March and 7 April 1997) into a point source and an extended source [29]. Claims of proper motion detection and intensity changes of the extended source [4] were not corroborated with a second HST observation on 4 September 1997, that detected the point source at the same position and the extended source at the same intensity [11].

GRB970402: This event was detected only with BeppoSAX; WFC located it to 3' accuracy and the NFIs were repointed. A very weak X-ray source was detected that was not observed during a second pointing [28]. No optical/radio counterpart was found for this source.

GRB970508: H. Bond [2] discovered the optical counterpart associated with GRB970508 using the 36" telescope at Kitt Peak. It was subsequently observed with the Palomar [8] and with the Keck [23] telescopes; the latter observation led to the first limits for a GRB redshift, ie., $0.835 < z < 2.3$. The lower limit is associated with the detection of absorption lines in the optical spectrum of the counterpart; the upper limit is set from the absence of the Lyman-α absorption features in the spectrum [23]. HST observations of the source did not detect any nebulosity asssociated with the burst counterpart [27]. GRB970508 is the only event so far with an identification of a radio counterpart [10]. The VLA data show rapid fluctuations during the first two weeks, that decline thereafter. The simplest explanation of the fluctuations is that they are due to scintillation caused by the interstellar medium in our Galaxy [10]. The decline of their amplitude is a consequence of the increase of the angular size of the source; initial estimates of the source size are $\sim 3\mu$arcsec. Such a source size is consistent with the fireball expansion models that are prevalent in the field these days [22].

GRB970616: This is the first afterglow detected with the PCA scanning of a BATSE error box [21]. The RXTE/PCA source intensity was $\sim$0.5 mCrab and was located with 1' accuracy [21]; subsequent observations of the RXTE error box with ASCA revealed four X-ray sources within the PCA region [24]. ROSAT observations have shown that two of these are variable [13], which makes a choice for an X-ray counterpart uncertain. No optical/radio counterpart was detected for this event.

GRB970815: The burst was detected with the RXTE/All Sky Monitor(ASM) [30]; its error box was subsequently scanned with the PCA, but no X-ray afterglow was discovered. The ASM error box was observed with ASCA and ROSAT but no counterparts were found [25, 14].

GRB970828: This is the second burst that was detected with the ASM [31]. The source region was observed with ASCA and ROSAT, which detected a declining X-ray afterglow within its error box [26, 15]. No optical/radio counterpart was found to coincide with the ASCA/ROSAT source [18].

TABLE 1. Summary of search results of GRB counterparts

GRB	Prompt X-rays (Crab)	γ-rays erg/cm^2	Secondary X-rays erg/cm^2 s	Optical	Radio mJy	Comments
970111	4.00^a	$6 \times 10^{-5,b}$	—	—	—	Very intense GRB
970228	2.30^a	$10^{-6,c}$	$3 \times 10^{-12,d}$	V= 21.3, I= 20.6	—	First optical counterpart with nebulosity
970402	0.46^a	?	$2 \times 10^{-13,d}$	V> 22.5, R> 21.0	—	Very weak NFI source
970508	1.00^a	$3 \times 10^{-6,b}$	$6 \times 10^{-13,d}$	V= 20.5, R= 19.8	0.43^e	First redshift. Radio source size: $< 10^{17}$ cm
970616	?	$4 \times 10^{-5,b}$	$1 \times 10^{-11,f}$	V> 25.5	—	Uncertain X-ray counterpart
970815	2.00^g	$10^{-5,b}$	$< 10^{-13,h}$	V> 21.5, R> 23.0	—	
970828	$\sim 0.8^g$	$7 \times 10^{-5,b}$	$1 \times 10^{-11,f}$	R> 25.0	—	Very intense GRB

^{a} Detected with the Wide Field Cameras (WFC) on BeppoSAX ^{b} Detected with the Burst And Transient Source Experiment (BATSE) on the Compton Gamma-Ray Observatory (CGRO) ^{c} Detected with TGRS on WIND ^{d} Detected with the Narrow Field Instruments (NFI) on BeppoSAX ^{e} Detected with VLA ^{f} Detected with the Proportional Counter Array (PCA) on the Rossi X-Ray Timing Explorer (RXTE) ^{g} Detected with the All Sky Monitor (ASM) on RXTE ^{h} Detected with ASCA

3. SUMMARY

Table 1 summarizes the GRB counterpart search results so far. The questions that are mostly asked today are: why some events do not have counterparts, what is the relation between detections in different wavelengths, what characteristics in GRB are correlated with the optical/X-ray afterglows. Although we have no definitive answers to any of these questions, simple explanations, such as large absorption from *circumstellar* dust in high to moderate redshifts could explain the lack of optical afterglows in several cases. Many more detections are needed, however, before we can establish a connection between GRB sources and their parent population.

References

1. Boella, G. et al. (1997) BeppoSAX, the wide band mission for X-ray astronomy, Astron. Astrophys. Suppl. Ser., 122, 299–399
2. Bond, H.E. (1997) IAU Circular No. 6654
3. Butler, R.G. et al. (1997) IAU Circular No. 6539
4. Caraveo, P. (1997) Talk presented at the 4th Compton Symposium in Williamsburg, VA, 27-30 April, 1997
5. Costa, E. et al., (1997a) IAU Circular No. 6533
6. Costa, E. et al., (1997b) IAU Circular No. 6572
7. Costa, E. et al. (1997c) Discovery of an X-ray afterglow associated with the γ-ray burst of 28 February 1997, Nature, 387, 783–785
8. Djorgovski, S.G. et al. (1997) The optical counterpart to the γ-ray burst GRB970508, Nature, 387, 876–878
9. Frail, D.A. et al. (1997a) Radio monitoring of the 1997 January 11 Gamma-Ray Burst, Astrophys. J., 483, L91–L94
10. Frail, D.A., Kulkarni, S.R., Nicastro, L., Feroci, M., & Taylor, G.B. (1997b) The radio afterglow from the γ-ray burst of 8 May 1997, Nature, 389, 261–263
11. Fruchter, A. et al., (1997) IAU Circular No. 6747
12. Galama, T.J. et al. (1997) Radio and optical follow-up observations and improved interplanetary network position of GRB970111, Astrophys. J., 486, L5–L9
13. Greiner, J. et al., (1997a) IAU Circular No. 6722
14. Greiner, J. (1997b) IAU Circular No. 6742
15. Greiner, J. et al., (1997c) IAU Circular No. 6757
16. Groot, P. et al., (1997a) IAU Circular No. 6584
17. Groot, P. et al., (1997b) IAU Circular No. 6588
18. Groot, P. et al., (1997c) Astrophys. J. Letters, in press
19. Hurley, K. et al., (1997) IAU Circular No. 6594
20. in't Zand, J. et al. (1997) IAU Circular No. 6569
21. Marshall, F.E. et al. (1997) IAU Circular No. 6683
22. Meszaros, P. & Rees, M. (1997) Optical and long-wavelength afterglow from gamma-ray bursts, Astrophys. J., 476, 232–237
23. Metzger, M.R. et al. (1997) Spectral constraints on the redshift of the optical counterpart to the γ-ray burst of 8 May 1997, Nature, 387, 878–880
24. Murakami, T., Fujimoto, R., Ueda, Y., & Shibata, R. (1997a) IAU Circular No. 6687
25. Murakami, T., Ueda, Y., Ishida, M., & Fujimoto, R. (1997b) IAU Circular No. 6722
26. Murakami, T. et al. (1997c) IAU Circular No. 6732
27. Pian, E. et al. (1997) Astrophys. J., submitted
28. Piro, L. et al., (1997) IAU Circular No. 6617
29. Sahu, K. et al. (1997) The optical counterpart to γ-ray burst GRB970228 observed using the Hubble Space Telescope, Nature, 387, 476–478
30. Smith, D.A, Levine, A., Morgan, E.H., & Wood, A. (1997a) IAU Circular No. 6718
31. Smith, D.A, Levine, A., Remillard, R., & Wood, A. (1997b) IAU Circular No. 6728
32. Voges, W., Boller, T., & Greiner, J. (1997) IAU Circular No. 6539
33. Van Paradijs, J. et al. (1997) Transient optical emission from the error box of the gamma-ray burst of 28 February 1997, Nature, 387, 686–689

QUICK OBSERVATIONS OF THE FADING X-RAYS FROM GAMMA-RAY BURSTS WITH ASCA

T. MURAKAMI, Y. UEDA, R. FUJIMOTO, M. ISHIDA, R. SHIBATA,
S. UNO, F. NAGASE AND ISAS TEAM
Institute of Space and Astronautical Science
3-1-1, Yoshinodai, Sagamihara, Kanagawa 229, Japan

A. YOSHIDA, N. KAWAI, F. TOKANI, C. OTANI AND RIKEN TEAM
Institute of Physical and Chemical Research
2-1, Hirosawa, Wako, Saitama 351, Japan

AND

F.E. MARSHALL, R.H.D. CORBET, J.H. SWANK, T. TAKESHIMA,
D.A. SMITH, A. LEVINE, R.A. REMILLARD, R. VANDERSPEK,
(RXTE TEAM), C.R. ROBINSON, C. KOUVELIOTOU, C. MEEGAN,
V. CONNAUGHTON, R.M. KIPPEN, (BATSE TEAM), K. HURLEY,
(UCB), S.D. BARTHELMY (GCN), L. PIRO, E. COSTA, J. HEISE,
F. FIORE, (SAX TEAM), J.V. PARADIJS, Y. TANAKA, (UOA) AND
J. GREINER(AIP)
The Large International Collaborations

1. Introduction

Since the discovery of fading X-rays from Gamma-Ray Bursts (GRBs) with BeppoSAX (Piro et al. 1997, Costa et al. 1997), world-wide follow-up observations in optical band have achieved the fruitful results. The case of GRB 970228, there was an optical transient, coincides with the BeppoSAX position and faded (Paradijs et al. 1997, Sahu et al. 1997). These optical observations also confirmed the extended component, which was associated with the optical transient. The new transient are fading with a power-law function in time and the later observation of HST confirmed the extended emission is stable (Fruchter et al. 1997). This extended object seems to be a distant galaxy and strongly suggests to be the host.
In the case of GRB 970508, the optical transient showed the lines of absorptions which were consistent with z=0.835, but there was no extended

emission confirmed. The existence of the absorption and later emission at the same z suggest the transient at z=0.835 but again no galaxy was confirmed yet (Djorgovski et al. 1997, Metzger et al. 1997). Based on these two cases, most are now believing that the origin of GRBs is from distant galaxies probably not from the nuclei but in the arm. However, the examples are too few to firmly convince that all GRBs come from the distant galaxies. In fact, many GRBs described below have shown no optical transient. We hope to confirm the scenario observing much local GRBs clearly associate with a galaxy. ASCA was out of this scope, because of no capability of detecting GRBs but the quick informations of SAX enabled us to monitor the fading X-rays of GRB 970228, GRB 970402 (Yoshida et al. 1997a, Feroci et al. 1997). Recently RXTE has successfully localized GRBs responding to the LOCBURST and informed their locations to us within several hours after the detections. This also enables us to observe the fading X-rays within 2 days. We have done observations of GRB 970616, GRB 970815 and GRB 970828 following the RXTE detections (Connaughton et al. 1997, Marshall et al. 1997, Smith et al. 1997, Murakami et al. 1997a, b and c). TM reviews the ASCA efforts briefly in this report.

2. Observations

2.1. GRB 970228

We have done monitor of this burst 1 week after the burst and detected the flux. The observed flux was almost our detection limit of 7.2 ± 2.1(1 sigma)$\times 10^{-14}$ erg sec^{-1} cm^{-2} in 2-10 keV. We have detected the flux but could not say anything about the variability and the spectrum. However the detected flux was fully consistent with the power-law decay in time even one week after the burst (Costa et al. 1997, Yoshida et al. 1997a).

2.2. GRB 970402 AND GRB 970616

We have started the monitor of GRB 970402, 2.8 days after the burst on April 5.74 (U.T.) but could not detect any flux. The 90% upper limit was about 1×10^{-13} erg sec^{-1} cm^{-2} in 2-10 keV.

The observation of GRB 970616 was also done 3.5 days after the burst. At least four X-ray sources were in the SIS FOV. Two were in the reported IPN error and one was slightly outside but consistent with the IPN considering the ASCA error (Murakami et al. 1997a, Kevin et al. 1997a). One X-ray source labeled A1 in the ROSAT IAUC was very variable but fading during our observation (Greiner et al., 1997). The most probable source; A1 faded in factor of 22 between the observations of ROSAT and ASCA and A2 and A3 were clearly out of the IPN error. However A4 in the IPN error was not detected. Based on the informations above, we could not conclude that the A1 source was the X-ray counterpart. However, the fading during the ASCA observation strongly suggests that A1 X-ray source might be the fading X-ray counterpart of this GRB. If this were true, the fading was not simple, the X-rays faded in power-law in time in average but in short time scale, the flux was very variable. The time scale of variability was several hours. There were no optical transient for both cases.

2.3. GRB 970815 AND GRB 970828

We have done observation of GRB 970815, 3.2 days after the burst. Although there was an X-ray source in the FOV, which was consistent with the new IPN but not in the RXTE/ASM error (Smith et al. 1997, Kevin et al. 1997b). The intensity of this source did not show a fading and looked stable, so we rejected the possibility of an X-ray counterpart (Murakami et al. 1997b).

Soon after this burst during the IAU symposium at Kyoto, GRB 970828 was observed. We successfully reached to the source within 1.17 days after the burst. We have carried out the ASCA maneuver at the Kyoto International Hole and also from the Kyoto university through the network not from the operation room at ISAS. The precise location of the X-ray source was reported two days after the detection but there was no optical transient discovered (Murakami et al. 1997c). The detail of the ASCA observation was reported by Yoshida at the GRB workshop in Huntsville (Yoshida et al. 1997b). This was fading during our observation, so there was no doubt to be the fading X-ray counterpart but no optical counterpart discovered.

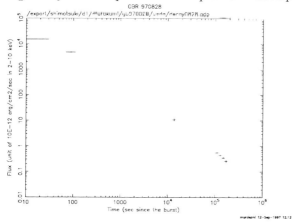

Figure 1. ASCA X-ray lightcurve of GRB 970828 together with the reported RXTE fluxes. The power-law decay index in time, between the last RXTE flux and the ASCA was about 1.44 is much faster than the case of GRB 970228.

3. Discussion

Our interest is to add another case of an optical transient and hope to find a galaxy associated with a fading X-ray counterpart. However, unfortunately there was no so far for the ASCA observations. Only the two SAX positions showed the optical transient but the four of the ASCA positions resulted no optical transient. Many ground-based observers such as Keck and Palomar have tried to find an optical transient following the ASCA notices but non. What were the differences between GRB 970228, GRB 970508 and others which were no optical transient? Let us make figure 1 of the X-ray lightcurve of GRB 970828 in the same scale of GRB 970228 (figure 3 of Costa et al. 1997). The cases without optical transient faded much faster than the cases with the optical transient. However this is still a small sampling. It is clear that the size of the bursts such as the peak flux and/or the fluency did not relate to the existence of the optical transient. What comes from this difference? We do not know the reason why. Schaefer mentioned no host galaxy is common to his deep HST observations (Schaefer et al. 1997). In any case, we require further detections of a burst which shows a slower decay and brighter flux (probably local) in future to make the origin of the optical transient clear.

Acknowledgments: This work was done by the large international collaboration. We cannot specify the contributions of each person and are afraid of missing your name. TM is responsible for that case and thanks to all the ASCA member who allowed us to carry frequent ToO observations.

References

Connaughton, V. et al., (1997) *IAUC*, **6683**
Costa, E. et al., (1997), *Astroph*, **no. 9706065**
Djorgovski, S. G. et al., (1997), *Nature*, **Vol. 387**, 876
Feroci, M. et al., (1997) *IAUC*, **6610**
Fruchter, A. et al., (1997), *IAUC*, **6747**
Greiner, J. et al., (1997), *IAUC*, **6722**
Hurley, K. et al., (1997a) *IAUC*, **6687**
Hurley, K. et al., (1997b) private communication
Marshall, F. et al., (1997) *IAUC*, **6727**
Metzger, M. R. et al., (1997), *Nature*, **Vol. 387**, 878
Murakami, T. et al., (1997a,b,c) *IAUC*, **6687, 6722, 6729**
Paradijs, van J. et al., (1997), *Nature*, **Vol. 386**, 686
Piro, L. et al., (1997), *Astroph*, **no. 9707215**
Remillard, R. et al., (1997) *IAUC*, **6726**
Sahu, K. C. et al., (1997), *Nature*, **Vol. 387**, 476
Schaefer, B. et al., (1997), *Astroph*, **no 9704278**
Smith, D. et al., (1997), *IAUC*, **6718, 6728**
Yoshida, A. et al., (1997) *IAUC*, **6593**
Yoshida, A. et al., (1997) *Proc. of Gamma-Ray Bursts*, **AIP** in press at Huntsville meeting. Ed. C. Meegan.

Session 4: Large Scale Hot Plasmas and Their Relation with Dark Matter

X-RAY LARGE SCALE STRUCTURE AND XMM

M. PIERRE
Service d'Astrophysique & XMM Survey Science Center
CEA/DSM/DAPNIA/SAp
CE Saclay
F-91191 Gif sur Yvette

1. Introduction

The formation of Large Scale Structures (LSS) in the universe was first studied at optical wavelengths as the galaxy spatial distribution appeared to be far from homogeneous. Considerable effort has been invested in semi-analytical approaches and in numerical simulations (DM + hot gas) to explain the observed structures, given some set of initial conditions and using additional constraints provided by the COBE results. It is now clear however, that these two extreme data set are not sufficient to discriminate between the possible remaining cosmological scenarios. It is thus timely to investigate LSS at a much higher redshift than the present survey limits *both* in the optical and in other wavebands. In this context, the X-ray band will certainly become a hot field with the advent of the XMM observatory. The next section briefly summarizes what is known about LSS from optical wavelengths and simulations. Sect. 3 reviews the particular points that can be addressed in the X-ray band. Last section presents realistic prospects for mapping LSS with XMM.
Throughout the paper we assume $H_o = 50$ km/s/Mpc and $q_o = 1/2$.

2. Present knowledge

Optical - up to what "scale"?
Some 10 years ago, the CfA redshift survey revealed the "bubbly" structure of the local universe ($z_{max} \sim 0.03$) [1]. The Las Campanas Redshift Survey - the largest galaxy survey to date - largely confirms this topology out to $z_{max} \sim 0.2$ [2]. Its power spectrum indicates a characteristic scale of ~ 100 Mpc, which is reminiscent of the "periodicity" of 128 Mpc found in deep

pencil beam surveys at the galactic poles [3]. A similar scale (120 Mpc) was also recently found in the galaxy cluster distribution [4].

N-body simulations - down to what "depth"?

Constraining the initial spectrum to be compatible with the COBE results leaves still room for "tilted" CDM models (e.g. Λ CDM, SDM, ODM, τ [5]). Large N-body simulations are quite succcessful in reproducing the characteristics of the local observed network. Moreover, they clearly show model dependent identifiable features at $z \sim 3$ and suggest that it will still be possible to discriminate among the various models at $z \sim 1$ but certainly not later. More precisely, hydrodynamical simulations show that filaments begin to form in both the dark matter and the gas by $z \sim 4$, with material flowing over large distances (> 10 Mpc) along the filaments to reach forming clusters [6]. These timescales are in global agreement with independent recent observations of galaxies [7] and thus, with the overall evolutionary picture in "bottom-up" models where accretion or merging is an integral part of formation processes on all scales.

A realistic goal for the next generation of LSS surveys should be to investigate scales of the order of 100 Mpc at $z \sim 1$

3. X-ray implications for the LSS

High galactic latitude fields observed at medium sensitivity ($\sim 10^{-14}$ erg/s/cm^2 in the [2-10] keV band) contain basically two types of objects: clusters of galaxies (extended) and AGNs (pointlike). In all that follows it is assumed that coordinated optical (photometry & spectroscopy) observations are available.

Galaxy clusters

Clusters are the largest bound entities in the universe, they are located at the intersection of filaments and, thus, constitute key objects for investigating LSS. Systematic cluster searches in the optical ($22 < I < 25$) are a difficult task as the cluster galaxy density contrast above field galaxy counts becomes marginal beyond $z \sim 1$ [8]. Fine-tuned multicolour photometry (on 8-10 m class telescopes) seems to be efficient but requires large amounts of telescope time. On the other hand, the X-ray band is ideally suited to the detection of high redshift clusters, as it is free from projection effects. Moreover, X-ray data readily provide information about the cluster potential depth and relaxation state (L_X, T_X, morphology). Mapping large areas in X-ray should, therefore, provide a unique view of the topology of the deep potential wells of the universe out to $z \sim 2$.

Filaments

Hydrodynamical simulations predict the existence of a "cool", low density gas trapped in the filaments connecting clusters. This tenuous component has so far not been detected unambiguously and a proper measurement would be of prime interest for our understanding of the formation of LSS. Moreover, combining X-ray observation of filaments with the analysis of weak lensing in the optical (caused by the underlying DM [9]) would provide invaluable information on bias mechanisms in low density structures.

AGNs and QSOs

QSOs are known to be strongly correlated [10], but the origin of the signal and what fraction of the amplitude is due e.g. to lensing are still unclear [11]. AGNs and QSOs will constitute by far the largest source population for XMM at medium sensitivity (up to 90%). As their spatial distribution will appear unambiguously on top of the 3-D cluster network, it will be possible to isolate "true" QSO clustering, quantitatively study the AGN formation process and its relation to high density fluctuations (e.g. BL-LAC in clusters) or, alternatively, with the presence of filaments or voids.

4. Investigating LSS with XMM [12]

XMM ([0.1-10] keV) is one of ESA's cornerstones for the next millenium and is expected to be launched by the end of 1999. With a collecting area of some 6500 cm^2 at 1.5 keV it is by far the most sensitive X-ray telescope of the next generation, and is thus best suited for large scale investigations. It has a field of view of 30', an on-axis FWHM of $\sim$ 15" and, in imaging mode, a spectral resolution of 10 % at 1.5 keV [13].

Taking a canonical value of 500 kpc for cluster characteristic sizes ($R_c = 250$ kpc), XMM will readily flag clusters as extended sources out to $z \sim 1-2$ ($R_c = 12$"), provided enough photons are detected. We have performed detailed simulations in order to estimate the cluster population that can be seen by XMM as a function of sensitivity assuming: (1) the local $F(L_X)$ and $T_X - L_X$ relationship hold out to $z = 2$; (2) a King luminosity profile for clusters with $\beta = 0.75$ and $R_c = 250$ kpc; (3) Raymond-Smith type spectra, with $N_H = 5\ 10^{20}$ cm^{-2}; (4) a folding of the spectra with XMM response; (5) a detection limit set to nσ, for a given energy band and exposure time, within $1.5R_c$ (60 % of the flux) with respect to the background.

Predictions at medium sensitivity are displayed in Fig.1. Clusters are best detected in the soft energy band; this is easily understandable since low luminosity (i.e "cool") objects are the most numerous and, in addition, are seen with redshifted spectra. In a 5 × 5 sq. degree area (perfectly adapted for probing the 100 Mpc scale at $z \sim 1$) some 320 clusters are expected out to $z = 2$. Out of these, about 1/4 will be beyond $z > 1$, which, in addition, should provide strong constraints on any cluter evolution theory.

Preliminary calculations seem to indicate that it will not be possible to detect filaments with XMM, despite its large collecting area; the truly dif-

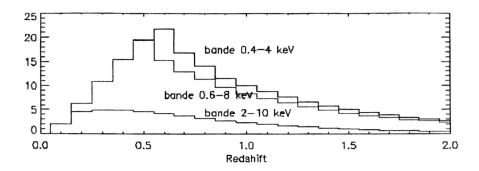

Figure 1. Expected cluster population seen by XMM, for a 15 sq. deg. area and 20 ks exposure. Detection at the 5σ level ($\sim 10^{-14}$ erg/s/cm^2). For exposure times of ~ 100 ks, the distribution peaks around $z = 0.8$, with ~ 1.5 times as many clusters.

fuse medium is expected to have a mean temperature well below the XMM range and to be too tenuous (e.g. [14]) for producing a significant signal. Hydrodynamical simulations are, however improving significantly and, in the near future, they will be able to resolve the small galaxy groups supposed to be embedded within the filaments, a population that will be detectable by XMM. This will enable a detailed confrontation with theory.

References

[1] De Lapparent V., Geller M.J., Huchra J.P., 1986, ApJ Let 302, 1L
[2] Landy S. D. et al 1996, ApJ Let, 456, L1
[3] Broadhurst T.J., Ellis R.S., Koo D.C., Szalay A.S., 1990, Nature 343, 726
[4] Einasto M., Tago E., Jaaniste J., Einsato J., Andernach H., 1997 A&AS, 123, 119
[5] White S.D.M. 1997, in *The Early Universe with the VLT*, p 199, Springer, Ed. J. Bergeron
[6] Katz N., & White S.D.M., 1993, ApJ 412, 455
[7] Madau P., Fergusson H.C., Dickinson M.E., Giavalisco M., Steidel C.C., Fruchter A., 1996, MNRAS 283, 1388
[8] Dickinson M.E, 1997 in *The Early Universe with the VLT*, p 274, Springer, Ed. J. Bergeron
[9] Mellier Y., Fort B., 1997 in *The Early Universe with the VLT*, p 189, Springer, Ed. J. Bergeron
[10] Shanks T., Boyle B.J., 1994 MNRAS, 271, 753
[11] Wu X.P., Fang L.Z., 1996 ApJ Let, 461, 5L
[12] Information on the **XMM Medium Deep Survey** can be found at:
http://www-dapnia.cea.fr/Phys/Sap/Activites/Science/Structures/XMDS.html
[13] http://astro.estec.esa.nl/XMM/xmm.html
[14] Bryan G., Cen R., Norman M., Ostriker J., Stones J.M., 1994, ApJ 428, 405

HIERARCHICAL STRUCTURE IN DARK MATTER DISTRIBUTIONS

K. MAKISHIMA
Department of Physics, University of Tokyo
7-3-1 Hongo, Bunkyo-ku, Tokyo, Japan 113

1. Introduction and Summary

Galaxy observers argue for the dark halo surrounding each galaxy, while cluster investigators discuss the dominance of dark matter (DM) in the intra-cluster space. Then, what is the relation between the cluster DM and the DM associated with member galaxies of the cluster?

ASCA studies of cD clusters (§2) have shown that, while a cluster has a massive dark halo, the cD galaxy also has its own smaller-scale dark halo, thus making a hierarchical halo-in-halo structure. Such a nested potential structure has also been discovered around several X-ray bright non-cD elliptical galaxies (§3).

2. Clusters with cD Galaxies

As has long been known (e.g. Jones & Forman 1984), some clusters of galaxies, particularly those with cD galaxies, exhibit centrally peaked X-ray surface brightness profiles with small (e.g. < 100 kpc) X-ray core radii; others exhibit flatter X-ray profiles with larger core radii.

In the former type of objects, the ICM is thought to be cooling significantly and becoming very dense in the central regions, because of the short cooling time. The centrally peaked X-ray emissivities observed from many cD clusters are usually explained in terms of such cooling effects.

As an alternative possibility, the X-ray emissivity may be enhanced in the core region of a cD cluster because the gravitational potential associated with the cD galaxy attracts excess ICM. Although this simple-minded interpretation so far received little attention, this has actually been confirmed to be the case with many cD clusters through *ASCA* studies.

TABLE 1. Parameters of the central excess X-ray emission ($h_0 = 75$).

object	kT [a] (keV)	R_{excess} [b] (kpc)	M_{excess} [c] ($M_\odot$)	$L_{excess}^{0.5-10}$ (erg s^{-1})	$L_{cool}^{0.5-10}$ (erg s^{-1})	M/L
NGC 4636	0.9	25	6×10^{11}	1.5×10^{41}	—	20
Fornax	1.2	60	2×10^{12}	1.6×10^{42}	—	25
Centaurus	3.8	70	3×10^{12}	1×10^{43}	4×10^{42}	60
Hydra-A	3.5	60	5×10^{12}	8×10^{43}	3×10^{42}	45
A1795	6.1	100	1.5×10^{13}	2×10^{44}	6×10^{43}	50

a) Average ICM temperature excluding the central region.
b) The radius within which the excess X-ray emission is seen.
b) The excess total gravitating mass inside R_{excess}.

2.1. THE FORNAX CLUSTER

The Fornax cluster is the first example in which the central excess X-ray emission was confirmed to arise due to the potential well of the cD galaxy, NGC 1399 (Ikebe et al. 1996). Details are summarized as follows.

   a. The X-ray surface brightness profile exhibits a clear excess around the cD galaxy, above a single β-model fitted to the outer region.
   b. The overall X-ray emission is approximately isothermal within $\sim 20\%$.
   c. The X-ray profile can be described by a sum of two projected β-model components, which cross over at a radius R_{excess} given in Table 1.
   d. The X-ray volume emissivity can be modeled as a quadrature sum of the two deprojected β-model components.
   e. The spherically-integrated total mass curve $M(R)$ exhibits a clear hierarchy, corresponding to the cD galaxy and the cluster.
   f. The hierarchy exists not only in the total mass but also in the DM mass, estimated by removing the stellar and ICM masses from $M(R)$.

2.2. THE HYDRA-A CLUSTER AND A1795

Similar results as described above have been obtained from two prototypical "cooling flow" clusters; the Hydra-A cluster by Ikebe et al. (1997a), and A1795 by Xu et al. (1997). Using the *ROSAT* and *ASCA* data, these authors revealed that the prominent central excess X-ray emission of these clusters are detectable not only in soft X-rays but also in hard X-rays up to 10 keV. Items a.–f. listed in §2.1 apply to these two clusters as well. Furthermore, they commonly exhibit the following two additional properties.

   g. An additional cool emission is seen near the cluster center, but its luminosity L_{cool} does not account for the excess emission (Table 1).

h. The cooling-flow rates estimated with *ASCA* fall several times short of those estimated previously without the knowledge of hard X-ray data.

Based on N-body simulations, Navarro et al. (1996) propose a universal shape of the DM distribution, which exhibits a central cusp as compared to the classical King-type potential. However, the central excess emission of A1795 is too strong to be explained by the "universal halo" profile.

2.3. THE CENTAURUS CLUSTER

The Centaurus cluster shows a very strong central cool emission, and the most prominent central metallicity increase ever measured from clusters (Fukazawa et al. 1994; Ikebe et al. 1997b). However even in this case, the central cluster volume is mostly filled with the hot ICM in an apparent two-phase configuration, and the central excess is seen in hard X-ray continuum as well. Therefore the centrally-peaked X-ray profile of this cluster must be caused, at least partially, by the central excess potential associated with the cD galaxy, NGC 4696 (Ikebe et al. 1997b).

Within $\sim$ 50 kpc of the center, there take place the three outstanding phenomena of the Centaurus cluster; the cool emission component, the metallicity increase, and the excess hard X-ray emissivity. All these phenomena are probably caused by the cD galaxy, in such a way that the metal-enriched cool inter-stellar medium of NGC 4696 is confined by its own potential plus the external ICM pressure.

3. Elliptical galaxies

Matsushita (1997) and Matsushita et al. (1997a) have discovered that elliptical galaxies can be classified into two subgroups in terms of their X-ray properties. One (the other) class of objects are characterized by a high (low) X-ray luminosity, a high (low) ISM metallicity, and the X-ray emission extending (not extending) beyond the optical galaxy.

The giant elliptical galaxy NGC 4636 is the best example of the X-ray extended class of objects. From a deep (200 ks) *ASCA* exposure, the radial X-ray brightness profile of NGC 4636 was found to consist of two β-model components (Matsushita 1997; Matsushita et al. 1997b). The subsequent story (Fig.1) is essentially the same as that of the Fornax cluster (§2.1). However in the case of NGC 4636, the mass component corresponding to the larger β-model component, with $M/L \sim 200$, has no clear counterpart in the visible light. It is hence suggested that NGC 4636 sits at the bottom of a group of galaxies that is optically invisible. Other X-ray luminous non-cD elliptical galaxies are also inferred to signal the presence of a group-size DM concentration, with little optical indication.

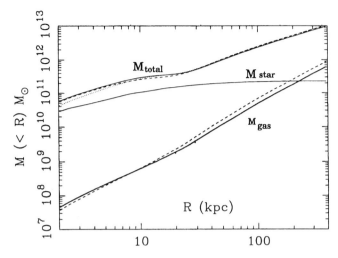

Figure 1. The spherically integrated mass profiles of NGC 4636 (Matsushita 1997).

4. Discussion and Implication

When the X-ray emitting hot gas confined in a self-gravitating system exhibits a deviation from a single-β distribution, there is a high probability that the underlying gravitational potential has a nested halo-in-halo structure, which in turn reflects a hierarchy in the dark matter distribution. These effects are seen over a wide range of system richness, including ellipticals (§3), groups (Mulchaey & Zabludoff 1997), and clusters (§2).

The double-β model has been found to be very useful; its application to the X-ray data analysis is strongly encouraged. When a single-β profile is forced to fit a brightness profile that is intrinsically of double-β nature, the core radius and β both tend to get smaller. This effect can nicely account for the apparent positive correlation between these two quantities observed from a large *ROSAT* sample of clusters (Pownall & Stewart 1995).

References

Fukazawa, Y., et al. (1994), *Publ. Astr. Soc. Japan* **46**, L55
Pownall, H. R., & Stewart, G. C. (1996), in *Röntgenstrahlung from the Universe.*
Ikebe, Y., et al. (1996), *Nature*, **379**, 427
Ikebe, Y., et al. (1997a), *Astrophys. J.*, **481**, 660
Ikebe, Y., et al. (1997b), submitted to *Astrophys. J.*; see also this volume.
Jones, C., & Forman, W. (1984), *Astrophys. J.*, **276**, 38
Mulchaey, J. S., & Zabludoff, A. I. (1997), *Astrophys. J.*, in press
Matsushita, K. (1997), PhD Thesis, University of Tokyo
Matsushita, K., et al. (1997a), *Astrophys. J.*, in press; see also this volume.
Matsushita, K., et al. (1997b), submitted to *Nature*; see also this volume.
Navarro, J. F, Frenk, C. S., & White, S. D. M. (1996), *Astrophys. J.*, **462**, 563
Xu, H., et al. (1997), submitted to *Astrophys. J.*; see also this volume.

DIFFUSE EUV EMISSION FROM CLUSTERS OF GALAXIES

S. BOWYER
Space Sciences Laboratory
University of California, Berkeley, CA 94720-7450, USA

R. LIEU
Department of Physics
University of Alabama, Huntsville, AL 35988, USA

AND

J. P. MITTAZ
Mullard Space Science Laboratory
Holmbury St. Mary, Dorking, Surrey, UK

1. Introduction and Survey of Results

Diffuse EUV emission has been detected in five clusters of galaxies: Virgo (Lieu et al., 1996a), Coma (Lieu et al., 1996b), Abell 1795, Abell 2199, and Abell 4038. These results were obtained with the Deep Survey Telescope of the Extreme Ultraviolet Explorer (Bowyer and Malina, 1991) using a Lex/B filter which covers the 60 to 170 Å band. The extent of this diffuse emission from these clusters ranges from 20' to 40' in diameter. The statistical significance of these results varies from 8 to 50 standard deviations. Some EUV emission would be expected from the low-energy tail of the well-studied X-ray cluster emission. However, the EUV emission detected is far greater than the expected emission from the X-ray-emitting gas. Marginal signatures of this "soft excess" are often present in the lowest energy resolution band of the ROSAT PSPC where it is typically less than a 20% effect. In the case of Abell 2199, no low energy excess is present in the ROSAT data. In the data taken with EUVE, however, the excesses range from 70% to 600%.

Although other suggestions as to the source of this emission have been recently advanced, most workers have interpreted these results in terms of an additional thermal gas component in the cluster. In this work the EUV and X-ray data are first fitted to the know X-ray emitting hot cluster gas.

The excess EUV emission is then fitted by an additional thermal gas; a much lower temperature gas is required.

2. Criticisms and Concerns

We address a number of concerns that have been repeatedly raised regarding the validity of this data, and on alternate suggestions that these data may be the result of detailed characteristics of the Galactic ISM.

A major concern has been that holes in the interstellar medium in our Galaxy would allow excess EUV emission from the X-ray emitting gas through to our location in the Galaxy. However, we have obtained new NRAO twenty-one centimeter data with 21-minute arc resolution, oversampled by a factor of three, for the Virgo, Coma, and Abell 1795 clusters, and similar data with the Effelsberg telescope on the Virgo cluster, with nine arc-minute resolution. All these data show the ISM is smoothly distributed. The distribution of dust in this region was examined with IRAS data, which showed a smooth distribution with four arc-minute resolution. Finally, hydrogen columns have been obtained by absorption in QSO spectra and these give the same hydrogen columns as those obtained by high-resolution radio observations. We conclude that holes in the Galactic ISM cannot be the explanation for these results.

Fabian (1996) has suggested that ionization of neutral helium in the ISM would allow sufficient EUV flux from the low-energy tail of the X-rays to reach Earth and account for the observations. Bowyer et al. (1996) showed that while this effect could work in principle, the required ionization was vastly at variance with the overall ionization state of helium in the Galaxy.

Concerns have been raised that the effects observed may be artifacts due to the presence of the North Polar Spur in the direction of the Virgo Cluster and Abell 1795 which could produce an uneven background which would distort the observations. Detailed study of the background show this is not a problem, but independent of this analysis, this explanation certainly cannot in any way affect the observations of the Coma Cluster, Abell 2199, and Abell 4038. Hence, this effect cannot be the explanation of what is being observed.

Many critics have questioned whether supposed instabilities in the EUVE detectors may have produced the effects observed. These concerns may be related to the fact that the ROSAT detectors do have variable backgrounds in the lowest energy channels, and the Wide Field Camera underwent substantial changes in its sensitivity over time. However, the EUVE detectors are calibrated every three months through observations of hot white dwarfs. No changes in sensitivity have been observed to a level of a few percent, which is the statistical uncertainty in the observations. A few pixels were destroyed when a bright EUV source was observed. However, these dead pixels cover only a few arcseconds and the extended cluster sources typically have twenty to forty arcminute diameters. So this certainly cannot be the explanation for the observations.

It has been suggested that the effect is due to an X-ray leak in the EUVE detector. This concern may arise from the fact that there is an X-ray leak in the Wide Field Camera. However, detailed analyses of the response of the EUVE deep survey telescope shows that this cannot be the case. Independent of the details of this technical analysis, we can eliminate this concern with reference to the data obtained on Abell 1795. In this cluster, the X-ray emission is peaked in the center while the fractional excess EUV emission increases radially outward. If an X-ray leak were producing the effects observed, we would see the maximum pseudo-EUV response in the center of the cluster, but we observe the opposite. This provides confirmation that the EUVE deep-survey telescope detector does not have an X-ray leak which affects these observations.

It has been suggested that an ultraviolet leak in the EUVE deep-survey telescope may be producing these results. This concern may be motivated by the fact that there is a UV leak in the ROSAT PSPC. EUVE, however, does not have a UV leak, and detailed examinations of a large number of stars show that only an eighth-magnitude star or brighter shows any spurious UV signal. There are no O-stars or B-stars even in the general region of any of these clusters.

3. Theoretical Suggestions

Most analyses have assumed that the diffuse EUV emission is the result of cooler ($\sim$ 1 million K) gas in these clusters. The problem is that gas at these temperatures are at the peak of the cooling curve and consequently rapidly loses energy. As one example, the cooling of the X-ray gas in the Virgo cluster produces about ten solar masses per year of million-degree gas. However, to produce the observed EUV flux, we need 340 solar masses per year. Hence, if the source of emission is a low-temperature plasma, some mechanism must be present to continually inject energy into this gas.

One suggestion is that this energy could be supplied by energy from the member galaxies of the cluster . The problem with this suggestion is exemplified by Abell 1795 where the total energy requirement is over a Hubble time is $\sim 10^{65} ergs$, which is 500 times larger than the total energy of all of the supernovae in the central part of the cluster.

Another suggestion (Fabian, 1997) is that the hot X-ray emitting cluster gas contains many cold (10^4 K) clouds. This increases the EUV emission of the gas by some two orders of magnitude. Unfortunately, cooling is so rapid that this process could maintain the existing emission for only one percent of the Hubble time.

Mittaz, et al (1997) have suggested that accretion from the intergalactic medium may be the energy source. This concept works in principle, but for Abell 1795, the total mass accreted over a Hubble time is 4×10^{15} solar masses per year. Briel and Henry (1996) obtained this mass for the total virial mass of the cluster, which leaves no room for dark matter, which everyone knows must be present in some exotic form, rather than as a simple plasma.

Several authors (Hwang, 1997; Bowyer and Berghoefer, 1997; and Ensslin and Bierman, 1997) have explored the possibility that high-energy electrons, known to be present from their syncrotron emission, may produce the observed effect via inverse Compton-scattering against the 3° K Black Body background.

4. Conclusions

We conclude that diffuse EUV emission is present in some clusters of galaxies, and this emission is not due to spurious instrumental effects or effects of the Galactic ISM. Only five clusters have been examined to date, and from this small number it is hard to establish the underlying mechanism producing this emission. More observations of a variety of types of clusters are clearly needed. Unfortunately, for the next several years the Extreme Ultraviolet Explorer may be the only observatory available to study this effect.

References

S. Bowyer and and R. F. Malina, in *EUV Astronomy*, ed. R. F. Malina and S. Bowyer. New York: Pergamon, 1991, 397.
S. Bowyer, M. Lampton, and R. Lieu, *Science*, **274**, 1338, 1996.
S. Bowyer and T. Berghoefer, 1997, in preparation.
U. G. Briel and J. P. Henry, *Astrophys. J.*, **472**, 131, 1996.
Ensslin and Bierman, 1997, in preparation.
A. C. Fabian, *Science*, **271**, 1244, 1996.
A. C. Fabian, *Science*, **275**, 48, 1997.
C.-Y. Hwang, 1997, in preparation.
R. Lieu, J. P. D. Mittaz, S. Bowyer, F. J. Lockman, C.-Y. Hwang, and J. H. M. M. Schmitt, *Astrophys. J.*, **458**, L5-7, 1996a.
R. Lieu, J. P. D. Mittaz, S. Bowyer, J. O. Breen, J. O. Lockman, E. M. Murphy, and C.-Y. Hwang, *Science*, **274**, 1335, 1996b.
J. P. M. Mittaz, R. Lieu, and F. J. Lockman, *Nature*, submitted.
C. L. Sarazin, *Rev. Mod. Phys.*, **58**, 1, 1986.

HOT ELECTRONS AND COLD PHOTONS: GALAXY CLUSTERS AND THE SUNYAEV-ZEL'DOVICH EFFECT

J.P. HUGHES
Rutgers University
Department of Physics and Astronomy
P.O. Box 849, Piscataway, NJ 08855-0849 USA

1. Introduction

The hot gas in clusters of galaxies emits thermal bremsstrahlung emission that can be probed directly through measurements in the X-ray band. Another probe of this gas comes from its effect on the cosmic microwave background radiation (CMBR): the hot cluster electrons inverse Compton scatter the CMBR photons and thereby distort the background radiation from its blackbody spectral form. Although this, the Sunyaev-Zel'dovich (SZ) effect, is quite small, heroic efforts during the 1980's resulted in its detection in three moderately distant clusters of galaxies: A665, A2218, and CL 0016+16. It is well known that one of the purposes of conducting such measurements is to determine the Hubble constant. The technique has generated considerable interest because it is independent of all other rungs of the cosmic distance ladder and is effective over a wide range of redshifts: ~ 0.02 to ~ 1.

In the last few years, the development of sensitive new instruments for measuring the SZ effect in clusters has sparked a revolution in the field. Current radio interferometric arrays can now detect and map the SZ effect in even distant clusters ($z \sim 1$). Another important development in this field was the launch of the *ASCA* satellite with its broadband X-ray imaging and spectroscopy that allows, for the first time, accurate determination of gas temperatures in distant galaxy clusters. This information is critically important to the interpretation of the SZ effect, since the determination of H_0 depends on the square of the cluster gas temperature. In the following I report on the progress that has been made in determining the cosmic distance scale from the SZ effect and I highlight what has been learned about galaxy clusters from these investigations.

TABLE 1. Summary of X-Ray/SZ Effect H_0 Measurements

Cluster	z	H_0 (km s^{-1} Mpc^{-1})	Reference
Coma	0.0232	64^{+25}_{-21}	Herbig et al. 1996
Abell 2256	0.0581	68^{+21}_{-18}	Myers et al. 1997
Abell 478	0.0881	30^{+17}_{-13}	Myers et al. 1997
Abell 2142	0.0899	46^{+41}_{-28}	Myers et al. 1997
Abell 2218	0.171	59 ± 23	Birkinshaw & Hughes 1994
Abell 2218	0.171	35^{+16}_{-15}	Jones 1995
Abell 665	0.182	46 ± 16	Hughes & Birkinshaw 1998
Abell 2163	0.201	56^{+39}_{-22}	Holzapfel et al. 1997
CL 0016+16	0.5455	47^{+23}_{-15}	Hughes & Birkinshaw 1997

2. Current Results

There are eight galaxy clusters with published measurements of the SZ effect based on single-dish radiometry, infrared bolometry, or radio interferometry. Table 1 summarizes the derived H_0 values (and 68% confidence level errors) from the several clusters, ordered by increasing redshift and determined under the following assumptions:

1. spherical symmetry
2. gas density distribution given by $n_e = n_{e0}[1 + (\theta/\theta_C)^2]^{-3\beta/2}$
3. isothermal gas distribution
4. unclumped
5. $\Omega_0 = 2q_0 = 0.2$

Relativistic corrections for the CMBR intensity change and the X-ray bremsstrahlung spectral emissivity, which result in reductions of order 10% in the derived H_0 values, are both included.

The average value of these nine measurements weighted by the individual errors is $H_0 = 48.5 \pm 6.5$ km s^{-1} Mpc^{-1}. However, this value is potentially quite strongly biased by systematic effects, as I discuss below.

3. Systematic Uncertainties

Birkinshaw & Hughes (1994) and Holzapfel et al. (1997) allowed for large scale radial temperature gradients when analyzing the SZ effect and X-ray data for A2218 and A2163, respectively. Both groups found that, for temperature profiles that fell with radius, the value of H_0 derived under an isothermal assumption would underestimate the true H_0 value by 20%–30%.

If cluster gas is clumped, then X-ray emissivity will be increased relative to SZ by a factor greater than unity. In this case the value of H_0 derived assuming an unclumped gas distribution will be an upper limit to the true H_0 value. Holzapfel et al. (1997) used X-ray spectral fits to constrain the amount of isobaric clumping in A2163 and found that a reduction in H_0 of only ~10% from the unclumped case was allowed.

The peculiar motion of clusters relative to the Hubble flow introduces an additional distortion to the CMBR spectrum usually referred to as the "kinematic" SZ effect. For a cluster with a peculiar velocity of $1000\,\mathrm{km\,s^{-1}}$ and temperature of 10 keV the strength of the kinematic SZ effect would be 9% of the thermal effect in the Rayleigh-Jeans portion of the CMBR spectrum. Since the SZ effect intensity enters as a square in the equation determining H_0, the kinematic SZ effect could introduce up to a $\sim\pm20\%$ correction. Peculiar velocities are unlikely to be correlated for clusters that are widely distributed in redshift and position, so this effect would result in an additional random uncertainty in H_0 for any single cluster.

It is now clear based on *ROSAT* observations that many, if not most, clusters show evidence for complex surface brightness distributions. In recent work Hughes & Birkinshaw (1997) analyze CL 0016+16, a distant cluster that displays strong ellipticity (see the left panel of Fig. 1). They fit elliptical isothermal-β models to the X-ray image and deproject under the assumption that the three-dimensional structure of the cluster is axisymmetric, either prolate or oblate. If the symmetry axis is assumed to lie in the plane of the sky, then the different assumptions about the shape of the gas distribution yield values for H_0 that differ by 17%, which is the percentage difference between the major and minor axis lengths of the cluster. As the symmetry axis of the ellipsoid is allowed to vary toward the line-of-sight, then the intrinsic cluster ellipticity (defined as the ratio of major to minor axis lengths) grows increasingly larger, as does the uncertainty on H_0, which is shown graphically in the right panel of Fig. 1. In order to bound the uncertainty, Hughes & Birkinshaw argue that it is unlikely for a cluster to have an intrinsic ellipticity greater than about 1.5, based on observations of other galaxy clusters. This limit results in a uncertainty in H_0 of $\sim\pm20\%$ for CL 0016+16 from morphology alone.

To ensure that the effects of unknown geometry and arbitrary inclination are uncorrelated from cluster to cluster, it is essential that the cluster sample for determining H_0 be selected properly. For example, as pointed out by Birkinshaw, Hughes, & Arnaud (1991), it is important that clusters *not* be selected based on the strength of their SZ effect signal or central X-ray surface brightness, since this would result naturally in a bias toward prolate clusters with their long axes aligned to the line-of-sight. Recent cluster samples for H_0 determination have been selected based on the strength of their integrated X-ray flux from surveys by the *Einstein Observatory* or *ROSAT*. These samples should be relatively unbiased.

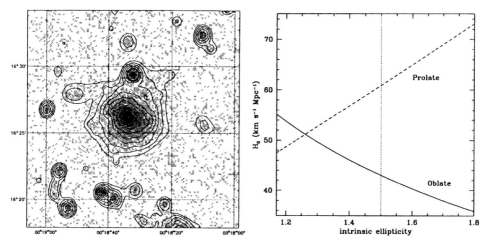

Figure 1. (Left panel) *ROSAT* PSPC X-ray map of CL 0016+16. (Right panel) Variation of the derived value of the Hubble constant, H_0, with the intrinsic ellipticity of CL 0016+16 for oblate and prolate geometries.

4. A Value for the Hubble Constant

When known systematic uncertainties are included, the best estimate of the Hubble constant becomes

$$H_0 = 44 - 64 \text{ km s}^{-1} \text{Mpc}^{-1} \pm 17\%,$$

where the range accounts for biases from temperature gradients (+30%) and clumped gas (−10%). The quoted random error includes observational errors combined in quadrature with the random systematic errors from peculiar velocities (±7%) and geometry/inclination (±7%), which have been reduced from the values given in §3 by $1/\sqrt{N}$ where N is the number of clusters. Future observational efforts should be directed toward measuring the large scale temperature gradients in galaxy clusters since this is the single largest uncertainty in the determination of H_0 from the SZ effect.

References

Birkinshaw, M. & Hughes, J.P. (1994), *Ap.J.*, **420**, 33.
Birkinshaw, M., Hughes, J.P., & Arnaud, K. A. (1991), *Ap.J.*, **379**, 466.
Herbig, T., et al. (1995), *Ap.J.*, **449**, L5.
Holzapfel, W.L., et al. (1997), *Ap.J.*, **480**, 449.
Hughes, J.P. & Birkinshaw, M. (1997), *Ap.J.*, submitted.
Hughes, J.P. & Birkinshaw, M. (1998), *Ap.J.*, in preparation.
Jones, M. (1995), *Astrophy. Lett. Comm.*, **6**, 347. *Ap.J.*, **447**, 8.
Myers, S.T., et al. (1997), *Ap.J.*, **485**, 1.

This research was partially supported by NASA LTSA Grant NAG5-3432.

X-RAY OBSERVATIONS OF THE HOT INTERGALACTIC MEDIUM

Q. DANIEL WANG
Dearborn Observatory, Northwestern University
2131 Sheridan Road, Evanston, IL 60208-2900
e-mail: wqd@nwu.edu

1. Introduction

A definite prediction from recent N-body/hydro simulations of the structure formation of the universe is the presence of a diffuse hot intergalactic medium (HIGM; e.g., Ostriker & Cen 1996). The filamentary structure of the today's universe, as seen in various galaxies surveys, is thought to be a result of the gravitational collapse of materials from a more-or-less uniform and isotropic early universe. During the collapse, shock-heating can naturally raise gas temperature to a range of $10^5 - 10^7$ K. Feedbacks from stars may also be an important heating source and may chemically enrich the HIGM. The understanding of the heating and chemical enrichment of the IGM is critical for studying the structure and evolution of clusters of galaxies, which are nearly virialized systems (e.g., Kaiser 1991; David, Jones,& Forman 1996). Most importantly, the HIGM may explain much of the missing baryon content required by the Big Bang nucleosynthesis theories (e.g., Copi, Schramm, & Turner 1995); the total visible mass in galaxies and in the hot intracluster medium together is known to account for $\lesssim 10\%$ of the baryon content (e.g., Persic & Salucci 1992).

What are the probable observational signatures of the HIGM? First, the HIGM may contribute considerably to the soft X-ray background. Various emission lines of heavy elements (O, Ne, and Fe) may collectively form a distinct spectral bump at ~ 0.7 keV (Cen et al. 1995). This bump has the potential as a diagnostic of the physical and chemical properties of the HIGM. Second, the density and temperature of the HIGM can be greatly enhanced in filamentary superstructures near rich clusters of galaxies because of deep gravitational potential. In these superstructures, X-ray emission from the HIGM can be greatly enhanced.

Here I review lines of observational evidence for the HIGM. These results appear to be consistent with the predictions from the simulations, and demonstrate the potential of X-ray observations as a powerful tool to study the large-scale structure of the universe.

2. Point-like versus Diffuse X-ray Components

While $\gtrsim 75\%$ of the background in the 1 - 2 keV range is apparently due to point-like sources (AGNs; Hasinger et al. 1993), the point-like contribution to the background decreases considerably at lower energies. Dick McCray and I have conducted an auto-correlation function (ACF) analysis of the X-ray background, based on *ROSAT* observations (Wang & McCray 1993). Using a multi-energy band ACF analysis technique, we characterized the mean spectrum of point-like sources, below a source detection limit of $S(0.5-2 \text{ keV}) \sim 1 \times 10^{-14}$ ergs s^{-2} cm^{-2}, as a power law with $\alpha = 0.7 \pm 0.2$. This spectrum is significantly flatter than that of the observed background. We find that point-like sources cannot account for more than about 60% of the background in the M-band ($\sim 0.4 - 1$ keV) without exceeding the total observed flux in higher energy bands. To explain the background spectrum, an additional thermal component of a characteristic temperature of $\sim 2 \times 10^6$ K is required, in addition to the well-constrained local 10^6 K component.

3. Spectroscopic Evidence for a Thermal Component

Direct evidence for a *thermal* M-band component of the background comes from the *ASCA* X-ray Observatory, which provides spectroscopic capability between 0.4-10 keV. The *ASCA* spectrum of the X-ray background clearly shows an excess in the M-band range above the extrapolation of the power law ($\alpha \approx 0.4$) that fits well to the 1-10 keV spectrum. The excess also shows signs of OVII and OVIII lines, suggesting that at least part of it is thermal in origin (Gendreau et al. 1995). There are, however, some significant discrepancies (up to $\sim 30\%$) between soft X-ray background normalizations derived from different instruments (McCammon & Sanders 1990; Wu et al. 1991; Garmire et al. 1992; Gendreau et al. 1995; Snowden et al. 1995), making it hard to compare different measurements of absolute background intensities. Nevertheless, both the *ROSAT* ACF analysis and the *ASCA* spectrum do seem to suggest that a considerable fraction of the M-band background arises in diffuse thermal gas. The question is where the gas is located: Galactic or extragalactic?

4. Galactic Foreground versus Extragalactic Background

While the extragalactic nature of the hard X-ray background in the 1 - 100 keV range is now generally accepted, a considerable Galactic contribution is expected at lower energies. X-ray shadowing studies are probably the only viable approach to separate the Galactic foreground from the extragalactic background. Such studies have yielded various estimates of the extragalactic background intensity in the range of 26 - 62 keV s^{-1} cm^{-2} keV^{-1} sr^{-1} at ~ 0.25 keV (e.g., Snowden et al. 1994; Cui et al. 1996; Barber, Roberts, & Warwick 1996). The uncertainty in these estimates is still too large, however, to place a useful constraint on the diffuse HIGM contribution.

We have recently measured the extragalactic background at ~ 0.7 keV by observing the X-ray shadowing of a neutral gas cloud in the Magellanic Bridge region (Wang & Ye 1996). The cloud is at a distance of ~ 60 kpc and has a peak HI column density of a few times 10^{21} cm^{-2}. From the anti-correlation between the observed background intensity and the HI column density of the cloud, we derived an unabsorbed extragalactic background intensity as ~ 28 keV s^{-1} cm^{-2} keV^{-1} sr^{-1} at ~ 0.7 keV, with the 95% confidence lower limit as 18 keV s^{-1} cm^{-2} keV^{-1} sr^{-1}. Part of this extragalactic emission must come from point-like sources such as AGNs. But the average spectrum of point-like sources is significantly flatter than that of the background and apparently flattens with decreasing source fluxes, as shown in our ACF analysis. Using the total observed background intensity in the 1-2 keV band, we obtain an upper limit to the source contribution as $\lesssim 14$ keV s^{-1} cm^{-2} keV^{-1} sr^{-1}, which is smaller than the 95% confidence lower limit of the measured extragalactic background (i.e., $18 - 14 = 4$ keV s^{-1} cm^{-2} keV^{-1} sr^{-1}). Although the actual intensity of the diffuse component is still greatly uncertain, our measurement is consistent with the HIGM contribution to the M-band background as predicted with popular cosmological models (Cen et al. 1995).

5. HIGM Features near Rich Clusters

As illustrated in the simulations, one may also expect to detect enhanced X-ray-emitting filamentary superstructures near rich clusters of galaxies. In a pilot study of the environs of intermediate redshift clusters, A. Connolly, R. Brunner, and I (1997) have discovered that A2125 is within a filamentary complex consisting of various extended X-ray-emitting features. We have further made a multi-color optical survey of galaxies in the field, using the Kitt Peak 4-m telescope. The color distribution of galaxies in the field suggests that this complex represents a hierarchical superstructure spanning $\sim 11 h_{50}^{-1}$ Mpc at redshift ~ 0.247. The multi-peak X-ray morphology of A2125 suggests that the cluster is an ongoing coalescence of at least three major subunits. The dynamic youth of this cluster is consistent with its large fraction of blue galaxies observed by Butcher & Oemler (1984). The complex also contains two additional clusters. But the most interesting feature is the large-scale low-surface-brightness X-ray emission from a moderate galaxy concentration associated with the complex.

This hierarchical complex, morphologically and spectrally very similar to superstructures seen in various N-body/hydrodynamic simulations (e.g., CDM+Λ universe; Cen & Ostriker 1994). We find that such superstructures can naturally explain the positive cross-correlation function between nearby Abell clusters and the X-ray background surface brightness at ~ 1 keV, as detected by Soltan et al. (1996).

While these results have offered us a first glimpse of the HIGM, observations with upcoming X-ray observatories such as AXAF, XMM, and Astro-E, will enable us to greatly improve the measurements. Detailed comparisons with the simulations will become possible, providing unique information on various physical and chemical processes of the structure formation of the universe.

6. References

Barber, C. R., Roberts, T. P., & Warwick, R. C. 1996, MNRAS, 282, 157
Butcher, H., & Oemler, Jr. A. 1984, ApJ, 285, 426
Cen, R., & Ostriker, J. P. 1994, ApJ, 429, 4
Cen, R., Kang, H., Ostriker, J. P., & Ryu, D. 1995, ApJ, 451, 43
Copi, C. J., Schramm, D. N., & Turner, M. S. 1995, Science, 267, 192
Cui, W., et al. 1996, ApJ, 468, 117
David, L. P., Jones, C., & Forman, W. 1996, ApJ, 473, 692
Garmire, G. P., et al. 1992, ApJ, 399, 694
Gendreau, K. C., et al. 1995, PASJ, 47, L5
Hasinger, G., et al. 1993, A&A, 275, 1
Kaiser, N. 1995, ApJ, 383, 104
McCammon, D., & Sanders, W. T. 1990, ARA&A, 28, 657
Ostriker, J. P., & Cen, R. 1996, ApJ, 464, 27
Persic, M., & Salucci, P. 1992, MNRAS, 258, 14p
Snowden, S. L., et al., 1994, ApJ, 430, 601
Snowden, S. L., et al., 1995, ApJ, 454, 643
Soltan, A. M., et al. 1996, A&A, 305, 17
Wang, Q. D., Connelly, A., & Brunner, R. 1997, ApJL, 487, 13
Wang, Q. D., & McCray, R. 1993, ApJL, 409, 37
Wang, Q. D., & Ye, T. 1996, New Astronomy, 1, 245
Wu, X.-Y., Hamilton, T. T., Helfand, D. J., & Wang, Q. D. 1991, ApJ, 37

ASCA SKY SURVEY OBSERVATIONS AND THE COSMIC X-RAY BACKGROUND IN 2–10 KEV

H. INOUE, T. TAKAHASHI, Y. UEDA AND A. YAMASHITA
Institute of Space and Astronautical Science, Kanagawa 229

Y. ISHISAKI
Tokyo Metropolitan University, Tokyo 192-03

AND

Y. OGASAKA
NASA Godard Space Flight Center, MD 20771

1. Introduction

The X-ray background in the energy range above 2 keV is highly uniform except for an excess component along the Galactic plane. The excess along the plane is considered to be associated with our Galaxy, whereas the rest of the emission is believed to be of extragalactic origin. In this paper, the X-ray background at high Galactic latitude is discussed and is designated as the CXB (cosmic X-ray background) to distinguish it from the Galactic origin.

The most likely explanation of the CXB is that the observed CXB flux is supplied by a collection of unresolved, weak discrete sources. The deepest sky survey of X-ray sources in 0.5 – 2 keV was done by ROSAT (Hasinger et al. 1993; 1994). The contribution to the 1–2 keV CXB from sources with flux above 2.5×10^{-15} erg cm^{-2} s^{-1} in 0.5 – 2 keV is estimated to be about 60 %. The discrete source contribution to the CXB in 2 – 10 keV were so far estimated through the deep survey observations with HEAO-1 A2 (Piccinotti et al. 1982) and with Ginga (Kondo 1992; Hayashida 1990). These results show that the contribution to the CXB in 2–10 keV from sources with flux above 10^{-12} erg cm^{-2} s^{-1} is $\sim$10 % and that the averaged spectral slope is still significantly steeper than that of the CXB.

ASCA carries nested thin-foil X-ray mirrors with an spatial resolution of about an arcminute even in the 2 – 10 keV band. This enables us to observe very faint sources with flux down to several times 10^{-14} erg cm^{-2} s^{-1} in 2 – 10 keV for the first time and to resolve a larger fraction of the CXB into a number of faint sources than previously possible.

2. ASCA sky survey

Two kinds of sky survey observations have been being done with ASCA.
Large sky survey (LSS): 76 pointing observations with mean exposure of about 30 ks were performed to cover a fairly wide field of sky ($\sim$ 5.4 Sq. Deg.) near the north Galactic pole.
Deep sky survey (DSS): ASCA was pointed to some sky fields (the SA 57 field, the Lockman Hole field, the Lynx field and so on) for a hundred to several hundreds ks.

In parallel to the LSS and DSS observations, a serendipitous source survey is being done for the ASCA public archive data and it is called as the ASCA medium sensitivity sky survey (MSS). Form the public archive data before Aug. 1995, 481 pointed fields satisfying some criteria ($|b| > 10°$, net exposure > 10 ks, the primary target < 10 c/s/GIS) were selected and 992 sources were serendipitously detected in the GIS fields. The catalogue of these sources will soon be open to public as the GIS Source Catalogue (Ver.1) via WWW (Ishisaki et al. 1997).

3. log N – log S relation in 2 – 10 keV

Figure 1 shows the log N – log S relations in 2 – 10 keV obtained from the DSS mainly in the SA57 field (Ogasaka et al. 1997), and from the LSS (Ueda et al. 1997). The results are consistent with an extrapolation of a N $\propto$ S$^{-3/2}$ relation from the results of the Ginga fluctuation analysis (Hayashida 1990), the Ginga high-latitude survey (Kondo 1992) and the HEAO-1 A2 survey (Piccinotti et al. 1982). If we integrate source-fluxes down to the ASCA DSS limit of $\sim 4\times 10^{-14}$ erg cm^{-2} s^{-1}, it is 40% of the CXB at most.

The 2–10 keV log N – log S relation is also obtained from the MSS and the result is consistent with those from the LSS and DSS (Ishisaki et al. 1997). Furthermore, the analysis of the angular fluctuation of the CXB is in progress. Although the result is still preliminary, we can say that the non-fluctuating component in the CXB is 20% at most and that the discrete source contribution to the CXB should be more than 80%.

Fig.2 shows the average of the spectral photon indices of sources in 2 – 10 keV in the flux range of $(1 - 4) \times 10^{-13}$ erg cm^{-2} s^{-1} detected in the LSS as a function of the source flux, in comparison with that obtained in the higher flux range by the fluctuation analysis of the Ginga background (Hayashida 1990). The average spectral index of sources detected in the

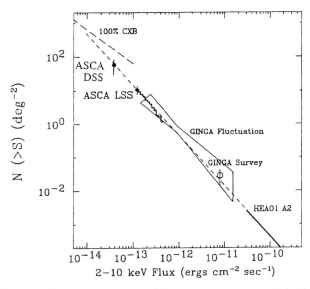

Figure 1. The log N – log S relation of X-ray sources in 2 – 10 keV obtained from the DSS observation (Ogasaka et al. 1997), and from the LSS observation (Ueda et al. 1997). The dimmer and brighter points are from the DSS and from the LSS, respectively. A dash-dotted line is an extrapolation of a $N \propto S^{-3/2}$ relation from the results of Ginga fluctuation analysis (Hayashida 1990), Ginga high-latitude survey (Kondo 1992) and HEAO-1 A2 survey (Piccinotti et al. 1982).

ASCA LSS is significantly harder and closer to the CXB index than that obtained from the previous measurements in the much higher flux range in this energy band. This result implies that a population of sources with an average index similar to the CXB begins to dominate in the flux range around 10^{-13} erg cm^{-2} s^{-1} in 2 – 10 keV.

4. What are the dim, hard sources?

Some DSS observations were done in fields overlapped with fields in which deep ROSAT observations and the optical follow-up observations were done (Lockman Hole: Hasinger et al. 1997; Schmidt et al. 1997. GSGP4, QSF3, F855: Georgantopoulos et al. 1997; Boyle et al. 1997).

In the Lockman Hole observations, sources detected with ASCA in 2 – 10 keV in the field overlapped with the ROSAT field are almost all identified with ROSAT sources to which optical counterparts were found. Almost all the optical counterparts (11/12) are classified as AGNs by Schmidt et al. (1997) and their luminosities and the cosmological redshifts distribute 10^{43-45} erg cm^{-2} s^{-1} and 0.5 – 2.

The ratios of the 0.5 – 2 keV ROSAT flux to the 2 – 10 keV ASCA flux largely scatter by an order of magnitude and several sources have the

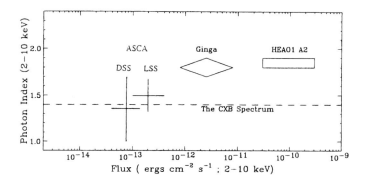

Figure 2. The average spectral photon index of sources detected in the lowest flux range of the LSS (Ueda et al. 1997) and of the DSS (Ogasaka et al. 1997), in comparison with those obtained in the higher flux range: the results of the fluctuation analysis of the Ginga background (Hayashida 1990) and of the Piccinotti sample by HEAO-1 A2. The spectral index of the cosmic X-ray background is also indicated.

ratio as low as 0.1. The low 0.5 – 2 keV flux relative to the 2 – 10 keV flux suggests that those AGNs would suffer from heavy obscuration. In fact, some of them show no optical broad line (Schmidt et al. 1997).

Other optical follow-up observations of the ASCA DSS or LSS sources are also in progress. An observation of a ASCA source in the SA57 field discovered a type-2 quasar at $z \simeq 0.9$ (Ohta et al. 1996). Another optical observation also found a type-2 Seyfert galaxy at the position of a very hard source in the LSS field (Sakano et al. 1997; Akiyama et al. 1997). These all seem to support the idea that most of the dim, hard sources would be highly obscured AGNs at the cosmological distance.

The authors would like to thank G. Hasinger for his providing us the results of the ROSAT Lockman Hole observation.

References

Akiyama, M., et al., 1997, this Symposium.
Boyle, B.J., et al., 1997, preprint.
Georgantopoulos, I., et al., 1997, preprint.
Hasinger, H., et al., 1993, A&Ap 275, 1.
Hasinger, H., et al., 1994, A&Ap 291, 348(Erratum).
Hasinger, H., et al., 1997, preprint.
Hayashida, K., 1990, Ph.D.Thesis, Univ. of Tokyo.
Ishisaki, Y., et al., 1997, this Symposium.
Kondo. H, 1992, Ph.D.Thesis, Univ. of Tokyo.
Ogasaka, Y., et al. 1997, in preparation.
Ohta, K., 1996, Ap.J., 458, L57.
Sakano, M., et al., 1997, in preparation.
Schmidt, M., et al., 1997, preprint.
Ueda, Y., et al. 1997, preprint.

Contributed Papers

Session 1: Plasma and Fresh Nucleosynthesis Phenomena

1-1. Sun and Stars

AN ATTEMPT TO CLASSIFY SOLAR MICROWAVE-BURSTS BY SOURCE LOCALIZATION CHARACTERISTICS AND DYNAMICS OF FLARE-ENERGY RELEASE

A. KRÜGER, B. KLIEM, J. HILDEBRANDT,
Astrophysical Institute Potsdam,
An der Sternwarte 16, 14482 Potsdam, Germany

AND

V.P. NEFEDEV, B.V. AGALAKOV, G.YA. SMOLKOV
Institute of Solar-Terrestrial Physics, P.O. Box 4026,
664033 Irkutsk, Russia

Abstract

An overview of spatially resolved observations of solar radio bursts obtained by the Siberian Solar Radio Telescope at 5.8 GHz during the last ten years reveals the occurrence of different classes of burst emission defined by their source localization characteristics. Four major classes of bursts according to the source position relative to sunspots, the source size and structure, and the source height, could be tentatively distinguished and compared with burst spectral characteristics as well as with soft X-ray emission oberved by YOHKOH. These findings are in favour of a magnetic origin of the underlying flare process.

1. Introduction

It is anticipated that the consideration of topological microwave-burst source characteristics may help to get a deeper insight into the related physical processes of coronal energy release which, due to the lack of observations with sufficient spatial resolution, was hindered for many years before. The present study makes an attempt to include geometrical source characteristics in a burst classification scheme based on spatially resolved observations.

2. Observations

We used the observations of the large cross-type Siberian Solar Radio Telescope (SSRT) operating at 5.2 cm wavelength (5.8 GHz) (cf., e. g., Smolkov et al., 1986). For the present purpose daily observations obtained by the E-W arm of the instrument have been considered. The time and space resolution at local noon (about 05 UT) are 2 min 15 sec and $17''$, respectively.

3. Topological classes of microwave bursts

A systematic inspection of daily observations of the SSRT during 1984–1994 revealed regularities of the burst sources stimulating the attempt of a classification scheme of microwave bursts according to the source topology and related properties determining four groups of bursts:

Group 1 comprises rather strong microwave bursts situated in the vicinity of large sunspots (distance d to sunspots less than about $20''$). These events are signatures of strong energy dissipation in the deep corona.

Group 2 ($d \approx 1'$) signifies moderate energy release in higher reaching flare loops (altitude h up to about 100 000 km) lasting a few hours.

Group 3 is indicative for energy storage in between main sunspots with timescales up to several hours occurring typically before the onset of large flare events ('preheating').

Group 4 comprises smaller, compact microwave sources in active regions characterized by no or only weak sunspots.

4. Conclusions

The groups of microwave bursts given above indicate a certain connection between the burst characteristics and the position of the burst source relative to the site of sunspot-magnetic fields. These findings support the magnetic origin of the released flare energy similar as found, e. g., by Nishio et al. (1997). A more detailed description will be published elsewhere.

Acknowledgement

The present work was supported by the Deutsche Agentur für Raumfahrtangelegenheiten (DARA) under contracts 50 QL 9602 and 9208.

References

Nishiio, M., Yaji, K., Kosugi, T., Nakajima, H., Sakurai, T,: 1997, Ap. J., in press.
Smolkov, G.Ya. and 5 coauthors: 1986, Astrophys. Space Sci. 119, 1.

RESISTIVE PROCESSES IN THE PREFLARE PHASE OF ERUPTIVE FLARES

The Role of the Perpendicular Magnetic Fields

T. MAGARA
Univ. of Kyoto, Faculty of Science
Sakyo-ku, Kyoto 606, Japan

AND

K. SHIBATA
National Astronomical Observator
Mitaka, Tokyo 181, Japan

Abstract

In this study, we perform 2.5-dimensional MHD simulations and clarify the role of *perpendicular magnetic fields* (which are perpendicular to the 2D plane) in a preflare current sheet of solar flares. At the first stage, a current sheet formed within a coronal magnetic structure is filled with the perpendicular fields (force-free structure). Then this sheet begins to be dissipated through the tearing instability under a uniform resistivity. As the instability proceeds, the distribution of the perpendicular fields vary in such a way that most of them gather around O-point (magnetic island) instead of X-point. Therefore, the magnetic pressure of these fields weaken in the vicinity of X-point so that they no longer suppress the inflows toward this point. These flows then make the current sheet thinner and thinner, which implies that the current density around X-point becomes high enough to cause an anomalous resistivity whose value is much larger than that of the normal collisional resistivity. In this way, the transition from a uniform resistivity to a locally-enhanced one occurs, which can make the violent energy release observed in solar flares.

References

Magara, T., Shibata, K., and Yokoyama, T. (1997), *ApJ*, **Vol.487**, pp.437
Ohyama, M. and Shibata, K. (1997), *PASJ*, **Vol.49**, pp.249

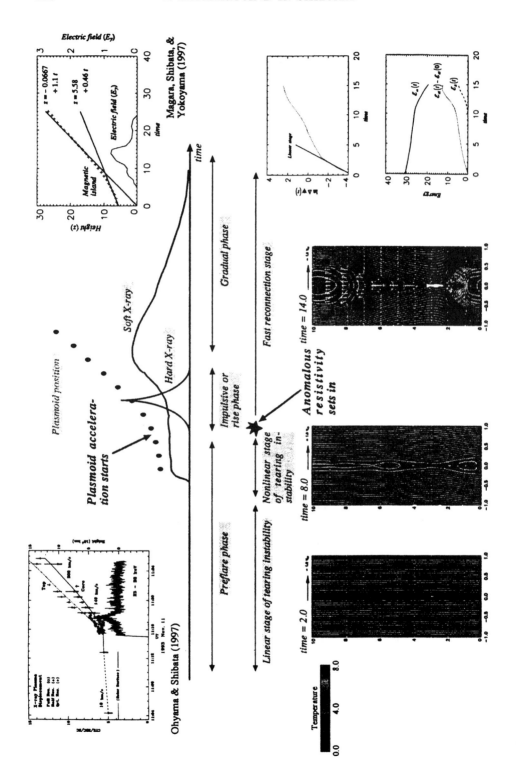

THREE-DIMENSIONAL SIMULATION STUDY OF PLASMOID INJECTION INTO MAGNETIZED PLASMA

Y. SUZUKI, T.-H. WATANABE, A. KAGEYAMA, T. SATO AND
T. HAYASHI
Theory and computer simulation center,
National Institute for Fusion Science
Toki 509-52, Japan

1. Introduction

Resent observations suggest that, during solar flares, plasmoids are injected into the interplanetary medium (Stewart et al., 1982). It has also been pointed out that solar wind irregularities modeled as plasmoids are penetrated into the magnetosphere (Lemaire, 1977). These plasmoid injections are considered to be an important process because they transfer mass, momentum, and energy into such magnetized plasma regions. Our objective is to investigate the dynamics of a plasmoid, which is injected into a magnetized plasma region and to reveal mechanisms to transfer them. To achieve this, we carried out three-dimensional magnetohydrodynamic (MHD) simulations.

2. 3D Dynamics of Injected Plasmoid

The simulation model is the same as that in Suzuki et al. (1997). The simulation region is composed of two cylinders which connect with each other. One with a smaller radius corresponds to a injection region in which the plasmoid is initially located, and the other with a larger radius is a magnetized plasma region where the magnetic field is almost unidirectional and perpendicular to the injection direction. The plasmoid is accelerated in the injection region into the magnetized plasma region.

The simulation result is shown in Figure 1. We can see that at $t = 20$ the plasmoid enters the magnetized plasma region, suffering from the tilting instability. Also, at this time, the plasmoid field reconnects with the field

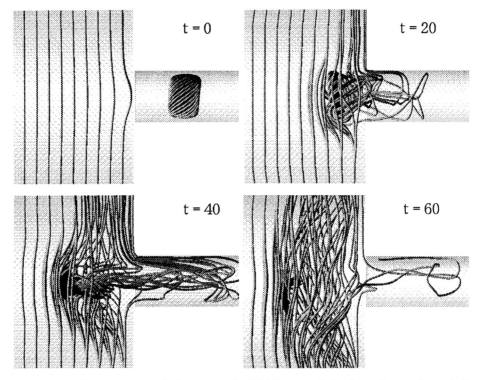

Figure 1. The structure of the magnetic field lines and the iso-value surface of the high density plasma (black region) at $t = 0, 20, 40, 60$. The time is normalized by the characteristic Alfvén transit time.

in the magnetized plasma region. As time goes, the magnetic reconnection process successively proceeds ($t = 40$). Finally the magnetic configuration of the plasmoid is disrupted ($t = 60$). As a result, the high density plasma confined in the plasmoid is supplied in the magnetized plasma region. From these results we can see that the plasmoid injection is an important process especially to transfer the plasmoid mass into the magnetized plasma region, where the magnetic reconnection is a key mechanism.

References

Stewart, R.T., Dulk, G.A., Sheridan, K.V., House, L.L., Wanger, W.J., Sawyer, C.J., and Illing, R., (1982) Visible Light Observations of a Dense Plasmoid Associated with a Moving Type IV Solar Radio Burst, *Astron. Astrophys.*, **116.**, pp. 217–223.

Lemaire, J., (1977) Impulsive Penetration of a Filamentary Plasma Element into the Magnetosphere of the Earth and Jupiter, *Planet. Space Sci.*, **25.**, pp. 887–890.

Suzuki, Y., Watanabe, T.-H., Kageyama, A., Sato, T., and Hayashi, T. (1997) to be published *in Proc. of the Joint Conference of 11th International Stellarator Conference & 8th International Toki Conference on Plasma Physics and Controlled Nuclear Fusion*, ed. Y.Ueda JPFR Series 1.

IMPULSIVE FLARE PLASMA ENERGIZATION IN THE LIGHT OF YOHKOH DISCOVERIES

YURII M. VOITENKO
Department of Space Plasma Physics, Main Astronomical Observatory, Holosiiv, Kyiv-22, 252650, Ukraine

Flare scenario, based on *Yohkoh* observations and magnetic reconnection (MR) hypotesis, has been elaborated recently (Tsuneta, 1996; Shibata, 1997). Using theoretical prediction that the fast MR results in a high-speed outflows from reconnection site, the reconnection downflow has been supposed to collide with the top of underlying magnetic loop, producing superhot plasma and high-energy particles.

We take into account another part of the reconnection outflow, diverting along field lines close to separatrix, and forming so-called separatrix jets (Strachan and Priest, 1994). These sepatatrix jets set up warm beams of ≥ 0.1 MeV protons (PBs), streaming down along just-reconnected field lines through steady underlying plasma (Voitenko, 1996a;b). Note that there are many observations indicating that during initial phase the bulk flare energy is carried by ≥ 0.1 MeV proton beams (Simnett, 1995). The principal significance for the solar flares theory has a question of the further transformation of PB kinetic energy in flaring loops (FLs), overlying well-known SXR loops.

We study **excitation of kinetic Alfvén waves (KAWs) by PB**, keeping effects of induced by PB large-scale electric field $\mathbf{E}_0 \| \mathbf{B}_0$ along the legs of flaring loop (Voitenko, 1996a). Taking into account high growth rate ($\sim 10^4 \text{s}^{-1}$), short relaxation distance ($\sim 10^6$cm), and energy flux partition between waves and PB after relaxation ($P_{KAW}/P_{PB} \sim 1$), we conclude that PB-driven KAW instability is an efficient mechanism of flare energy conversion in FL. The quasi-linear spectra of excited waves $\sim k_\perp^2$ in long-wavelength domain, and $\sim k_\perp$ in short-wavelength domain. Spectral energy concentration at largest wavenumbers below upper cut-off, where nonlinear three-wave interaction is strongest, give rise to nonlinear modification of the quasilinear spectra and to the formation of turbulent spectra $\sim k_\perp^{-5}$ in long-wavelength domain, and $\sim k_\perp^{-3}$ in short-wavelength domain.

Flare plasma energization by KAW turbulence may be responsible for various observed phenomena with rise time and time delays of about 5-10 s, corresponding to PB/KAW flux propagation time. Impulsive heating of plasma near injection site by a strong KAW flux results in time dependence of electron temperature in the upper part of FL legs (Voitenko, 1996b):

$$T_{e(7)} = (1.4 \times 10^4 B_{in(2)} t + T_0^{5/2})^{2/5}, \qquad (1)$$

where $T_{0(7)}$ is initial temperature, and subscripts in () denote the order (e.g., $T_{e(7)} = T_e/10^7$, etc.).

1992 Jan. 13 flare (Masuda et al., 1995). Taking plasma number density $n_{0(9)} = 2.5$ in legs and $n_{0(9)} = 10$ at the top, FL half-length $L_{(9)} = 2$, $n_{b(9)}(\approx n_{in(9)}) = 1$, $B_{0(2)}(\approx B_{in(2)}) = 0.57$, $T_{0(7)} = 0.6$, and PB injection time $t_b = 3$ s, we obtain high instability growth rate, $\gamma_k = 3 \times 10^4$ s^{-1}, short relaxation distance 7×10^5 cm, temperature of heated region $T_e = 6.7 \times 10^7$, spreading velocity $\geq 4 \times 10^8$ cm s^{-1} and flux of escaping (> 20 keV) electrons 10^{17} el. cm^{-2}s^{-1}. Hence the fast plasma heating by KAWs provides temperature and flux of escaping electrons, enough to produce observed (thermal synchrotron) microwave emission from the loop leg, and HXR bremsstrahlung from footpoints and loop top.

1992 Feb. 21 LDE flare (Tsuneta, 1996). In inflow region $n_{in(9)} = 1$, $T_{e(7)} = 0.5$; in FL legs $n_{0(9)} = 5$, $T_{e(7)} = 1.3$; heat conduction loss from legs $P_{th(9)} = 1$. Then PB number density $n_{b(9)} \approx 1$ and (super-Alfvén) velocity $V_{0b}/V_A \approx 2.2$. Assuming that the heating of FL legs by KAW flux is balanced by P_{th}, we get reasonable estimation for flare energy flux $P_{fl(9)} \approx P_{0b(9)} = 2.5$ and reconnecting field $B_{0(2)} = 0.2$.

Acknowledgements. Useful discussions of flare models with K. Shibata and S. Tsuneta are gratefully acknowledged. This work was supported in part by grant 2.4/1003 of the Ukrainian Fund for Fundamental Research; the presentation at the IAU Symposium No. 188 became possible due to IAU travel grant.

References

Masuda, S., Kosugi, T., Hara, H., Sakao, T., Shibata, K., and Tsuneta, S.: 1995, *P.A.S.J.* **47**, 677
Shibata, K.: 1997, in T. Watanabe, T. Kosugi, and A. C. Sterling (eds) *Proc. Yohkoh 5th Anniversary Symposium*, in press.
Simnett, G. M.: 1995, *Space Sci. Rev.* **73**, 387.
Strachan, N. R. and Priest, E. R.: 1994, *Geophys. Astrophys. Fluid Dynamics* **74**, 245-274
Tsuneta, S.: 1996, *Astrophys. J.* **456**, 840.
Voitenko, Yu. M.: 1996a, *Solar Phys.* **168**, 219.
Voitenko, Yu. M.: 1996b, *ASP Conference Series* **111**, 312.

MHD SIMULATION OF CHROMOSPHERIC EVAPORATION IN A SOLAR FLARE BASED ON MAGNETIC RECONNECTION MODEL

T. YOKOYAMA AND K. SHIBATA
National Astronomical Observatory of Japan
Mitaka, Tokyo, 181 Japan

Two-dimensional magnetohydrodynamic simulation of a solar flare is performed using a newly developed MHD code including nonlinear anisotropic heat conduction effect (Fig. 1; Yokoyama & Shibata 1997a). The numerical simulation starts with a vertical current sheet which is line-tied at one end to a dense chromosphere. The flare energy is released by the magnetic reconnection mechanism stimulated initially by the resistivity perturbation in the corona. The released thermal energy is transported into the chromosphere by heat conduction and drives chromospheric evaporation. Owing to the heat conduction effect, the adiabatic slow mode MHD shocks emanated from the neutral point are dissociated into conduction fronts and isothermal shocks (Yokoyama & Shibata 1997b). Temperature and derived soft X-ray distributions are similar to the cusp-like structure of long-duration-event (LDE) flares observed by the soft X-ray telescope aboard Yohkoh satellite. On the other hand density and radio maps show a simple loop configuration which is consistent with the observation with Nobeyama Radio Heliograph. Two interesting new features are found. One is a pair of *high density humps* on the evaporated plasma loops formed at the collision site between the reconnection jet and the evaporation flow. The other is the *loop-top blob* behind the fast-mode MHD shock.

References

Yokoyama, T. and Shibata K. (1997a) Two-Dimensional Magnetohydrodynamic Simulation of Chromospheric Evaporation in Solar Flare Based on Magnetic Reconnection Model, submitted to ApJ Letters.

Yokoyama, T. and Shibata K. (1997b) Magnetic Reconnection Coupled with Heat Conduction, *Astrophys. J.*, **474**, L61.

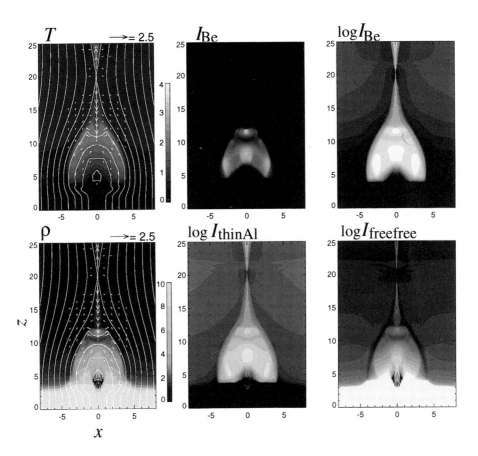

Figure 1. Results of the simulation. Snapshots at $t = 25$ of temperature, density, X-ray and radio maps derived from the simulation results are shown. The arrows show the velocity, and lines show the magnetic field lines. The unit of length, velocity, time, temperature, and density is 3000 km, 170 km s^{-1}, 18 s, 2×10^6 K, and 10^9 cm^{-3}, respectively. For the X-ray map, the response of the filter attached to the soft X-ray telescope on board Yohkoh is taken into account. I_{Be} and I_{thinAl} are the intensity through the beryllium and the thin aluminum filters. The radio map $I_{freefree}$ is derived under the assumption that the emission is by the thermal free-free mechanism. The unit of intensity is arbitrary.

CORONAL VARIABILITY AND FLARING OF THE RS CVN BINARIES σ^2 CRB AND HR1099

A. BROWN, R. A. OSTEN
CASA, U. of Colorado, Boulder, CO 80309-0389, USA

S. A. DRAKE
HEASARC, NASA/GSFC, Greenbelt, MD 20771, USA

K. L. JONES
Dept. of Physics, U. of Queensland, Brisbane, Australia

AND

R. A. STERN
Lockheed-Martin SAL, Palo Alto, CA 94304, USA

We present new X-ray, EUV, and radio observations of the coronal emission from the RS CVn binaries HR1099 (V711 Tau) and σ^2 CrB. RS CVn systems possess coronae that display extreme activity levels and frequent flaring. Our observations provide multiwavelength records of coronal variability and flaring over more than two binary orbits for each system. While the EUV and X-ray spectra show the flare response of the thermal plasma, the radio data give the corresponding information for the nonthermal electron population. The HR1099 data contain one of the most energetic flare outbursts observed from HR1099 that lasts for over 3 days. Coronal flaring is common in these systems and, in fact, is the normal condition.

σ^2 CrB was observed with the XTE and ASCA X-ray satellites and the VLA radio array on 1997 March 11-13. The radio observations show large variations at 3, 6 and 20 cm; including a flare rise seen on March 11 and a complex flare outburst that lasted for most of March 12. Polarised 20 cm bursts were seen associated with flare rises; those asssociated with the flares at March 11.7 and 12.3 were **left** circularly polarised, while that at March 12.6 was **right** circularly polarised. A weak excess of right CP was seen during the March 12.5 subflare at 6 cm. The rapid radio decay at 3 and 6 cm suggests that the flare region was expanding rapidly. Outside the flares CP was absent at all three frequencies. At least three flares were detected by XTE, with two of them having radio counterparts.

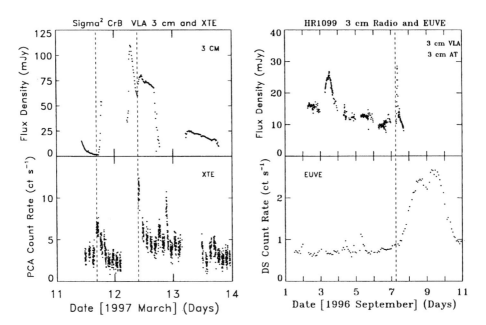

Figure 1. **Left** Coronal flaring from σ^2 CrB in the X-ray and radio. **Right** Variations of HR1099 in the radio and EUV spectral regions. The vertical dashed lines mark the onset of large flares.

Over the period 1996 September 1-11 we observed HR1099 with the EUVE and XTE satellites, and the VLA and the AT radio arrays. While the EUVE Deep Survey (DS) photometer shows a mildly varying signal between September 1 and 7, the 3cm flux density is steadily falling with a modest flare lasting 16 hours seen on September 3 without any obvious EUV counterpart. At about 6 UT on September 7 a sharp radio flare indicates the initiation of an extremely large coronal flare outburst. The EUV flare lasted over 86 hours with an integrated energy (80-180 Å) of $4.1 \; 10^{34}$ ergs. The e-folding times of the rise and decay were 23.5 and 25 hours respectively. XTE detected a small flare at 20 UT on September 4, which corresponds to a weak enhancement in the EUV and a gap in the radio coverage.

Our extensive multi-spectral-region studies of HR1099 over the last five years have demonstrated in a systematic way the ubiquitousness of coronal flaring and the extremely transitory coronal conditions present at any instant.

This work is supported by NASA grants NAG5-4735, NAG5-4534, NAG5-2259, and NAG5-3226 to the University of Colorado.

HGMN STARS AS APPARENT X-RAY EMITTERS

S. HUBRIG
University of Potsdam, Am Neuen Palais 10, D-14469 Potsdam, Germany

AND

T. W. BERGHÖFER
Space Sciences Laboratory, University of California, Berkeley, CA 94720-7450, USA

1. Introduction

In the ROSAT all-sky survey 11 HgMn stars were detected as soft X-ray emitters (Berghöfer, Schmitt & Cassinelli 1996). Prior to ROSAT, X-ray observations with the *Einstein Observatory* had suggested that stars in the spectral range B5–A7 are devoid of X-ray emission. Since there is no X-ray emitting mechanism available for these stars (also not for HgMn stars), the usual argument in the case of an X-ray detected star of this spectral type is the existence of an unseen low-mass companion which is responsible for the X-ray emission. However, this hypothesis is not easily testable. Based on high resolution X-ray images taken with the ROSAT HRI, Berghöfer & Schmitt (1994) showed that known visual late-type companions can be disregarded in this context. In almost all cases studied so far (including two HgMn stars in our sample) the X-ray emission is associated with the primary B star.

The purpose of the present work is to use all available data for our sample of X-ray detected HgMn stars and conclude on the nature of possible spectroscopic companions. We emphasize that some of our sample stars consist of two nearly equal B stars. The observed X-ray emission in these systems is also inconsistent with the secondary star and, thus, a third component must exist to explain the X-ray emission by a low-mass companion. Some of our sample stars show X-ray luminosities that exceed the X-ray output of normal late-type stars and, therefore, an active pre-main sequence companion (PMS) is required. This hypothesis is supported by the fact that a significant fraction of the HgMn stars found in the ROSAT survey belong to rather young stellar groups like the Pleiades supercluster or the Sco-Cen association.

Here we describe our method to conclude on the nature of possible low-mass companions. In a first step stellar masses and ages were derived for our sample of HgMn stars. For this we utilized the stellar model grids provided by Schaller et al. (1992); the stellar distances were taken from the recently released Hipparcos catalog and the effective temperatures were compiled from the literature. We then assumed that the absence of a secondary in the optical spectrum implies a mass ration of $M_1/M_2 \geq 1.5$ for the two binary components and all systems are formed coeval. A further criterion was the saturation limit of $log(L_x/L_{Bol}) \approx -3$ known for late-type star X-ray emission (cf. Schmitt 1997); for the observed X-ray luminosities this relation provides upper limits for the bolometric luminosities of the possible secondaries. Together with these limits for the companions masses, luminosities, and ages, we utilized the pre-main-sequence evolutionary tracks provided by D'Antona and Mazzitelli (1994) to limit the range of possible companions of the 11 HgMn stars.

2. Results and Conclusions

For all of our sample HgMn stars detected in the ROSAT all-sky survey we find that a late-type companion can provide a natural explanation for the observed X-ray emission. In 7 cases (HD 32964, HD 33904, HD 35497, HD 75333, HD 110073, HD 141556, and HD 173524) the detected X-ray emission can be explained by a main-sequence late-type star, whereas for the stars HD 27295, HD 27376, HD 29589, and HD 221507 a PMS star is required. Further investigations by means of radial velocity studies and high-resolution imaging (e.g., in the near IR) are needed to detect the predicted companions. According to the mass lower limits derived for possible companions in our sample of HgMn stars spectral types are in the range late K-M4. It is remarkable that in many cases when a spectroscopic binary with a late-B primary has a third, distant companion, the SB primary is a HgMn star (e.g., Hubrig & Mathys 1995).

References

Berghöfer, T.W., Schmitt, J.H.M.M.: 1994, A&A 292, L5
Berghöfer, T.W., Schmitt, J.H.M.M., Cassinelli, J.P.: 1996, A&AS 118, 481
D'Antona F., Mazzitelli I.: 1994, ApJ Suppl. 90, 467
Hubrig S., Mathys G.: 1995, Comm. Astroph. 18, 167
Schaller G., Schaerer D., Meynet G., Maeder A.: 1992, A&AS 96, 269
Schmitt, J.H.M.M.: 1997, A&A 318, 215

VARIABLE HIGH VELOCITY JETS IN THE SYMBIOTIC STAR CH CYGNI

T. IIJIMA
Astronomical Observatory of Padova
Osservatorio Astrofisico, I-36012 Asiago (Vicenza) Italy
e-mail: iijima@astras.pd.astro.it

Highly blue or red shifted broad emission components of H I lines have been noticed in some recent spectra of the symbiotic star CH Cyg. Their intensities, profiles, and displacements from the narrow component have changed with time scale of days. For example, a blue-shifted (-1545 km s^{-1}) intense emission component and a red-shifted weak ($+757$ km s^{-1}) emission were seen on 1994 September 9. One day after, the blue-shift and red-shift became -2076 km s^{-1} and $+1285$ km s^{-1}, respectively.

The profile of the broad emissions could be classified in some groups:

1: Main part is seen in the blue side of the narrow component, and a weak tail is found in the red side. The blue-shift of the main peak from the narrow component is roughly -1000 km s^{-1}.

2: Main part is seen in the red side. The red-shift of the main peak is roughly 1000 km s^{-1}.

3: Two emission components are seen in the blue and red sides of the narrow component. The blue-shift sometimes exceeds -2000 km s^{-1}, while the red shift is roughly 1000 km s^{-1}.

4: Broad emission component is merged in the narrow component.

5: Weak but very broad and roughly symmetric emission tail is seen. The full width at zero intensity is about 8000 km s^{-1}.

6: P Cygni type absorption reverse is superimposed on one of the above profiles.

The behaviour of the broad emission components suggests that there is a strong variable gas outflow in this system. The properties of the broad components are rather similar to those of broad emission components of H I lines of some Seyfert galaxies, except for the time scale of variation.

X-RAY EMISSION FROM DISTANT STELLAR CLUSTERS

Mean Luminosity versus Age in Late Type PMS Stars

N.S.SCHULZ, J.H.KASTNER
Massachusetts Institute of Technology
Cambridge MA 02143, USA

1. Introduction

Observations with the *Einstein* Observatory indicated that stellar X-ray activity diminishes in clusters older than 70 Myr (Pleiades). *ROSAT* observations of older clusters also support this result (see Caillault 1995 and references therein). The timescales over which young stars diminish in X-ray luminosity depends on spectral type (Randich et al. 1996), leading to the conclusion that X-ray activity in late type PMS depends on age and stellar mass. F and G-stars approach the main sequence much faster and the diminishing rates of X-ray activity from F to M stars start to differ considerably. Kastner et al. (1997) observed that the mean of the ratio L_x/L_{bol} for K and M dwarf stars increases monotonically for low-mass stars from the very early T Tauri stage through the age of the Pleiades cluster, reflecting the contraction and spin-up of such stars during pre-main sequence evolution. This ratio then decreases towards middle aged stars, as late-type main sequence stars spin down. Here we extend this result by including more distant clusters that are younger overall than those considered by Kastner et al. and also including earlier spectral types.

2. Results

For very young stars we should be safe including F and G-types, for older ones the error grows significantly. Since we do not have much knowledge on the spectral types in most of the detected X-rays sources, we assume a random distribution from F up to early M-type stars for an estimate of bolometric luminosities. Detailed photometric identifications clearly are an outstanding issue for the future. In figure 1 the estimated values of L_x/L_{bol} for IC 1396 (Schulz et al. 1997), the Rosette Nebula, NGC 2362 (Berghöfer 1997, priv. comm.) and IC 348 (Preibisch et al. 1996) are added to the ones

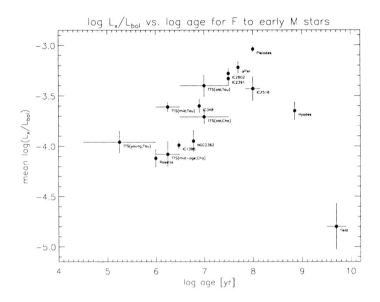

Figure 1. Mean of log of the ratio of X-ray to bolometric luminosity for late type PMS for F- up to early M-type stars. New entries are values for IC 1396, the Rosette Nebula, NGC 2362, IC 348 and IC 2516.

already shown in Kastner et al. (1997). For IC 348 figure 1 confines L_x/L_{bol} to an upper limit of -3.2 and thus allows us to set a mean L_x/L_{bol} of -3.6. For completeness an estimate for an older cluster in IC 2516 (Dachs et al. 1996) was also added.

The mean values for very young clusters fit nicely into the trend observed T Tauri stars and nearby, somewhat older clusters. It is apparent that from the diagram one may be able to estimate the mean L_x/L_{bol} for clusters for which we are unable to determine the X-ray luminosity function directly. If we exercise this for IC 1805, which due to its distance is near the *ROSAT* sensitivity limit, we would get an expected mean value between -3.5 and -4.4 for an age of 1 Myr. The same would apply to Tr 14, where no *ROSAT* pointing was found in the archive.

References

Caillault J.-P. (1996) , MPE report 263, ISSN 0178-0719, 67
Dachs J. , Hummel W. (1996), *A&A* 312, 818
Kastner J., Zuckerman B., Weintraub D.A., Forveille T., (1997), *Science* 277, 67
Preibisch T., Zinnecker H., Herbig G.H., (1996), *A&A* 310, 456
Randich S., Schmitt J.H.M.M., Prosser C.F., Stauffer J.R., (1996), *A&A* 300, 134
Schulz N.S., Berghöfer T., Zinnecker H., (1997), *A&A* 325, 1001

X-RAY CORONAE FROM SINGLE LATE-TYPE DWARF STARS

K.P. SINGH

Tata Institute of Fundamental Research
Homi Bhabha Road, Mumbai 400 005, INDIA

AND

S.A. DRAKE AND N.E. WHITE

LHEA, Goddard Space Flight Center/NASA
Greenbelt, MD 20771, U.S.A.

1. Introduction

Study of X-ray coronae from late-type stars with moderate resolution X-ray spectroscopy with *ASCA*, has led to the characterization of temperatures and measurements of elemental abundances in their coronae. Several RS CVn and Algol-type binary systems, and single late-type stars have been observed. We present here the results obtained from X-ray spectroscopy of *recently observed single F-G-K-M type dwarfs*. The sample observed with *ASCA* contains αCen (Mewe et al. 1997), π^1UMa (Drake et al. 1994), YY Gem, Speedy Mic, GJ 890 (Singh et al. 1997), EK Dra, HN Peg, κ^1Cet (Guedel et al. 1997), AB Dor (Mewe et al. 1996) and HD 35850 (Tagliaferri et al. 1997).

2. X-ray Activity vs. Rotation Period

The ratio of "quiescent" X-ray luminosity to the bolometric luminosity is an indicator of the X-ray activity level. Being independent of the radius and distances of the stars it allows us to combine stars of different spectral types. Augmenting the *ASCA* observed sample above with the *ROSAT* observed G dwarfs from Guedel et al., we find that the log L_x/L_{bol} shows a strong correlation with the period of rotation of stars and saturates near the value of -3 reached by stars rotating with periods of a day or faster.

3. X-ray Spectroscopy

3.1. DIFFERENTIAL EMISSION MEASURES (DEM)

Application of continuous emission measure polynomial method and using plasma emission models "MEKAL" (Mewe et al. 1995) shows that in the very rapidly rotating dwarfs e.g., GJ 890 and Speedy Mic, DEM is best represented by a bimodal temperature distribution, and shows the presence of a very hot component ($kT \geq 2$ keV). These stars show DEM characteristics similar to that observed in the RS CVn and Algol type binaries (Singh et al. 1995, Singh et al. 1996, Kaastra et al. 1996). During the frequently observed X-ray flares on active K and M dwarfs the temperature of the plasma and the emission measure in the hotter cpmponents increases significantly. Very hot components with $kT \geq 1$ keV are, however, not found in the Sun like stars e.g., αCen and π^1UMa.

3.2. CORONAL ABUNDANCES

The coronae of rapid rotators show extremely weak line emission and consequently a depletion (factor 3 to 8) of elemental abundances when compared to the solar photospheric abundances (Anders & Grevesse 1989). This underabandance appears to be more than that observed in Algol and AR Lac. The First Ionization Potential (FIP) effect as observed in the solar corona, where the low FIP elements have enhanced abundances compared to the high FIP elements has not been detected in the X-ray spectra of rapid rotators.

In single solar-types G stars, e.g. αCen (and π^1UMa), the corona is relatively cooler and the elemental abundances in them are closer to that in the solar corona, and thus indicative of the FIP effect as seen in the Sun.

References

Anders, E., & Grevesse, N., 1989, Geochimica et Cosmochimica Acta, 53, 197
Drake, S.A., Singh, K.P., White, N.E., & Simon, T. 1994, ApJ, 436, L87
Guedel, M., Guinan, E.F., & Skinner, S.L. 1997, ApJ, 483, 947.
Kaastra, J.S., Mewe, R., Liedahl, D.A., Singh, K.P., White, N.E., & Drake, S.A. 1996, A&A, 314, 547
Kaastra, J.S., & Mewe, R. 1993, Legacy, 3, 16
Mewe, R., Drake, S.A., Kaastra, J.S., Drake, J.J., Schmitt, J.H.M.M., Schrijver, C.J., Singh, K.P., & White, N.E., 1997, A&A, submitted
Mewe, R., Kaastra, J.S., White, S.M., & Pallavicini, R., 1996, A&A, 315, 170
Mewe, R., Kaastra, J.S., & Liedahl, D.A. 1995, Legacy, 6, 16
Singh, K.P., Drake, S.A., Gotthelf, E., & White, N.E. 1997, (in preparation)
Singh, K.P., Drake, S.A., & White, N.E. 1995, ApJ, 445, 840.
Singh, K.P., White, N.E., & Drake, S.A. 1996, ApJ, 456, 766.
Tagliaferri, G, Covino, S., Fleming, T.A., Gagne, M., Pallavicini, R., Haardt, F., & Uchida, Y. 1997, A&A, 321, 850.

CHEMICAL ABUNDANCES OF EARLY TYPE STARS

S. TANAKA, S. KITAMOTO, T. SUZUKI AND K. TORII
Osaka University, Department of Earth and Space Science, Osaka University 1-1, Machikaneyama-cho, Toyonaka, Osaka, 560, Japan

M.F. CORCORAN
Laboratory for High Energy Astrophysics, NASA/GSFC, Greenbelt, MD20771, USA

AND

W. WALDRON
Applied Research Corp., 8201 Corporate Dr., Landover MD 20785, USA

1. Introduction

X-rays from early-type stars are emitted by the corona or the stellar wind. The materials in the surface layer of early-type stars are not contaminated by nuclear reactions in the stellar inside. Therefore, abundance study of the early-type stars provides us an information of the abundances of the original gas. However, the X-ray observations indicate low-metallicity, which is about 0.3 times of cosmic abundances. This fact raises the problem on the *cosmic abundances*.

In this work, we obtained chemical abundances of six early-type stars, λOri, ζPup, τSco, ζOri, δOri and ζOph, from the *ASCA* observations. The abundances derived from simple isothermal optically-thin model fits for those stars are significantly smaller than cosmic abundances for certain elements. We show that if emitting region is substantially optically thick the abundances we derived are underestimated.

2. Analysis and Results

We fitted each spectrum of the six early type stars with an ionizition-equilibrium thin thermal plasma model. Absorption by neutral gas was taken into account. Two stars required an additional power-law component (Torii 1997, private communication). Observed spectra were well fitted by this single(or two)-component model.

We plotted the relative abundances to the cosmic value in figure 1. In comparison with the cosmic abundances, the result says all stars' abundances are small. And these six stars show similar tendencies each other, *i.e.* low O and Fe.

3. Discussion

To interpret this low matallicity, we considerd the effect of resonance scatterings in the gas. Since cross section of resonance scattering is large, photons which form resonance lines have to pass through long paths in order to escape the X-ray emitting plasma. In other words, for line photons optical depth becomes thicker than that of continuum photons. So line intensities will be suppressed, *i.e.* abundances will be apparently small.

We made rough estimation of cross sections for resonance scattering and photoelectric absorption for each element. Then we calculated effective optical thicknesses of the hot plasma as a function of the column density of the X-ray emitting plasma. We derived the expected abundances when we analyse the X-ray emission from the cosmic abundance plasma using an optically thin plasma model. The results were shown in figure 1 by solid lines.

We can see that the caluculations and observational data shows similar tendency, such that the abundance of Fe is low. The value of column density, $10^{22.5}$ H cm^{-2}, seems to be most plausible, and this column density is a reasonable value for the coronae of early type stars.

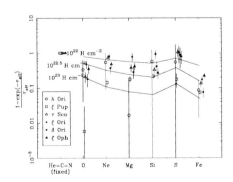

Figure 1: Derived abundances (marks) and expected apparent abundances for three values of column densities of X-ray emitting plasma (solid lines).

ASCA OBSERVATIONS OF 44I BOOTIS AND VW CEPHEI

CHUL-SUNG CHOI
Korea Astronomy Observatory, 36-1 Hwaam, Yusong, Taejon 305-348, Korea; cschoi@hanul.issa.re.kr.

AND

TADAYASU DOTANI
Institute of Space and Astronautical Science, 3-1-1 Yoshinodai, Sagamihara, Kanagawa 229, Japan; dotani@astro.isas.ac.jp.

Abstract.

We analyze X-ray archive data of the W UMa-type binaries 44i Boo and VW Cep, taken with ASCA on 1994 May 10 and 1993 November 5, respectively. By analyzing the light curve of VW Cep, we find a long-duration flare of ≈ 7.5 hrs with the peak luminosity of 1.2×10^{30} ergs s^{-1} (0.4–3.0 keV) for the assumed distance of 23.2 pc. We also find appreciable flux variations from the light curve of 44i Boo, and the variations are erratic and are not orbital phase dependent. From the spectral analysis of both data, we see that the spectra could be reproduced by the variable abundance plasma model with a combination of two different temperatures, kT = 0.64 – 0.65 keV and kT = 1.85 – 1.91 keV.

1. Results

From the light curve analysis of VW Cep observed with the SIS0 detector, we find a typical pattern of X-ray flare: a sudden increase of flux from 0.35 to about 1 counts s^{-1} during 1 – 1.5 hrs, followed by a decrease (almost exponentially) down to a quiescent level with a time scale of ≈ 6 hrs. The maximum flux level is ~ 2.5 times larger than the quiescent level, and the flux ratio is consistent with the previously detected X-ray flares (Vihlu et al. 1988; McGale et al. 1996). On the other hand, from the 44i Boo data, we find no significant flare-like event. Instead, the light curve shows erratic variations of X-ray flux with an amplitude of about a few tens percent, from ~ 1.0 to ~ 1.4 counts s^{-1}. We calculate χ^2-value to be χ^2/ν (degree of freedom) = 139.4/29 to the constant flux of 1.1 counts s^{-1}. This value implies that the flux fluctuation is substantial. We also investigate folded light curve of 44i Boo. However, we find no evidence that the flux variations are orbital-phase dependent.

To fit the energy spectrum, we use two-temperature "mekal" model with the elemental abundances which were allowed to be varied independently from the solar photospheric values (e.g., O, Ne, Mg, Si, Fe). From the spectral fitting, we see that both spectra are well explained by the combination of the two different temperatures, kT = 0.65 keV + 1.85 keV for 44i Boo and kT = 0.64 keV + 1.91 keV for VW Cep. When the results for VW Cep are compared with those of 44i Boo, we find that the elemental abundances for O, Mg, and Fe are almost the same as those of 44i Boo, while the abundances of Ne and Si are larger by a factor of 1.5 - 2. Details of VW Cep results are found in Choi and Dotani (1998).

2. Discussion

When the present flare is compared with the stellar X-ray flares observed by EXOSAT (Pallavicini et al. 1990), it is very similar to those of Algol and YY Gem in its general profile and the long decay time. From these analogies, the flare is considered to be "two-ribbon flare" similar to the one observed from the Sun. We did not observed any flare-like event from the 44i Boo system, but observed an X-ray flux variation which shows no orbital phase-dependence. This fact may indicate that coronal plasma are distributed inhomogeneously within the 44i Boo system.

We derived the coronal abundances of the elements O, Ne, Mg, Si, and Fe. The relatively high Ne and Si abundances of the VW Cep system are thought to be related with the flare, because an enhancement of the coronal abundances is inevitable during the flare activity. From the spectral analysis, we also found that the VW Cep spectrum is well reproduced by the variable abundance two-temperature model. The hotter component is believed to be associated with the flare. The 44i Boo observation shows no flare features, but its spectrum also required two-temperature model to be fitted acceptably. However, unlike the VW Cep case, the X-ray emission of 44i Boo is dominated by the cooler component.

References

Choi, C. S., & Dotani, T. 1998, *ApJ*, **492**, in press
McGale, P. A., Pye, J. P., & Hodgkin, S. T. 1996, *MNRAS*, **280**, 627
Pallavicini, R., Tagliaferri, G., & Stella, L. 1990, *A&A*, **228**, 403
Vilhu, O., Caillault, J.-P., & Heise, J. 1988, *ApJ*, **330**, 922

ASCA OBSERVATION OF NGC1333 STAR FORMING REGION

M. ITOH AND H. FUKUNAGA
Kobe University, Tsurukabuto, Nada, Kobe 657 Japan

K. KOYAMA AND Y. TSUBOI
Department of Physics, Graduate School of Science
Kyoto University, Sakyo-ku, Kyoto 606-01 Japan

S. YAMAUCHI
Faculty of Humanities and Social Sciences, Iwate University
Ueda, Morioka, Iwate 020 Japan

N. KOBAYASHI AND M. HAYASHI
National Astronomical Observatory
Osawa, Mitaka, Tokyo, 181 Japan

AND

S. UENO
Department of Physics and Astronomy, University of Leicester
Leicester, UK

The region south of the reflection nebula NGC1333 in Perseus is an active star forming region including numerous Herbig-Haro objects and at least 5 protostar candidates with molecular outflows and far-infrared emission. It has been actively studied in various wave bands (e.g. Aspin et al 1994 and references therein). We observed this region with *ASCA* with the primary objective to detect X-rays from the protostars embedded deep in the molecular cloud.

The X-ray images obtained by *ASCA* are shown in Fig.1. While the soft X-ray image is dominated by 4 or 5 point sources, about 15 X-ray sources were detected with the statistical significance higher than 5σ in the hard band. The positions of the protostar candidates are plotted in the figure. There are two cases of possible coincidence of the X-ray source position with the protostar candidate within the positional error circle: one is for SVS13(IRAS 3), and the other is for IRAS 2. However, these sources were detected by ROSAT/HRI observations with superior positional accuracy, and Preibisch (1997) identified them with other nearby sources.

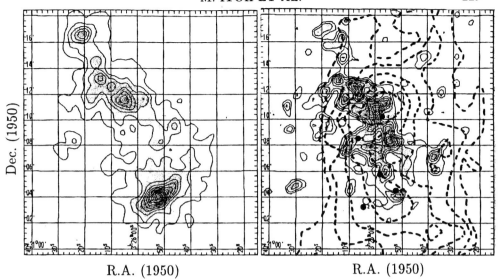

Figure 1. X-ray images of the NGC1333 south in soft (0.5-2 keV: left) and hard (4-10 keV: right) energy bands. In the hard band image, positions of the protostar candidates are indicated with filled circles, and the contours of the $C^{18}O$ column density (broken lines) are overlaid.

Overall distribution of the hard X-ray emission follows the high column density region traced by the $C^{18}O$ emission (Warin et al. 1996) as shown in Fig.1. Since it is difficult to extract spectrum of the individual sources with the point spread function of the *ASCA* telescope, we examined the X-ray spectrum of a region centered at $(3^h26^m10^s, 31°08'00'')$ with the radius of 2 arcmin which includes some of the hard sources and is free from the contamination from the sources visible in the soft band. The spectrum can be fit with thermal bremsstrahlung of $kT \sim 4 - 13 keV$ (90% confidence error) and the absorption of $N_H \sim 10^{22} cm^{-2}$. The X-ray luminosities of the detected hard sources are estimated to be $\sim 10^{30} erg \cdot sec^{-1}$ assuming the distance of 350 pc. These results imply that the *ASCA* sources are young stellar objects embedded in the molecular cloud. Significant time variation was observed for some of the detected sources, but no flare-like event was detected.

In conclusion, we did not detected X-rays from the known protostar candidates, but we detected X-rays from new young stellar objects. Evolutionary stages of these objects are to be studied by future observations.

References

Aspin, C., Sandell, G., Russel, A.P.G., 1994, A&AS 106,165
Cernis, K.,1990, ApSS, 166,315
Preibisch, T. 1997, A&A, 324, 690
Warin, S., et al, 1996, A&A, 306, 935

X-RAY OBSERVATIONS OF ORION OBIB ASSOCIATION

M. NAKANO
Oita University
Oita 870-11, Japan

1. Introduction

The signs of the active star formation in the Orion region are mainly found in the direction of the two giant molecular clouds - Ori A and Ori B - . Recent objective prism survey in the Orion region shows large number of Hα emission-line stars distributed outside of the giant molecular clouds (Nakano et al., 1995). Many weak-lined T Tauri star candidates are also discovered by the discrimination analysis of the X-ray sources found in the ROSAT all sky survey (RASS) (Sterzik et al., 1995). Although such huge number of pre-main sequence stars outside of the molecular cloud was not expected, their nature is still in controversial (Neuhäuser, 1997). To know the X-ray properties of these sources in the Orion region, we have carried out the ASCA observations.

2. Observations

We have obtained the broad-band X-ray images of the Orion OB1b association by ASCA. The obscuration effects are less important for detector with high sensitivity at high energies. We observed three areas in the OBIb association. A-region is between ε Ori and δ Ori, where relatively many emission-line stars were found. B-region is the east of ε Ori where no known young stars or emission-line stars. C-region is the interface region between Orion B giant molecular cloud (L1630) and HII region IC 434 excited by σ Ori, and where the horsehead nebula (B33) lies.

In total, more than forty faint X-ray sources were detected in three fields, and about half of them have likely optical counterparts in the Hubble Guide Star Catalog.

3. Results

In the C-region, several X-ray sources were identified with known possible low mass pre-main sequence stars. Some X-ray sources were also found in the chain of small reflection nebulosity in the A-region, which appears to be illuminated by OB association members.

Including the source with marginal detection, eight pre-main sequence stars were detected with ASCA observations - V469 Ori (marginal), V607 Ori, V615 Ori, Haro 5-36, NGC2023/S105, and #101, #105, #110 in our emission-line star survey (Wiramihardja et al., 1989). As a detailed analysis of the spectral properties of all sources is not feasible, we used the count ratio to estimate the X-ray temperature. We simulated a hot, optically thin plasma with interstellar absorption and calculated model count ratios for a grid of hydrogen absorption column densities in the range $N(H)=3\times10^{20} - 1\times10^{22} cm^{-2}$ and X-ray temperature kT = 1-10 keV. All of pre-main sequence stars with X-ray emission are rather X-ray luminous in our source list ($> 2\times10^{30}$ ergs/s) and count ratio for most sources are around 0.9 - 1.0, which mean the X-ray temperature kT = 1-2 keV.

New X-ray source in the Horsehead nebula is identified as B33-10 (Reipurth and Bouchet, 1984). Its NIR colors on the two-color diagram indicates that it has no NIR excess emission. The spectrum of the bright classical T-Tauri star V615 Ori is fitted by a Raymond-Smith model with a 0.3 solar abundance, kT=2.8 keV, and $N(H)=0.29\times10^{22} cm^{-2}$.

There are five high tempearture sources in the B-region. One of them is identified with a small nebulous object on the Palomar Sky Survey Print. Its position is RA=$5^h38^m36^s$, Dec=$-1°38'33''$ (2000). We tried to model the spectra with simple combination of absorbed thin hot plasma. Exteremely high X-ray temperature is needed to fit one temperature hot plasma model (kT > 10 keV) for the spectrum. We suggest that it might be an extragalactic source.

References

Nakano, M., Wiramihardja, S.D., and Kogure, T. (1995) Survey observations of emission-line stars in the Orion region V. The Outer regions, *Publ. Astron. Soc. Japan*, **47**, 418.

Neuhäuser, R. (1997) Low-mass pre-main sequence stars and their X-ray emission, *Science*, **276**, 1363.

Reipurth, B. and Bouchet, P. (1984) Star Formation in Bok globules and low-mass clouds II. A collimated flow in the Horsehead, *Astron. and Astrophys.*, **137**, L1.

Sterzik, M.F., Alcalá, J.M., Neuhäuser, R., and Schmitt, J.H.M.M (1995) The spatial distribution of X-ray selected T-Tauri stars: I. Orion, *Astron. and Astrophys.*, **297**, 889.

Wiramihardja, S.D.*et al.* (1989) Survey observations of emission-line stars in the Orion region I. The Kiso Area A-0904, *Publ. Astron. Soc. Japan*, **41**, 155.

FLARES AND MHD JETS IN PROTOSTAR

MITSURU HAYASHI
Japan Atomic Energy Research Institute(JAERI)
801-1 Mukouyama,Naka-machi,Naka-gun,
Ibaraki,311-01,Japan

KAZUNARI SHIBATA
National Astronomical Observatory
Mitaka, Tokyo 181, Japan

AND

RYOJI MATSUMOTO
Department of Physics, Faculty of Science, Chiba University
Inage-Ku, Chiba 263, Japan

1. Introduction

By using ASCA, Koyama et al. (1994,1996) carried out a systematic survey of hard X-ray sources in molecular clouds and revealed that protostars are strong hard X-ray emitting sources. Some of them show flare-like activities. Protostellar flares differ from solar flares in their total energy(10^{35-36} erg), size(several times the radius of protostar), and higher temperature(8keV). Protostellar flares are also observed in lower energy band by ROSAT in YLW15 (Grosso et al. 1997). By extending the model of solar flares associated with footpoint shearing motion, we proposed a model of protostellar flares in which the magnetic field connecting the protostar and the disk disrupt by twist injection from the rotating disk(Hayashi et al. 1996).

2. Numerical Models & Results

We solve the resistive MHD equations in clindrical coordinate by applying a modified Lax-Wendroff scheme with artificial viscosity. The effects of radiative cooling and rotation of the central star are neglected. Magnetic field lines connecting the central star and the disk are twisted by the rotation of the disk. As the magnetic twist accumulates, the magnetic loops expand quasi-statically in the early stage but later, they expand dynamically. Magnetic reconnection takes place in the current sheet formed inside the expanding loops. Outgoing magnetic island and post flare loops are formed after the reconnection (See Fig 1 in Hayashi et al. 1996).

We carried out longer time-scale simulations to investigate the evolution of the post flare loops created by reconnection event. We found that magnetic reconnection and plasmoid ejection occur intermittently.

3. Discussion

We found that the dipole magnetic fields of the protostar can become partially open by imposed twists within one rotation of the disk. We also showed that hot plasmoids are ejected in bipolar directions with velocity 2− 5 times the Keplarian rotation speed(v_{K0}) around the inner edge of the disk. Plasma heating occurs both by Joule heating in the current sheet and by shock waves created by magnetic reconnection. The plasma temperature corresponding to a reconnection flow speed(∼ Alfvén speed) of 2-5 v_{K0}, where v_{K0} is the Keplarian velocity at unit radius r=r_0=1, can be 10^7-10^8 K, consistent with the observed spectrum which extends up to 10 keV. Since magnetic reconnection accelerates electrons, we expect synchrotron radiation from ejected plasmoids. Recently, Hughes (1997) reported that in the young star forming region Cepheus A, nonthermal sources are separating with a relative transverse velocity(∼ 370 km/s). In T-Tauri region, Ray et al. (1997) found two lobes which emit radio emission exhibiting strong circular polarization of opposite helicity. These observations are consistent with our reconnection model of hot plasmoid ejections.

The size of the flaring loop is several times the stellar radius. In addition, recently ASCA detected X-ray flaring variability which repeats three times recursively (Tsuboi et al. 1997). This can be the first observatuon which supports intermittent flaring variability in star forming region. Such recurrent activities can be explained by intermittent magnetic reconnection in the disk-star connecting magnetic loop.

References

Grosso, N., Montmerle, T., Feigelson, E.D., André, P., Cassanova, S.,
   & Gregorio-Hetem, J. 1997, *Nature*, **385**, 56
Hayashi, M.R., Shibata, K. & Matsumoto, R. 1996, *ApJ*, **468**, L37
Hughes, V. A., 1997, *ApJ*, **481**, 857
Koyama, K., Hamaguchi, K., Ueno, S., Kobayashi, N.,
   & Feigelson, E. D. 1996, *PASJ*, **48**, L87
Koyama, K. et al. 1994, *PASJ*, **46**, L125
Ray, T. P., Muxlow, T. W. B., Axon, D. J., Brown, A., Corcoran, D., Dyson, J.,
   & Mundt, R. 1997, *Nature*, **385**, 415
Tsuboi, Y. et al. 1997 IAU Symposium No.188, "The Hot Universe"

X-RAY VALIABILITIES FROM PROTOSTARS IN THE RCRA MOLECULAR CLOUD

KENJI HAMAGUCHI, HIROSHI MURAKAMI AND KATSUJI KOYAMA
Department of Physics, Faculty of Science, Kyoto University
Sakyo-ku, Kyoto 606-01, Japan

AND

SHIRO UENO
Department of Physics & Astronomy, University of leicester
Leicester, LE1 7RH, U.K.

1. Introduction

R CrA molecular cloud, at a distance of 130pc(Marraco, Rydgren 1981), is one of the active low-to-middle mass star forming regions. The core of this cloud, named Coronet, contains five protostar candidates. In 1994, we observed the Coronet Cluster using the X-ray satellite ASCA, and discovered hard X-rays from protostar candidates(Koyama et al. 1996).

T-Tauri stars are known to exhibit the strong X-ray variablities with occasional rapid flares, whose emission mechanisms are similar to the solar-type magnetic reconnection process(Montmerle et al. 1983), and the emitting regions are thought to be as large as the diameter of T-Tauri stars. On the other hand, the origin of protostar variabilities is less clear.

In 1996, we observed this region again, and found that the fluxes of each protostar candidates are different from those in 1994. This is the first evidence for the long time variablities of protostar candidates.

2. Observations and Results

The quesient state images for the 4-10keV band in 1994 and 1996 are shown in figure 1. The gray scale peaks will correspond to the position of IR cataloged protostar sources. One of the prominent feature is that, at the region centered on the protostar candidate, R1(R.A=$18^h58^m32.7^s$, DEC

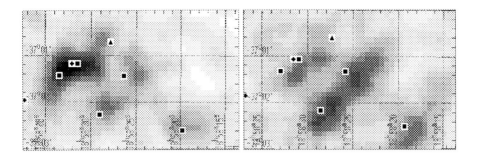

Figure 1. SIS(ASCA X-ray CCD Camera) images from the the Coronet cluster for the 4-10keV band in 1994(left side) and 1996(right side). Gray scales are normalized for the surface flux density. Squares are protostar candidates, crosses are Herbig Ae/Be stars and a triangle is a classical T-Tauri star

$= -37°01'39''$(1950), upper left on fig.1), the flux in 1996 was at least 2.5 times lower than that in 1994. On the other hand, the fluxes of other sources are increased by ~1.5.

In the observation of 1994, we detected a powerful X-ray flare from one of the sources, probably the embedded far-infrared source R1. The X-ray flux increased at least by a factor of 3-4 from a quiesent level. The flare spectrum is unusual, showing a broadend or double emission lines between 6.2 and 6.8 keV (Koyama et al. 1996).

3. Discussion

What is the emission mechanism of these time valiabilities?

The flux reduction of R1 from 1994 to 1996 may be the feature like FU-Ori outbursts, which occur by the change of accretion rates.

The flare in 1994 is probably due to a solar-type magnetic activity like those seen in T-Tauri stars, but the flare spectrum is mysterious because there is no candidate line near 6.2keV. One possibility is that these lines are doppler-shifted iron lines in a bipolar jet, whose speed is ~0.1c. However, no star forming theory predicts such a relativistic acceralation mechanism.

4. References

Marraco H G., Rydgren A.E. 1981 AJ 86, 62
Koyama K., Hamaguchi K., Ueno S., Kobayashi N., Eric D. Feigelson. 1996 PASJ 48, L87
Montmerle T., Koch-Miramond K., Falgarone E., Grindlay J.E. 1983, ApJ 269, 182

ASCA OBSERVATIONS OF CLASS I PROTOSTARS IN THE RHO OPH DARK CLOUD

Y. TSUBOI AND K. KOYAMA
Department of Physics, Graduate School of Science, Kyoto University, Sakyo-ku, Kyoto, 606-01, Japan

Y. KAMATA
Department of Astrophysics, School of Science, Nagoya University, Furo-cho, Chikusa-ku, Nagoya 464-01, Japan

AND

S. YAMAUCHI
Faculty of Humanities and Social Sciences, Iwate University, 3-18-34, Ueda, Morioka, Iwate, 020, Japan

1. Introduction

On 1993 August 20, we observed the Rho-Oph dark cloud and detected hard X-rays from Class I sources (Koyama et al.(1994), Kamata et al.(1997)). One of the sources (EL29) showed a flare-like variability, while another (WL6) exhibited sinusoidal variation with no large spectral change. The later would be due to a spin of the protostar. The sinusoidal period of about 1 day is shorter than spin periods of TTSs of ∼3–7 day.

From these X-ray emitting YSOs, we found bipolar flows with radio observations (Sekimoto et al.(1997)). This places the sources to be protostars at the dynamical mass accretion phase. The common feature of these out-flows is that the blue and red lobes are largely overlapped, suggesting nearly pole-on geometry. On the other hand, no significant X-ray has been reported from out-flow sources, VLA 1623 (Kamata et al.(1997)), L1551 IRS5 and L1551 NE, HL Tau (Carkner et al.(1996)), all these would be edge-on systems (André et al.(1990), Ohashi et al.(1996)). Consequently, we proposed a unified picture of protostars; every protostar emits X-rays, but the X-rays can only be detected from pole-on viewing angle, where X-rays are less absorbed by dense circumstellar disks.

To study the time variability of the X-ray emitting Class I sources, we re-observed this region deeply.

2. Observations & Results

On 1997 Mar 2–3, we observed the central region of the cloud with 100 ksec exposure, and found that WL6 and EL29 became very faint. Instead, new hard X-ray (> 2 keV) objects appeared at the positions of other Class I stars. The brightest was YLW15, on which a large flare had been detected with the ROSAT deep pointing observation (Grosso et al.(1997)). The outflow map of YLW15 (Bontemps et al.(1996)) has also a signature of pole-on configuration, confirming our unified picture.

In the hard band (> 2 keV), we detected three flares on the YLW15 with about 20 hours interval. The time interval is comparable to the spin period of the protostar WL6, which we suggested from the sinusoidal X-ray light curve. It is also comparable to the period of the inner-most Keplarian orbit (r = several $\times 10^{-2}$ AU). Accordingly, the quasi-periodicity of the flares would be related to the spin of the central star or/and rotation of the circumstellar disk.

References

André, P., Martin-Pintado, J., Montmerle, T.: 1990, *A&A*, 236, 180.
Bontemps, S., André, P., Terebey, S., Cabrit, S.: 1996, *A&A* 311 858.
Carkner, L., Feigelson, E.D., Koyama, K., et al.: 1996, *ApJ*, 464, 286.
Grosso, N., Montmerle, T., Feigelson, E. D., et al.: 1997, *Nature*, 387, 56.
Kamata, Y., Koyama, K., Tsuboi, Y., Yamauchi, S.: 1997, *PASJ*, 49, 461.
Koyama, K., Maeda, Y., Ozaki, M., et al.: 1994, *PASJ*, 46, L125.
Ohashi, N., Hayashi, M., Ho, P. T. P., et al.: 1996, *ApJ*, 466, 957.
Sekimoto, Y., Tatematsu, K., Umemoto, T., et al.: 1997, *A&AL*, in press.

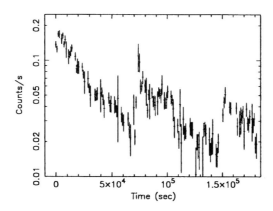

Figure 1. Lightcurve of YLW15 in 2–10 keV band (GIS 2 + 3). Each time bin width is 1024 sec.

Session 1: Plasma and Fresh Nucleosynthesis Phenomena

1-2. Supernovae, Supernova Remnants and Galactic Hot Plasma

THE HALF-LIFE OF TITANIUM 44 AND SN 1987A

YUKO S. MOCHIZUKI
*The Institute of Physical and Chemical Research (RIKEN)
Hirosawa 2-1, Wako, Saitama 351-01, JAPAN*

AND

SHIOMI KUMAGAI
*Department of Physics, College of Science and Technology, Nihon University
Kanda-Surugadai 1-8, Chiyoda-ku, Tokyo 101, JAPAN*

The radioactive isotopes, such as ^{44}Ti and ^{56}Ni, are synthesized as a result of rapid nucleosynthesis in supernova explosions. The gamma-ray photons coming out from the decay sequence of ^{44}Ti is now a strong candidate to explain the the late light curve of SN 1987A. It is noted here that the energy release from the ^{44}Ti decay depends strongly on its half-life. However, the published values for ^{44}Ti half-life display a large spread, ranging from $\sim$ 35 to $\sim$ 68 years. In this paper (Kumagai et al. 1997, see also Kumagai et al. 1993), we discuss the value of the half-life by comparing the theoretical light curves and the observations in SN 1987A. The unestablished half-life value is related to the ratio of the abundance of ^{44}Ti to that of ^{56}Ni.

In Figure 1, we show the ranges of $<^{44}$Ti/^{56}Ni$>$ which satisfy the observed luminosity integrated from the infrared to the ultraviolet wavelength as the function of the half-life of ^{44}Ti. In this figure, the region sandwiched between the two dotted lines accounts for the late luminosity of SN 1987A derived by the combined observations of CTIO and HST at 3600 days after the explosion (Suntzeff 1997). On the other hand, the wider region between the two dashed lines in Figure 1 satisfies the luminosity observed with CTIO at 1699 days after the explosion (Suntzeff 1997). Here, we define $<^{44}$Ti/^{56}Ni$>$ as the ratio of ^{44}Ti/^{56}Ni (in amounts) in the supernova remnant to ^{44}Ca/^{56}Fe in the solar neighborhood, i.e., $<^{44}$Ti/^{56}Ni$> \equiv [X(^{44}$Ti$)/X(^{56}$Ni$)]/[X(^{44}$Ca$)/X(^{56}$Fe$)]_\odot$. Note that ^{44}Ca and ^{56}Fe are the daughter nuclei of ^{44}Ti and ^{56}Ni, respectively.

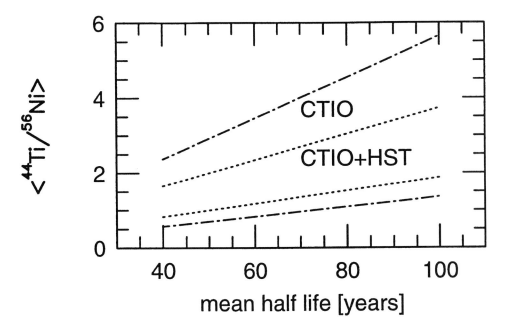

Figure 1. The ranges of $<^{44}\text{Ti}/^{56}\text{Ni}>$ which satisfy the late observed luminosity in SN 1987A are shown against the half-life of ^{44}Ti.

One sees in Figure 1 that the overall range of the published values of the half-life mentioned above is appropriate to explain the late observed luminosity (abbreviated by CTIO+HST), if $<^{44}\text{Ti}/^{56}\text{Ni}>$ is roughly in between 1 and 2. If we assume $<^{44}\text{Ti}/^{56}\text{Ni}> = 1$, Figure 1 shows that the real half-life value of ^{44}Ti is larger than ~ 45 years but is less than ~ 55 years, taking the $\sim 10\%$ uncertainty of the solar abundance measurements into account. Contrarily, it is quite interesting to note that we can deduce the value of $^{44}\text{Ti}/^{56}\text{Ni}$ in SN1987A from this figure, if the real half-life of ^{44}Ti is established experimentally.

References

Kumagai, S., Mochizuki, Y.S., Nomoto, K., and Tanihata, I. (1997) in preparation

Kumagai, S., Nomoto, K., Shigeyama, T., Hashimoto, M., and Itoh, M. (1993) "Detection of ^{57}Co γ rays from SN 1987A and prospect of X-ray observations of the pulsar with ASCA", *Astron. & Astrophys.*, **273**, pp. 153-159

Suntzeff, N.B. (1997) in SN1987A: Ten Years After, The Fifth CTIO/ESO/LCO Workshop, eds. M.M. Phillips and N.B. Suntzeff, in press

BLACK HOLE DISK ACCRETION IN SUPERNOVAE

H. NOMURA AND S. MINESHIGE
Department of Astronomy, Faculty of Science, Kyoto University, Sakyo-ku, Kyoto 606-01, Japan

M. HIROSE
Theoretical Physics, Astronomical Observatory of Japan, Japan

AND

K. NOMOTO AND T. SUZUKI
Department of Astronomy and Research Center for the Early Universe, School of Science, University of Tokyo, Japan

1. Introduction

Massive stars in a certain mass range ($20-40M_\odot$) may form low mass black holes after supernova explosions. In such massive stars, fall back of $\sim 0.1M_\odot$ materials onto a black hole is expected due to a deep gravitational potential or a reverse shock propagating back from the outer composition interface. We study hydrodynamical disk accretion onto a new-born low mass black hole in a supernova using the SPH (Smoothed Particle Hydrodynamics) method.

2. Results and Discussions

As for particular case, we apply the quantities of SN1987A; that is, the mass of the central object, $M_0 = 1.4 M_\odot$, the mass of the fallback matter, $M_{\rm fb} = 0.1 M_\odot$, the place where a reverse shock appears, $r_0 = 5 \times 10^{10}$cm, the sound velocity, $c_s = 3.45 \times 10^7$cm s^{-1}, the angular frequency, $\Omega = 10^{-3}$rad s^{-1} (for comparison, we also simulate cases with $\Omega = 0.0$), and the specific heat ratio $\gamma = 4/3$.

When the ambient gas has no angular momentum, it accretes toward the center with a free-fall velocity, whereas if the gas has a certain angular momentum, it first falls onto the equatorial plane, forming a rotating gas disk, and then accretes inward via viscosity (Figure 1).

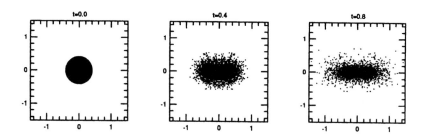

Figure 1. The time evolution of SPH particle distributions with the initial angular momentum. The length scale is normalized by $[L] = 1.22 \times 10^{11}$ cm and the times are in the unit of $[T] = 3.14 \times 10^3$ s.

Numerically derived mass accretion rate is roughly,

$$\dot{m} \sim 1.0 \times 10^5 (\alpha t/\text{yr})^{-1.35}$$

where $\dot{m} \equiv \dot{M}/\dot{M}_{\text{crit}}$, $\dot{M}_{\text{crit}} \equiv L_E/c^2$, L_E is the Eddington luminosity, and α is the viscous parameter.

The results thus indicate a hypercritical disk accretion. When $\dot{M}$ exceeds the critical rate, a disk becomes advection dominated and optically thick (so-called the slim disk), as long as shear-viscous tensor does depend on the radiation pressure.

This suggests the view on following two topics.

1. The luminosity of SN 1987A

The observed bolometric luminosity of SN 1987A is $\sim 10^{36}$ erg s^{-1}, which can be explained by the energy deposition from the ^{44}Ti decay. If standard disk accretion occurs, the disk luminosity should be, at least, of the order of the Eddington luminosity ($\sim 10^{38}$ erg s^{-1}), contrary to the observations. But in the advection-dominated, hypercritial accretion disk, that discrepancy is avoided due to advection of the radiation energy and photon trapping at the hottest part of the disk. (The photon trapping occurs inside the radius $\sim 10^{11}$ cm for $\dot{M}/\dot{M}_{\text{crit}} \sim 10^6$.)

2. Possible nucleosynthesis

The slim disk is hot and dense; for $\dot{M}/\dot{M}_{\text{crit}} \sim 10^6$, $T \sim 10^9 (\alpha/0.01)^{-1/4}$ K and $\rho \sim 10^3 (\alpha/0.01)^{-1}$ g cm^{-3}. If some hydrogen and helium have been mixed down to deeper layers and accreted, interesting nucleosynthesis processes via rapid proton and alpha captures on heavy elements would take place. The elements produced in this way might be advected inward and swallowed by the central black hole, but some of them could be ejected in a disk wind or a jet.

References

Mineshige, S., Nomura, H., Hirose, M., Nomoto, K., & Suzuki, T.(1997) *ApJ*, **489**, 227

THE X-RAY SPECTRUM OF SUPERNOVA SN1993J

its Long-Term Evolutions, and Shock Heating.

SHIN'ICHIRO UNO, KAZUHISA MITSUDA, TADAYUKI TAKAHASHI,
HAJIME INOUE AND FUMIYOSHI MAKINO
Institute of Space and Astronautical Science
3-1-1 Yoshino-dai Sagamihara-city Kanagawa 229 Japan

KAZUO MAKISHIMA
Department of Physics, the University of Tokyo
Bunkyo-ku Tokyo 113 Japan

YOSHITAKA ISHISAKI
Dept. of Physics, Tokyo Metropolitan University
Minami-Osawa 1-1, Hachioji, Tokyo 192-03, Japan

YOSHIKI KOHMURA
The Institute of Chemical and Physical Research
3-1, Hirosawa, Wako, Saitama 351-01 Japan

MASAYUKI ITOH
Faculty of Human Development, Kobe University
3-11 Tsurukabuto Nada-ku Kobe 657

AND

WALTER .H.G. LEWIN
Massachusetts Institute of Technology
Center for Space Research, Room 37-627, Cambridge, MA 02139

SN1993J is very unique object which was discovered in the nearby Sb galaxy M81 (NGC 3031) on March 28[1]. The first detection of the radio emission was at 22.5 GHz by the VLA only 5 days after the optical outburst[2]. Subsequently X-ray emission was detected by ROSAT and ASCA at 6 days and 8 days after the explosion respectively. These emissions are expected when the SN shock front sweeps out the circumstellar matter (CSM). The early detection of radio and X-ray emission implies the existence of high-density CSM in the vicinity of the supernova(e.g. [3][4]).

ASCA[5] observed SN1993J eleven times during 1993 April to 1995 October. Although the two bright X-ray sources, M81 X−5 and X−6 are only about 3 arc minutes and 1 arc minute from the supernova, we successfully separated the three sources for the SIS (Solidstate Imaging Spectometer) data utilizing a kind of 2-dimensional image fittings and obtained the spectra of the three sources.

The spectrum showed drastic softening with a power-law photon index of 0.4 to 4, while the X-ray intensity decreased from 0.03 to 0.008 counts/sec/SIS. The early-phase of spectra require two thermal emission components of different absorption columns, if they are fitted with thermal models. The temperatures of two emission components cannot be well constrained from ASCA contiuum spectra. However, the detection of iron K emission line with ASCA [4] and the hard X-ray spectra observed by OSSE imposes strong constraints on the temperatures.

The properties of the two emission components are consistent with those of emissions from the front and reverse shocks of the supernova explosion. The drastic softening of the X-ray spectra is explained by decrease of the absorption column density of the initially heavily- abosrbed reverse shock component; the dominant emission component in the soft X-ray band altered from the front shock to the revserse shock.

We investigated the emission line structure in detail. The central energy of the line is 6.70 ± 0.15 keV, and width of 0.4 ± 0.2 for the spectrum of 10 days after the explosion. The interpretation of the center energy and the line width involves, the geometry and the motion of the supernova shock.

The continuum X-ray spectra could be described with the two-shock scheme. In such models, the emission line is estimated to come from "reverse shock" and "cold shell". Considering bloadning and blue-shift by the expansion veolocity of the supernova, we fitted the emission line profile with three gaussian models, i.e. Hydrogen-like, Helium-like and neutral Iron. From the ratio of the line equivalent widths of 6.9 to 6.7 keV reflects the temperature of the plasma. The temperature on April 5 to be between 3 keV and 20 keV and the temperature of April 7 and April 16 to be lower than 12 keV. This result is consistent with the two shock scheme.

1. Garcia, F. et al. 1993, *IAUC*, 5731
2. Weiler, K. W. et al. 1993, *IAUC*, 5752
3. Zimmermann, H. U., et al. 1994, *nature*, **367**, 621–623
4. Kohmura, Y., et al., 1994, *Publ. Astron. Soc. Japan*, **46**, L157–L161
5. Tanaka, Y., et al., 1994, *Publ. Astron. Soc. Japan*, **46**, 37
6. Leising, M. D. et al., 1994, *Astrophys. J. Let.*, **431**, 95
7. Chevalier, R. A. 1982, *Astrophys. J.* , **259**, 302–310

NEUTRINO TRANSPORT IN TYPE II SUPERNOVAE : BOLTZMANN SOLVER VS MONTE CARLO METHOD

SHOICHI YAMADA AND HANS-THOMAS JANKA
Max-Planck-Institut für Astrophysik
Karl - Schwarzschild - Str. 1, D-85740 Garching, Germany

AND

HIDEYUKI SUZUKI
National Laboratory for High Energy Physics (KEK)
Oho, Tsukuba, Ibaraki 305, Japan

In this paper we discuss the results of extensive comparison of the Boltzmann solver recently developed by us and the Monte Carlo method coded by Janka. The aim of our project is to improve the treatment of the neutrino transfer and to study the neutrino reactions systematically, since the numerical treatment of neutrino transfer and neutrino reaction rates in the hot dense medium are two of the greatest uncertainties in the supernova simulations.

We assume spherical symmetry of the system. The fully general relativistic Boltzmann equation for the Misner-Sharp metric is finite differenced with respect to each coordinate, time t, baryon mass m, directional cosine of neutrino 3-momentum μ and neutrino energy ε_ν, and solve the resulting coupled equations. In the Monte Carlo simulations, on the other hand, $\sim 500,000$ particles are typically used. We used Wilson's realistic models for three different times after core bounce, and calculated the neutrino transport for this static backgrounds.

In the typical calculations, the Boltzmann solver used 105 spatial grids, 6 angular meshes in 180 degrees, 12 energy bins covering $0 \sim 110$ MeV. However, these numbers are varied systematically to see the effect of the resolution.

As shown in Figure 1, the number density and the number flux as well as the energy spectrum of neutrinos are quite well reproduced by the Boltzmann solver. Figure 2a, on the other hand, clearly shows the limitation of the Boltzmann solver. The energy integrated flux factor and Eddington fac-

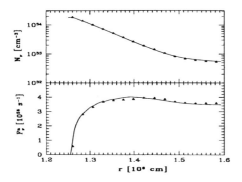 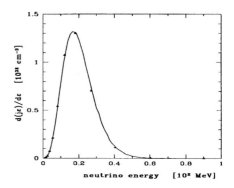

Figure 1. a. The number density (upper panel) and the number flux (lower panel) of the electron-type neutrino at $t = 3.32$ sec. Triangles are for Monte Carlo results. b. The electron-type anti-neutrino energy spectrum at $t = 3.32$ sec. Triangles are for Boltzmann results.

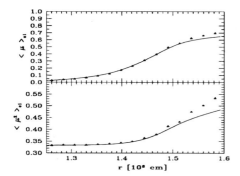 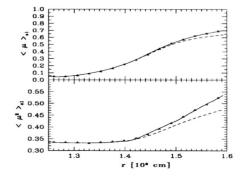

Figure 2. The flux factor and Eddington factor for the muon-type neutrino with (a) and without (b) a variable angular mesh method.

tor are considerably underestimated by the Boltzmann solver in the outer region where the neutrino angular distribution is forward peaked. This is entirely due to the poor angular resolution of the Boltzmann solver. This is a serious drawback of the Boltzmann solver in studying the neutrino heating mechanism behind shock where the very forward peaked angular distribution is expected. To improve the angular resolution without increasing the mesh number we implemented the variable angular mesh method in which the angular mesh is collected more in the forward direction depending on the time and position. As shown in Figure 2b both the flux factor and Eddington factor are improved remarkably.

RADIO EMISSION FROM EXTENDED SHELL-LIKE SNRS

A.I. ASVAROV
Institute of Physics, Azerbaijan Academy of Sciences,
Baku 370143, Azerbaijan Republic
e-mail: physic@lan.ab.az

1. Introduction

Observations of the soft X-Ray background and interstellar UV absorption lines have indicated that a large fraction of interstellar space is filled with a high temperature low density "coronal" gas. In such low density environments SNRs will expand up to 200 pc in radius without thin shell formation which occurs due to radiative cooling effects. Such SNRs can occupy a large fraction of volume of Galaxy and can be the main source of background emissions. In the present work we examine the evolution of the radio emission of shell-like SNR evolving in the hot ISM.

2. The model and results

The main assumptions made in present study are the same as in our previous papers [1,2] but now SNR is modeled by using an analytical approximation of [3] which follows the development of an adiabatic spherical blast wave in homogeneous ambient medium of finite pressure. At early times this approximation resembles the zero pressure Sedov similarity solution, but extends the range of investigation well into the regime in which the external pressure is significant.

At high Much numbers of the SNR's shock wave, M_s, emission highly concentrated directly behind the shock front but with decreasing of the shock strength the profile of distribution broadens. At $M_s \simeq 3$ the peak in profile of radial distribution separates from the shock front. SNRs at the end of their life (at $2 \leq M_s \leq 4$) have emissivity characteristics resembling the mean characteristics of Galactic background synchrotron emission, namely, $\alpha \simeq 0.7$ and $\varepsilon_{1GHz} \simeq 10^{-40} - 10^{-41}$ erg/cm^{2}s sr Hz. From the fact that the number of extended SNRs with small values of M_s is considerable more then the number of young SNRs we can expect that such extended remnants occupy most of the galactic volume and are able to form galactic background synchrotron emission. The comparison of the modeled $\Sigma - D$ tracks with the observational $\Sigma - D$ relation for the shell-type SNRs (Fig.1) leads us to the conclusion similar to one made in [4], namely, adiabatic shell-type SNRs evolve at nearly constant Σ followed by a steep decrease. In our model it begins when $M_s \leq 8 - 9$. The influence of the parameters of the ISM is more prominent then the parameters of SN explosion

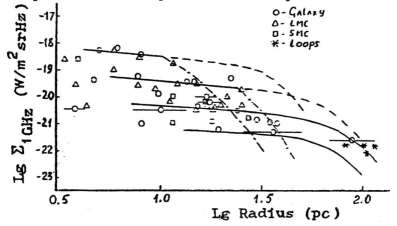

Fig.1. Σ-R (Radius) diagram at 1 GHz for shell-like SNRs. Data are taken from [5] and [6] for Galactic and MC SNRs, respectively. The positions of the giant loops are from [4]. Tracks are drawn (from bottom) for: (a) $n_{0e}=10^{-3}$ cm^{-3}, $P_0=5\cdot 10^3$ K cm^{-3}, $H_0=3$ μG; (b) $n_{0e}=5\cdot 10^{-3}$ cm^{-3}, $P_0=5\cdot 10^3$ K cm^{-3}, $H_0=5$ μG; (c) $n_{0e}=0.5$ cm^{-3}, $P_0=5\cdot 10^3$ K cm^{-3}, $H_0=5$ μG; (d) $n_{0e}=5$ cm^{-3}, $P_0=2\cdot 10^4$ K cm^{-3}, $H_0=10$ μG. The parts of tracks after beginning of radiative cooling are drawn by dashed and dot-dashed (no more acceleration) lines.

The main results are following: if the acceleration efficiency is such that one electron out $(2-3)10^3$ thermal electrons is subjected to acceleration, then typical for the ISM magnetic fields compressed by a shock wave to a factor of four are sufficient for an explanation of the observed radio fluxes. The mean value of the spectral index of shell-like SNRs, evolving without setting in the radiative cooling, at the end of their life remains bounded at the value of $0.70 - 0.74$. The extended SNRs can serve as the main source of galactic nonthermal background radio emission.

References: [1] Asvarov A.I., 1992, AZh, v.69, p.753; [2] Asvarov A.I., 1994, AZh, v.71, p.228; [3] Cox D.P. and Anderson P.R., 1982,ApJ,v.253, p.268; [4] Berkhuijsen E.M., 1986,AA,166,257; [5] Green D.A., 1996, "A Catalogue of Galactic SNRs (August ver.)", MRAO, UK [6] Mathewson D.S. et al. 1983, ApJ (Suppl),51,345.

DISCOVERY OF X-RAY EMISSION FROM THE RADIO SNR G352.7-0.1

K. KINUGASA, K. TORII, AND H. TSUNEMI
Graduate School of Science, Osaka University
1-1 Machikaneyama-cho, Toyonaka, Osaka 560, Japan

S. YAMAUCHI
Faculty of Humanities and Social Sciences, Iwate University

K. KOYAMA
Graduate School of Science, Kyoto University

AND

T. DOTANI
Institute of Space and Astronautical Science

1. Introduction

One major objective of our ASCA Galactic Plane Survey Project (AGPSP) is, utilizing the wide and high energy band (up to 10 keV) X-ray imaging capability and the high spectral resolving power of ASCA, to search possible X-ray SNRs in the Galactic inner disk. The observation of the field including G352.7-0.1 reported in this paper, was performed on 1996 March 14 during the first AO4 survey. We report on the X-ray SNR G352.7-0.1 found in AGPSP. G352.7-0.1 is one of the radio SNRs (Green 1996), and is classified as a shell-like SNR with the size of $8' \times 6'$.

2. Analysis and Results

At X-ray image, we clearly found an extended source with the size of $\sim 6'$. The center coordinate (J2000) of the source is $(\alpha, \delta) \sim (17^h27^m42^s, -35°07'20'')$. This position is consistent with that of the radio SNR G352.7-0.1. From the positional coincidence and the extended structure, we firmly conclude that the X-rays are attributable to the radio SNR. It is the first detection of X-rays from this SNR. We then overlaid SIS image on the VLA radio contours at 1465MHz (Dubner et al. 1993) in figure 1. The SIS image exhibits a shell-like structure which roughly coincides with the radio shell.

The spectra show two remarkable spectral features. One is a low energy turn-off which implies a large interstellar absorption, and the other is the existence of prominent emission lines which correspond to the Kα lines from He-like Si, S, and Ar. We fitted a non-equilibrium ionization (NEI) model (Masai 1984). The final results we obtained are in table 1. The observed and intrinsic fluxes in the 1-10 keV band are respectively estimated to be $\sim 3.0 \times 10^{-12}$ erg cm^{-2} s^{-1} and $\sim 1.2 \times 10^{-11}$ erg cm^{-2} s^{-1}.

3. Discussion

The observed $N_{\rm H}$ to G352.7−0.1 is the same to that of the hard component of the Galactic ridge emission, hence we reasonably assume that the distance to G352.7−0.1 is nearly equal to that to the Galactic Center (8.5 kpc). Then applying Sedov solution, we can estimate the age and the SN explosion energy to be $t_{\rm age} \sim 2200$ yr and $E_{\rm SN} \sim 2 \times 10^{50}$ erg, respectively. Taking into account the Galactic abundance gradient, the S abundance of G352.7−0.1 is consistent with the S abundance near the GC region.

Table 1. The best-fit parameters by the NEI model.

Parameters	Values
kT(keV)	2.0 (1.3 − 3.6)
log τ(cm^{-3} s)	11.0 (10.7 − 11.5)
Si*	3.7 (2.7 − 5.1)
S*	3.4 (2.3 − 5.0)
EM/4πD^2 (cm^{-5})	3 (2 − 6) × 10^{11}
$N_{\rm H}$ (10^{22}cm^{-2})	2.9 (2.4 − 3.5)
χ^2 / d.o.f	106.0 / 82

* Relative to solar abundance.

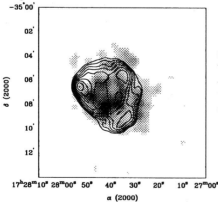

Figure 1. The SIS image (0.7–10keV) of G352.7–0.1 overlaid with the VLA radio contours at 1465MHz.

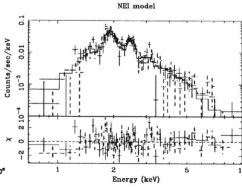

Figure 2. The energy spectra of G352.7–0.1 and the best-fit NEI model. Solid lines and dashed lines are those of SIS and GIS, respectively.

THE EFFECT OF DUST SPUTTERING IN SUPERNOVA REMNANT

TAKASHI KURINO, MASAYUKI FUJIMOTO AND ASAO HABE
Astrophysics Labolatory Division of Physics, Graduate School of Science, HOKKAIDO University
Kita 10 Nishi 8, Sapporo, Hokkaido 060, JAPAN

IRAS observations of supernova remnants (SNRs) reported that infrared luminosity from SNRs are very different, though some of them have same temperature. This is because infrared emittion depends on density of the dust grains.

We calculated infrared and X-ray emission from SNR just after supernova explosion, in case dust number density $n_i = 10^2, 10^5$, released energy $E = 10^{51} erg$. We used one dimensional Flux-Split method, uniform molecular cloud, and spherically symmetric physical parameters. To make our model more realistic, we include the effect of dust grain sputtering which is expected to decrease infrared emission.

This effect is expressed by decreasing the radii of the dust grains depending on surrounding hot gas density.

We found that it is in dense molecular cloud that infrared emittion become greater than X-ray emittion, in case dust sputtering model. The effect of dust grain cooling quicken the shift to cooling phase, and make more compact SNR.

References

Eli Dwek *The Infrared Diagnostic of a Dusty Plasma with Applications to Supernova Remnantos.* Apj,322,812-821

B.T. Draine and D.T. Woods *Supernova Remnants in Dense Clouds, I, Blast-Wave Dynamics and X-ray Irradiation.* Apj, 383,621-638

ASCA OBSERVATION OF THE CYGNUS LOOP SUPERNOVA REMNANT

E. MIYATA AND H. TSUNEMI
Department of Earth and Space Science,
Graduate School of Science, Osaka University, Japan
CREST, Japan Science and Technology Corporation (JST)

1. Introduction

The Cygnus Loop is the prototype shell-like supernova remnant (SNR) and one of the brighest SNRs in X-ray wavelength. We have observed the entire Cygnus Loop with the X-ray satellite, ASCA. Its large apparent size, high surface brightness, and low absorption features have made the Cygnus Loop to be an ideal target for the study of the spatially-resolved spectroscopic structure in detail. Part of this work was summarized in Miyata (1996). Here, we present the first X-ray image of the Cygnus Loop obtained with ASCA.

2. Observation

ASCA observations were performed from PV-phase (1993/04) to AO-5 (1997/06). The total number of observations is 30 and mean observing time is $\simeq$ 10ks. The preliminary mosaic image obtained with the ASCA GIS is shown in figure 1. We see some aritificial structures since the background component did not properly subtracted and vignetting effect was not corrected. Generally speaking, the limb-brightening structure can clearly be seen in this figures as previously shown in the Einstein image (Ku et al. 1984) or in the Rosat image (Aschenbach 1994). Comparing with these softband images, we can find moderately strong emission from inner part of the Cygnus Loop. The most significant difference is the bright X-ray compact source appeared at southern blow region of the Cygnus Loop. The flux of the source is 7.2×10^{-12} erg/s/cm^2 and the ASCA spectra were well fitted with the absorbed power-law with photon index of -2.1 ± 0.1 (Miyata et al. 1997a).

3. Si Map

By using the energy resolving power of the GIS, we investigated the extent of heavy elements. Since Si lines are well resolved in the GIS spectra, we extracted the narrow line images of Si (1.6–2.1 keV) as well as the continuum image between Mg lines and Si lines (1.4–1.6 keV). The equivalent width map of Si can be constructed by dividing the Si image with the continuum image shown in figure 2. Equivalent width depends on kT_e, $\log\tau$, and abundance of heavy elements. Based on the detail studies of equivalent widths of Si at the center portion, Miyata et al. (1997b) suggested that the maximum equivalent width of Si lines was $\simeq 0.9$ keV for the cosmic plasma. Therefore, figure 2 refrects the abundance distribution of Si. Based on figure 2, we can find that Si distributes from center portion toward the southern blow region. This suggests that ejecta is still confined inside the shell and has not yet mixed with the shocked interstellar medium.

At the southern blow region, the equivalent width is fairly high. Since the X-ray surface brightness of this region is quite low, the density of the interstellar medium might be low. This suggets that ejecta could be observable without any pollusion of the shocked interstellar medium.

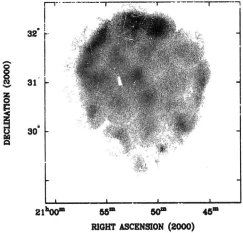

Figure 1. The X-ray image of the Cygnus Loop obtained with the ASCA/GIS

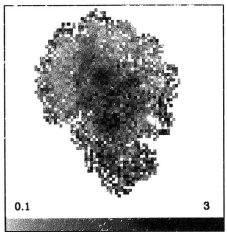

Figure 2. The equivalent width map of Si obtained with the ASCA/GIS in unit of keV

References

Aschenbach, B. 1994, New Horizon of X-ray Astronomy, p.103
Ku, W.H.-M. et al. 1984, ApJ, 278, 615
Miyata, E. 1996, Ph D. thesis, Osaka University
Miyata, E. et al. 1997a, PASJL submitted
Miyata, E. et al. 1997b, PASJ submitted

THERMAL AND NON-THERMAL X-RAYS FROM SN 1006 AND IC 443

M. OZAKI
The Institute of Space and Astronautical Science,
3-1-1, Yoshinodai, Sagamihara, Kanagawa, 229, Japan

AND

K. KOYAMA
Department of Physics, Faculty of Science, Kyoto University,
Kitashirakawa, Sakyo, Kyoto, 606-01, Japan

1. Introduction

From many Galactic supernova remnants (SNRs), X-ray emissions consisting of non-equilibrium ionization (NEI) plasma and additional hard component are detected. The hard emission have been usually interpreted as a high temperature plasma of ≥ 10 keV. However, the recent observation with ASCA made it clear that the hard components of some SNRs are of non-thermal origin. Here we report the ASCA results of SN 1006 and IC 443 observations as an example of such SNRs.

2. SN 1006

Koyama et al. (1995) found that SN 1006 has diffuse thermal component in its interior region in addition to the bright rims. Unfortunately, the region is strongly contaminated by the rim-emission leakage due to the loose XRT PSF: most of the photons above $\simeq 2.5$ keV come from bright rims, and even below $\simeq 2.5$ keV a significant fraction of the continuum is thought to be of rim origin. Fortunately, the leakage is mainly due to the distortion of the mirror surface so that we can expect that the leakage spectrum little changes position by position. Therefore, we used the off-source data as the background.

After being subtracted the Rim leakage, the resultant cannot be described by a single-component or two-component NEI plasma: they could not describe line profiles nor < 1 keV spectrum.

Nevertheless, strong Mg, Si and S lines implies that their abundances are much larger than the solar values. It suggests that SN 1006 is still in the free-expansion phase.

The bright rim spectrum can be described by the model of a power-law with an absorption which dominates above $\simeq 1$ keV and the interior thermal component. The best-fit photon index and the absorption column are 2.86 ± 0.05 and $(2.4 \pm 0.1) \times 10^{21}$ cm^{-2}, respectively (with 90% errors). The 2-10 keV flux of NE and SW rims are 1.9 and 1.5×10^{-11} erg s^{-1}cm^{-2}, respectively.

3. IC 443

In contrast to SN 1006, we fitted the IC 443 GIS data region by region and found that their thermal components can be described by 1-NEI model and the abundances in all regions are about 1 solar or lower, which suggests that IC 443 is in the adiabatic expansion phase. While the Einstein (Petre et al. 1988) and ROSAT (Asaoka and Aschenbach 1994) data implied that the southern part of the remnant has larger absorption, we could not find such a tendency.

In addition, we found that the hard (>4 keV) component forms a shell-like structure similar to radio, IR and optical emissions. Since it lies along the edge of the PV FOV, we could not judge how large the component extends. Nevertheless, we can expect that it extends from the FOV because the hard-component flux ($\simeq 1 \times 10^{-4}$ photons s^{-1}keV^{-1}cm^{-2} at 7 keV) is only about a half of that suggested by the Ginga observation.

Due to the poor statistics, we carried out spectral fitting for only two bright regions. Each region could be described by the model of a power-law and the thermal component of the nearby region. The photon indices of the hard components of two positions are $2.48^{+0.31}_{-0.28}$ and $2.35^{+0.65}_{-0.53}$, respectively (with 90% errors).

References

1. Asaoka, I. and Aschenbach, B. 1994, AA, 284, 573
2. Koyama, K., Petre, R., Gotthelf, E. V., Hwang, U., Matsuura, M., Ozaki, M. and Holt, S. S. 1995, Nature, 378, 255
3. Petre, R., Szymkowiak, A. E., Seward, F. D. and Willingale, R. 1988, ApJ, 335, 214

X-RAY STUDY OF CRAB-LIKE AND COMPOSITE SNRS

K. TORII, H. TSUNEMI
Osaka University

AND

P. SLANE
Harvard-Smithsonian Center for Astrophysics

1. Introduction

We have presented X-ray observations of Crab-like supernova remnants (SNRs) (*plerions*) and (non-thermal) composite SNRs. We have observed several objects including 3C58, CTA1, and G292.0+1.8. Since space is limited, here we summarize results on G292.0+1.8. For the other objects, please refer to each publication (Slane et al. 1997; Torii et al. 1997).

2. G292.0+1.8 (MSH11-54)

The composite nature of G292.0+1.8 has previously been in doubt. Although the center-filled radio morphology suggested an interior plerionic component, strong X-ray emission lines seemed to rule out such speculations. High resolution imaging with the *Einstein* HRI revealed its peculiar morphology (Tuohy et al. 1982) and X-ray emission was found to come from two distinct components, the central barlike feature and the ellipsoidal disk of approximately uniform surface brightness. Both the elemental abundances determined from X-ray spectra (Hughes et al. 1994) and the peculiar morphology suggested a Type II SN explosion.

We have found that the hard X-ray image (figure 1 right) shows a center-filled compact nebula in contrast to the extended morphology in the soft band (figure 1 left). The position of the hard source coincides the radio "ridge" (Braun et al. 1986) and it is displaced to the east south-east direction from the apparent emission center. We interpret this component as a synchrotron nebula embedded in the shock heated thermal plasma. Since the X-ray morphology is far from a standard limb-brightened shell, possibly suggesting the presence of circumstellar material, we have applied a self-similar solution for a point explosion expanding into a non-uniform density (Sedov 1993), $\rho \sim r^{-2}$, corresponding to a constant velocity stellar wind of the progenitor. With a single component non ionization equilibrium model, the temperature is obtained as $kT \sim 0.8$ keV. The age and the explosion energy are estimated to be $t_{age} \sim 4700 \pm 200$ yr and $E_{SN} \sim (1.25 \pm 0.07) \times 10^{50}$ ergs. Here we assumed the distance to the object of 4.8 kpc (Saken et al. 1992). From the estimated age, we derive the projected velocity of the putative pulsar born at the explosion center. If we adopt a nominal expansion center obtained by optical observations (Braun et al. 1983) and assume that the putative pulsar is at the center of the compact nebula, we obtain a reasonable value (Lyne & Lorimer 1994) for the velocity, $v_{trans} \sim 790 \pm 320\, d_{4.8 kpc} t_{age, 4700 yr}^{-1}$ km s^{-1}.

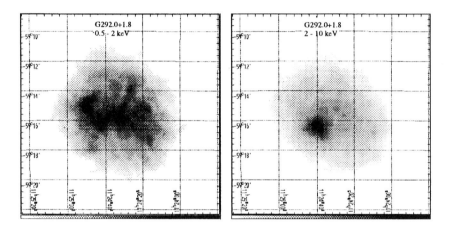

Figure 1. ASCA SIS images of G292.0+1.8. The images were deconvolved by the PSF of the telescope. Left: 0.5–2 keV, Right: 2–10 keV.

References

Braun, R., Goss, W.M., Danziger, I.J. and Boksenberg, A. (1983) *Supernova Remnants and Their X-ray Emissions*. Reidel, Dordrecht.
Braun, R., Goss, W.M. and Roger, R.S. (1986) *A&A*, 162, 259
Hughes, J.P. and Singh, K.P. (1994) *ApJ*, 422, 126
Lyne, A.G. and Lorimer, D.R. (1994) *Nature*, 369, 127
Saken, J.M., Fesen, R.A. and Shull, J.M. (1992) *ApJS*, 81, 715
Sedov, L.I. (1993) *Similarity and Dimensional Methods in Mechanics 10th Edition*. CRC Press, Boca Raton.
Slane, P., Seward, F.D., Bandiera, R., Torii, K. and Tsunemi, H. (1997) *ApJ*, 485, 221
Torii, K., Kinugasa, K., Hashimotodani, K., Tsunemi, H. and Slane, P.O. (1997) *PASJ*, submitted
Tuohy, I.R., Clark, D.H. and Burton, W.M. (1982) *ApJL*, 260, L65

THE MORPHOLOGIES OF SNRS AND THEIR ISM AND CSM

ZHENRU WANG
Dept. of Astronomy, Nanjing Univ., Nanjing 210093, PR China

Recently, many unusual morphologies of supernova remnants (SNRs) have been discovered: semicircular shell; center-brightened but without center point source; two-lobed; bipolar; irregular, etc. To understand these varieties of morphologies, we have made theoretical models. One of the main reasons of their unusualness is the multiphase structure of the interstellar medium (ISM) and circumstellar medium (CSM) where they evolved, and the CSM is connected with their progenitor.

The morphologies of SNRs were classified into three types—shell, plerion and composite types (Weiler 1985). It made progress in understanding SNRs.

We are interested in the irregular SNRs discovered recently that deviated from the above standard shapes. The most obvious example is the famous semicircular shell of CTB 109. It has been explained by our model of a SN explosion occured near the interface between a molecular cloud and diffuse ISM (Wang et al. 1992a). Another interesting example is IC443. Its morphology in soft X-xays is very different from that in the radio and optical (Petre et al. 1988). Its hard X-rays up to 20 kev has been observed by Ginga (Wang et al. 1992b). Considering IC443 expanding into a multiphase CSM and ISM, the very hard X-ray component is reasonably contributed from its western part. The age of IC443 was deduced to be in the range of 1000-1400 yrs, consistent with the explosion of the Tang Dynasty SN AD837 (Wang et al. 1992b, 1993). The joint ROSAT-Ginga observation further supported our above conclusion (Asaoka & Aschenbach 1994) because of the fact that the new SNR G189.6+3.3 discovered recently by ROSAT is not at the same distance as IC443.

Einstein, ROSAT and ASCA have discovered many center brightened SNRs without central point sources, e.g., G292.0+1.8, Kes 79, Kes 27, G18.9-1.1, HB9, G299.2-2.9, G272.2-3.2, RXJ 1713.7-3946 and W49B. W49B is center-brightened, but the other 8 SNRs are center-ring-brightened. It is obvious that they did not fit easily into the simple three-type classification scheme. White and Long (1991) considered SN evolving into a homogeneous intercloud medium with a uniform cloud distribution to explain its center-brightened shape. We (Chen, Liu & Wang 1995) have considered SNR expanding into an intercloud stellar wind bubble. The resulting thermal X-ray emission has a center-ring-brightened morphology. We conclude that the remnants of SN Ib/c can be center-ring-brightened even if no central point source is visible.

In the radio band, many SNRs have unusual morphologies. For example, G292.0+1.8 and G357.7-0.1 have center-brightened shapes with steep spectra but no central point sources, G76.9+1.0, Kes 79, G18.9-1.1, G93.7-0.3, and G65.7+1.2 have double-lobed or bipolar morphologies and so on.

The importance of the CSM to the evolution and non-thermal radio emission for the remnants of SN Ib/c and SN II is also obvious. We (Zhang, Wang & Chen 1996) have considered the evolution of SNRs in the circumstellar wind cavities. The freely expanding and shocked wind regions are the zones in this kind of SNRs that are most likely to be observed in the radio band. The magnetic structure of these two regions is likely to have toroidal configuration (Chevalier 1992). We expect that this orderly toroidal magnetic structure should affect the distribution of radio emission. Synchrotron radiation should be diminished in those parts of the remnant where the direction of the magnetic field route is parallel to the line of sight. As a result, the apparent morphologies will vary with aspect angles as follows (Zhang, Wang & Chen, 1996): center-brightened when observed in the equatorial plane; shell type when observed near the polar direction and two-lobed or bipolar when observed in the medial directions.

References

Asaoka, I. and Aschenbach, B. 1994, A&A, 284, 573
Chen, Y., Liu, N. and Wang, Z. R. 1995, ApJ, 446, 755
Chevalier, R. A. 1992, ApJ, 397, L39
Petre, R., Szymkoviak, E., Seward, F. D. and Willingale, R. 1988, ApJ, 335, 215
Wang, Z. R., Qu,Q. Y., Luo, D., McCray, R. and Mac Low, M. M., 1992a, ApJ, 388, 127
Wang, Z. R., Asaoka, I., Hayakawa, S. and Koyama, K. 1992b, PASJ, 44, 303
Wang, Z. R. 1993, in UV and X-ray Spectroscopy of Laboratory and Astrophysical Plasma, eds. Silver and Kahn (Cambridge Univ. Press), p. 407
Weiler, K. W. 1985, in The Crab Nebula and Related Supernova Remnants, eds. Kafatos and Henry, (Cambridge Univ. Press), p. 265
White, R. L. and Long, K. S. 1991, ApJ, 373, 543
Zhang, Q. C., Wang, Z. R. and Chen, Y, 1996, ApJ, 466, 808

THE AD393 GUEST STAR AND THE SNR RX J1713.7-3946

Z. R. WANG, Q. Y. QU & Y. CHEN
Dept. of Astronomy, Nanjing Univ., Nanjing 210093, PR China

RX J1713.7-3946 is a new and bright supernova remnant (SNR) in soft X-rays discovered by Pfeffermann & Aschenbach (1996, hereafter PA 1996). Its visual position as shown on its name is within the asterism Wei where the AD393 guest star occured (Wang et al. 1997).

The age of RX J1713.7-3946 was estimated about two thousand years old (PA 1996). It is nearly consistent with the explosion time of the AD393 guest star, 1.6 thousand years ago. The ASCA observation of RX J1713.7-3946 also supports it to be a young SNR (Koyama et al. 1997).

Since the morphology of RX J1713.7-3946 is central-ring-brightened shell-like in soft X-rays, we have reason to preferably consider it to be a remnant of SNIb/Ic (Chen et al. 1995). The peak luminosity of SNIb/Ic is much lower than SNIa (Nomoto et al. 1995). According to the estimated distance of 1.1 kpc for RX J1713.7-3946 (PA 1996), the visual magnitude of its corresponding SN is consistent with that of the AD393 guest star described in the ancient record (Wang et al. 1997). Therefore, the SNR RX J1713.7-3946 might be the remnant of the AD393 guest star.

Dr. Qu, Q. Y. gratefully acknowledges the support of K. C. Wong Education Foundation, Hong Kong.

References

Chen, Y., Liu, N., Wang, Z. R. 1995, ApJ, 446, 755
Koyama, K., Kinugasa, K., Matsuzaki, K. et al. 1997, PASJ, 49, L7
Nomoto, N., Iwamoto, K., Suzuki, T. 1995, Phys. Rep., 256, 173
Pfeffermann, E., Aschenbach, B. 1996, in Roentgenstrahlung from the Universe, International Conference on X-ray Astronomy & Astrophysics, MPE Report 263 (eds. H. Zimmermann, J. Trumper & H. Yorke), p.267
Wang, Z. R., Qu, Q. Y., Chen, Y. 1997, A&A, 318, L59

MULTIFREQUENCY SPECTRAL STUDIES OF SNRS

X.Z. ZHANG
Beijing Astronomical Observatory, CAS, Beijing 100080, P.R. China

X.J. WU
Beijing University, Beijing 100871, P.R. China

L.A. HIGGS AND T.L. LANDECKER
DRAO, P.O.Box 248, Penticton, B.C. V2A 6K3, Canada

D.A. GREEN
MRAO, Cambridge CB3 0HE, UK

AND

D.A. LEAHY
Department of Physics, The University of Calgary, T2N 1N4, Canada

1. Introduction

Radio supernova remnants(SNRs) of large angular diameter are obvious objects for multifrequency spectral studies from long wavelengths to short wavelengths, as resolutions at low frequency are usually about several arcminutes. An international cooperation[1] consists of MRAO, BAO, DRAO, MPI, and astronomers from Beijing University, Beijing Normal University, China and Calgary University, Canada. This international collaboration is also a part of the Panorramic Spectral Imaging of the Milky Way (PI R. Taylor). Some results on HB21, G78.2+2.1, and HB9 are presented in this paper.

2. Results

HB21
The integral flux density of HB21 at 232 MHz is about 390 ± 30 Jy measured from the new data. By combining this integral flux with that measured at 408 MHz (290 ± 20Jy) and 4750 MHz(110 ± 5Jy) from the maps of Tatematsu(1990) and

---

[1] The project is supported by the National Natural Science Foundation of China.

Reich(1983), integral spectral indices of $-0.41 \pm .02$ and $-0.39 \pm .02$ at frequency pairs of 232-4750 MHz and 408-4750 MHz are obtained respectively. A good linear fitting with the three flux densities together can be obtained.

T-T plot method was used to derive further the integral spectral index and spectral variations over the remnant. This method is independent of base level and is sensitive only to the variation of the temperature of object. The integral spectral index derived by this method is $-0.43 \pm .02$ between 232 - 4750 MHz and $-0.44 \pm .02$ between 408 - 4750 MHz. All maps used here were convolved to resolution of $5.2' \times 4.7'$.

The mean spectral index is 2.55 at southeast, falling to 2.35 at northwest. A ring-shape structure of steeper spectra is found from the calculated results. It surrounds the central flat spectrum area which contains the X-ray emission area. This is an interesting result to be discussed some where else.

G78.2+2.1

By using the same T-T plot method, the integral spectral indices of G78.2+2.1 and the mean indices in 12 boxes within the remnant were calculated respectively. From these results, two conclusions can be derived. The first is that spectral indices in each box in frequency range 232 - 2695 MHz are almost same as that in same box in frequency range 408 - 4750 MHz. The maximum diffusion of the spectral indices to their mean value is 0.12 in the box 9, while the mean-diffusion is 0.05.

The numbers suggest that the indices measured by this method are true, because of the small diffusion. The second conclusion is that, for the most case, spectral indices of frequency pairs 232 - 1420 MHz and 1420 - 4750 MHz in each box show flat-steep-relationships. Also the variations of spectral indices in low frequency range, 232 - 1420 MHz, show some distribution which is more far from the galactic plane, more flat spectra are found, whereas the spectral indices in frequency range 1420 - 4750 MHz do not show the tendency. Same phenomenon is also found in the field of HB21. The reason may be same. That is more far from the galactic plane more less the contribution of the galactic background emission which has property of steep spectrum.

HB9

Spectral index distribution of HB9 were calculated in three frequency pairs, i.e. 151 - 1420 MHz, 232 - 2695 MHz, and 408 - 4750 MHz. The maps used here has been convolved to $8' \times 8'$ beam size and maps were divided into 16 boxes which are same to that in the paper of Leahy D.A. and Roger R.(1991). Spectrum of HB9 in the boxes J,K,P have different property from other region. In the east and south of the remnant spectrum is steeper than that in west and north part. The diffusion of spectral indices of the three frequency pairs are also large in west and north region of the remnant.

References

Leahy, D.A. et al., 1991, AJ, 101, 1033
Reich, W., Furst, E. and Sieber, 1983, IAU Symposium 101, 377
Tatematsu, K., Fukui, Y., Landecker, T.L., and Roger, R.S., 1990, A&AP,1990, 237, 189
Wendker, H.J. et al., 1991, A&A, 241, 551

NEW DETECTION OF X-RAY PULSAR NEBULAE BY ASCA

N. KAWAI
The Institute of Physical and Chemical Research (RIKEN)
2-1 Hirosawa, Wako, Saitama 351-01, Japan

KEISUKE TAMURA
Venture Business Laboratory, Nagoya University,
Chikusa-ku, Nagoya 464-01, Japan

AND

S. SHIBATA
Department of Physics, Yamagata University,
1-4-12 Kojirokawa, Yamagata 990, Japan

1. Observations

X-ray images of rotation-powered pulsars were examined using ASCA Gas Imaging Spectrometer (GIS). The data sets are taken from those available in the ASCA public archive in the performance verification (PV) phase and the guest-observing (GO) phase 1. We detected diffuse X-ray sources in the vicinity of nine pulsars including five new detections. There are large variety in their morphology and spatial size. The high probability of finding such diffuse sources around pulsars suggests that they exist universally for all the active pulsars, and that they are powered by the pulsars. We propose that the pulsar-powered nebula is a good probe to measure the otherwise invisible energy flux dissipating from a pulsar into the surrounding space.

2. Energy Spectra and Luminosity

The spectra of these nebulae are generally hard, and characterized by the power-law photon index of $1.5 \sim 2.0$. The X-ray luminosities of these pulsar nebulae are derived based on the best-fit power-law models. The fraction of their luminosities in the total rotation energy losses range from 10^{-4} to 10^{-2}. These luminosities are plotted against the spin-down power of the pulsars $\dot{E}_{rot}\left(= I\Omega\dot{\Omega}\right)$ in Fig 1.

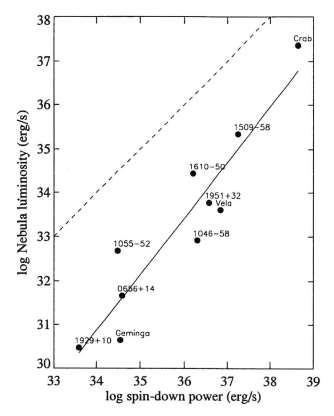

Figure 1. The X-ray pulsar nebula luminosity in the 0.7–10 keV band $L_{nebula}(0.7$–10 keV) plotted as a function of the spin-down power of the pulsars $\dot{E}_{rot}$ $(= I\Omega\dot{\Omega})$. The empirical relation obtained with least-square fit is shown by a solid line. The dashed line indicates $L_{nebula} = \dot{E}_{rot}$.

The least-square fit in the logarithmic space of the ten points here gives an empirical relation:

$$\log L_{nebula} = (33.42 \pm 0.20) + (1.27 \pm 0.17) \cdot \log\left(\frac{\dot{E}_{rot}}{10^{36}}\right), \quad (1)$$

Here the unit of the power and the luminosity is ergs, and the 1-σ uncertainties of the parameters shown in the formula reflect the scatter of the data points rather than the statistical uncertainties of the estimated luminosities. The theoretical interpretation of this relation is given in a separate paper.

PULSAR NEBULAE:

Relativistic Hot Clouds of Electron-Positron Pairs

SHINPEI SHIBATA
Department of Physcis, Yamagata University,
Kojirakawa, Yamagata 990, Japan

NOBUYUKI KAWAI
The Institute of Physical and Chemical Research,
2-1 Hirosawa Wako, Saitama 351-01, Japan

AND

KEISUKE TAMURA
Venture Business Laboratory, Nagoya University,
Chikusa-ku, Nagoya, 464-01, Japan

1. Introduction

Recent observation with ASCA reveals that radio pulsars own X-ray diffuse nebulae around them. 11 such nebulae have been reported (Harrus, Hughes & Helfand 1996, Shibata et al. 1997, Kawai, Tamura & Shibata 1997). It is most likely that these nebulae are synchrotron nebuale powered by the pulsar wind, the outflow of relativistic particles. Kawai, Tamura and Shibata (1997) reduced an empirical law for the nebula luminosity L_x as a function of the rotation power $\dot{E}_\mathrm{rot}$ of the central pulsar: $L_\mathrm{x}/\dot{E}_\mathrm{rot} = -12.3 + 0.27 \log \dot{E}_\mathrm{rot}$.

2. Pulsar Wind

A pulsar radiates its rotational energy at the rate $\dot{E}_\mathrm{rot} = \Im\Omega\dot{\Omega}$, where Ω and $\dot{\Omega}$ are the angular velocity and its time derivative, and $\Im$ is the moment of inertia of the neutron star. Since we have an reasonable estimate of $\Im$, the rotation power $\dot{E}_\mathrm{rot}$ is practically observable. However, most of this power is unseen. The pulsed luminosity is only a tiny fraction of $\dot{E}_\mathrm{rot}$. Studies of the Crab Nebula suggest that the pulsar wind carries off most of the rotation power.

The pulsar wind interacts with surrounding matter to form a shock. The shocked wind radiates in synchrotron radiation and is observed as a synchrotron nebula, whose luminosity is still a small fraction of $\dot{E}_{\rm rot}$.

The energy flux of the pulsar wind is composed of the electromagnetic part $\dot{E}_{\rm EM}$ and the kinetic energy part in bulk motion $\dot{E}_{\rm KE}$ (thermal energy is negligible). Near the star, $\dot{E}_{\rm EM} \gg \dot{E}_{\rm KE}$, and the pulsar magnetosphere is such a machine that $\dot{E}_{\rm EM}$ is transferred into $\dot{E}_{\rm KE}$. Therefore, the ratio $\sigma \equiv \dot{E}_{\rm EM}/\dot{E}_{\rm KE}$ (the magnetization parameter) at the preshock region defines the acceleration efficiency and has great importance in the theory of the relativistic wind.

Another important point in the pulsar nebula is its morphology. Recent ROSAT and HST images (Hester et al. 1995) shows a ring and a pair of bipolar jets. The efficiency of the wind acceleration is closely related to the wind structure.

3. Pulsar Wind as a Calorimeter: Results

Nebulae luminosity and spectrum depend on properties of the wind and the pressure confining the pulsar. Following the Kennel-Coroniti model (1984a, b) for the Crab, we have developed a scheme in which nebulae are used as calorimeter to detect wind paramters (Shibata, Kawai and Tamura 1997).

We have three model parameters: the magnetization parameter σ, the wind Lorentz factor $\gamma_{\rm w}$, and the nebula magnetic field B (magnetic pressure would roughly equal to the confining pressure). The model is aplied to the empirical luminosity curve, and we find $\gamma_{\rm w} \approx 6 \times 10^7 (\sigma/0.003)(B/10\mu{\rm G})^{-2}$ for the middle-aged pulsars, the spindown power of which is of order of 10^{35}erg/sec. Combining this result with the ASCA spectrum, we suggest that the wind Lorentz factor increases with spindown in spite of decrease of the electromotive force. The magnetization parameter σ is found to stay much smaller than unity as was found for the Crab pulsar wind. It is also found that the nebula pressure relaxes toward the pressure of the general interstellar medium in time scale of 10^5yr.

Harrus, I. M., Hughes, J. P., Helfand, D. J., 1996, ApJ., 464, L161
Hester, J.J., et al. 1995, ApJ, 448,240
Kawai, N., Tamura, K., Shibata, S., 1997, submitted to ApJ
Kennel, C.F., and Coroniti, F.V., 1984a, ApJ, 283, 694
Kennel, C.F., and Coroniti, F.V., 1984b, ApJ, 283, 710
Shibata, S., Kawai, N., Tamura, K., 1997, submitted to ApJ
Shibata, S., Sugawara, T., Gunji, S., Sano, S., Tukahara, H., Sakurai, H., S., Kawai, N., Dotani, T., Greiveldinger, H., Ogelman, H, 1997, ApJ, 483, 843

MAGNETIC RECONNECTION AS THE ORIGIN OF GALACTIC RIDGE X-RAY EMISSION

2D Numerical Simulation of Reconnection driven by Parker Instability and Supernova

S. TANUMA
Department of Astronomy, School of Science, University of Tokyo
7-3-1 Hongo, Bunkyo-ku, Tokyo 113, Japan

T. YOKOYAMA, T.KUDOH AND K. SHIBATA
National Astronomical Observatory of Japan
2-21-1 Osawa, Mitaka-shi, Tokyo 181, Japan

R. MATSUMOTO
Department of Physics, Faculty of Science, Chiba University

AND

K. MAKISHIMA
Department of Physics, School of Science, University of Tokyo1-
33 Yayoi-cho, Inage-ku, Chiba 263, Japan
7-3-1 Hongo, Bunkyo-ku, Tokyo 113, Japan

We present a scenario for the origin of the hot plasma in our Galaxy, as a model of a strong X-ray emission ($L_X(2-10\text{keV}) \sim 10^{38}$ erg s^{-1}), called Galactic Ridge X-ray Emission (GRXE), which has been observed near the Galactic plane. GRXE is thermal emission from hot component (~ 7 keV) and cool component (~ 0.8 keV). Observations suggest that the hot component is diffuse, and is not escaping away freely. Both what heats the hot component and what confines it in the Galactic ridge are still remained puzzling, while the cool component is believed to be made by supernovae. We propose a new scenario: the hot component of GRXE plasma is heated by magnetic reconnection, and confined in the helical magnetic field produced by magnetic reconnection or in the current sheet and magnetic field. We solved also the 2-dimensional magnetohydrodynamic (MHD) equations numerically to study how the magnetic reconnection creates hot plasmas and magnetic islands (helical tubes), and how the magnetic islands confine the hot plasmas in Galaxy. We conclude that the magnetic reconnection is

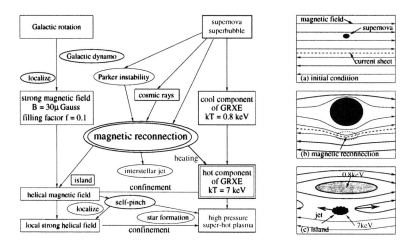

Figure 1. The schematic diagram showing possible routes for the Galactic Ridge X-ray Emission.

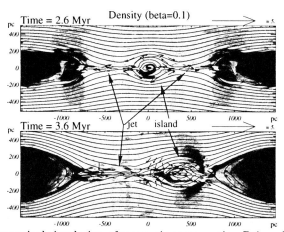

Figure 2. 2D numerical simulation of magnetic reconnection Driven by Supernova.

able to heat up the cool component to hot component of GRXE plasma if the magnetic field is localized into intense flux tube with $B_{\text{local}} \sim 30$ μG (the volume filling factor of $f \sim 0.1$).

References

Kaneda, H., Makishima, K., Yamauchi, S., Matsuzaki, K. & Yamasaki, N., (1997), *ApJ*, in press (11/1).
Tanuma, S., Yokomama, T., Kudoh, T., Matsumoto, R., Shibata, K. & Makishima, K. (1992), *ApJ*, submitted
Yokoyama, T. & Shibata, K., (1995), *Nature*, **375**, 42

OBSERVATIONS OF INTERSTELLAR O VI ABSORPTION AT 3 KM/S RESOLUTION

E.B. JENKINS
Princeton University Observatory
Princeton, NJ 08544-1001, USA

U.J. SOFIA
Dept. Astron. & Astrophys.
Villanova University
Villanova, PA 19085, USA

AND

G. SONNEBORN
Code 681
NASA Goddard Space Fight Center
Greenbelt, MD 20771 USA

1. Introduction

For studies of diffuse gases in the temperature range $10^5 - 10^6$K, observations of O VI absorption in the spectra of background stars provide an important supplement to information from surveys of soft x-ray emission. We report here the first observations of O VI absorption recorded at a resolution of 3 km s^{-1} for the stars 15 Mon and HD 64760. Both stars are behind regions that are suspected to be old supernova remnants. The observations were made with the Interstellar Medium Absorption Profile Spectrograph (IMAPS) during its flight on the ORFEUS-SPAS II mission in late 1996.

2. 15 Mon

15 Mon is 1.2 kpc away from us, and it appears within the boundary of the diffuse x-ray emission of the Monogem Ring (Plucinsky, *et al.*, 1996). The spectrum of 15 Mon in the vicinity of the O VI 1032Å feature is shown in Fig. 1. Velocities less than about -100 km s^{-1} could not be registered for

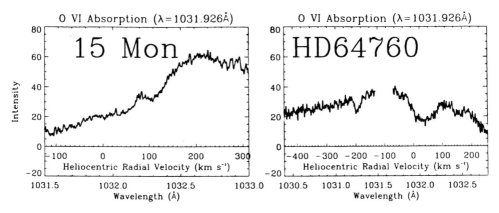

Figure 1. IMAPS observations of O VI absorption in the spectra of 15 Mon and HD 64760

this star because the stellar flux was too low. There is a very broad O VI feature centered at a heliocentric radial velocity of about $+35$ km s^{-1} and a stronger, asymmetric one with a peak absorption at $v = +105$ km s^{-1}. Possibly related to the high-velocity O VI feature are Mg II absorption features at $v = +57$ and $+86$ km s^{-1} that can be seen a spectrum in the HST archive.

3. HD 64760

HD 64760 (distance = 1 kpc) appears well inside the borders of the Gum Nebula, a structure that may be an old supernova remnant that is about about 2×10^6 yr old (Leahy, et al., 1992). Fig. 1 shows a very strong, symmetric O VI feature centered at $v = +35$ km s^{-1} situated between two high velocity components – one at $v = -195$ km s^{-1} and another at $v = +155$ km s^{-1}. The depth (45%) and width (80 km s^{-1} FWHM) of our O VI feature centered at $v = +35$ km s^{-1} are both greater than those of the model prediction of Slavin & Cox (1992) (30%, 35 km s^{-1} FWHM) for a SNR of this age. As with the features in the spectrum of 15 Mon, the broad feature does not seem to be composed of a random superposition of many narrower features.

This research was supported by NASA Grant NAG5–616 to Princeton University.

References

Leahy, D. A., Nousek, J., & Garmire, G. 1992, *Ap. J.*, **385**, 561
Plucinsky, P. P., Snowden, S. L., Aschenbach, B., Egger, R., Edgar, R. J., & McCammon, D. 1996, *Ap. J.*, **463**, 224
Slavin, J. D., & Cox, D. P. 1992, *Ap. J.*, **392**, 131

CAN GALACTIC γ-RAY BACKGROUND BE DUE TO SUPERPOSITION OF γ-RAYS FROM MILLISECOND PULSARS?

V.B. BHATIA, S. MISHRA, N. PANCHAPAKESAN
Department of Physics and Astrophysics
University of Delhi.
Delhi 110 007, India
$< vbb@ducos.ernet.in >$

The SAS 2 and COS B observations have established the existence of diffuse γ-rays in our Galaxy in various energy ranges. The diffuse radiation is attributed to the interaction of cosmic ray nuclei and electrons with the particles of interstellar atomic and molecular gas (via the decay of pions and bremsstrahlung, respectively). Inverse Compton scattering of interstellar photons by the high energy electrons of cosmic rays may also be contributing to this background. In addition some contribution may come from discrete sources of γ-rays.

We investigate if the whole of the diffuse γ-radiation in the energy range 300–5000 Mev be due to discrete sources. The discrete sources we have in mind are the millisecond pulsars which are found in large numbers in the Galaxy. Most millisecond pulsars have magnetic fields between 3 to 4 orders of magnitude smaller than that of canonical pulsars. But by virtue of their rotational velocities the charged particles in their magnetospheres could be accelerated to energies with Lorentz factor as high as

$$\gamma_e = 1.23 \times 10^7 R_6^{3/4} P_{ms}^{-1/4} B_8^{1/4}. \tag{1}$$

The critical energy of the curvature radiation photon becomes

$$E_c = 1.7 \times 10^{-2} P_{ms}^{-1/4} R_6^{18/8} B_8^{3/4}. \text{ ergs} \tag{2}$$

The luminosity of a millisecond pulsar between energies E_1 and E_2 can be shown to be

$$\dot{N}_\gamma = 9.7 \times 10^{36} B_8^{5/4} P_{ms}^{-9/4} R_6^{15/4} \int_{E_1}^{E_2} \frac{F(E)}{E} dE \text{ photons/sec} \tag{3}$$

For an estimate of the total contribution from all millisecond pulsars in the disk and the globular clusters of the Galaxy we require their period and space distributions. For the millisecond pulsars in the disk we have taken distributions already derived in the literature. For the millisecond pulsars in the globular clusters we have modeled their space distribution after the distribution of globular clusters themselves. Their period distribution was taken from literature. The integrated flux of γ-rays was calculated from the expression given in terms of the galacto-centric coordinates,

$$F_\gamma = \int\int\int q\phi(R,z,\theta)\,dR\,dz\,d\theta \tag{4}$$

where $\phi(R,z,\theta)$ is the space distribution of pulsars and q is given by

$$q = \frac{1}{4\pi}\int_{P_{min}}^{P_{max}} \dot{N}_\gamma \rho_P(P)\,dP \quad \text{photon/sec/sr} \tag{5}$$

The expression for the flux is transformed to galactic longitude and latitude and is evaluated for the energy range 300–5000 Mev. It is found that γ-rays from millisecond pulsars alone do not agree with the observations. However, in combination with those from the interaction of cosmic rays with the interstellar gas they could explain the observed flux quite satisfactorily. But for this the millisecond pulsars in the disk of the Galaxy must be $< 5x10^3$ and those in the globular clusters must not be > 200 per globular cluster. These constraints seem reasonable in view of some observations and can be a guide to the future pulsar surveys. (For detailed discussion, see Bhatia et al., 1997, Ap. J. **476**, 238.)

TRANSITION PROBABILITIES FOR ¹H IN STRONG MAGNETIC FIELDS

J. M. BENKŐ
Konkoly Observatory of the Hungarian Academy of Sciences,
P. O. Box 67, 1525 Budapest, Hungary,[‡]

AND

K. BALLA
Computer and Automation Institute, Hungarian Academy of Sciences, P. O. Box 63, 1518 Budapest, Hungary,[§]

When computing synthetic spectra we have to be aware of the strength of the lines. The determination of the required dipole strengths, oscillator strengths and transition probabilities is based on the evaluation of the dipole matrix elements. These quadratic functionals are defined by integrals over the whole space composed of the eigenfunctions Ψ of the atomic system belonging to the eigenvalues E_m and E_n, respectively:

$$\int \Psi^*(E_m)\mathbf{r}\Psi(E_n)d\mathbf{r} = \mathbf{p}. \tag{1}$$

The traditional way of computing (1) blocks the practical error estimates. An inaccuracy in the eigenfunctions of the time-dependent Schrödinger equation appears typically and it is amplified afterwards by both the weight function $\mathbf{r}$ and the numerical integration algorithm.

We describe here the general framework of an alternative method that works in a wide class of non-separable cases. Splitting the improper integral over the halfspace into two parts by fixing a plane at $z = z_c$, we assume

$$\int_0^{z_c} \int \int \Psi^*(E_m)\mathbf{r}\Psi(E_n)dxdydz = \mathbf{c}_m^{1iT}(z_c)K_{mn}^1(z_c)\mathbf{c}_n^{1j}(z_c) \tag{2}$$

[‡]e-mail: benko@buda.konkoly.hu. JMB acknowledges the support of the LOC of the 23rd General Assembly of IAU and that of PhD School of L. Eötvös University.

[§]e-mail: balla@sztaki.hu. The work of KB was partially supported by Hungarian Scientific Foundation, Grant No. T019460

$$\int_{z_c}^{\infty}\int\int \Psi^*(E_m)\mathbf{r}\Psi(E_n)dxdydz = -\mathbf{c}_m^{rT}(z_c)K_{mn}^r(z_c)\mathbf{c}_n^r(z_c) \tag{3}$$

where i, j refer to the z-parities, l (left) or r (right) refer to the directions of the integrations of the equation

$$\frac{dK_{mn}^p}{dz} - K_{mn}^p(Y_m^T Y_m)^{-1} Y_m^T \mathcal{P}_m Y_m - Y_n^T \mathcal{P}_n^T Y_n (Y_n^T Y_n)^{-1} K_{mn}^p - Y_n^T \mathcal{S}_2 Y_m = 0. \tag{4}$$

with initial values $K_{mn}^l(0) = 0$ and $K_{mn}^r(\infty) = 0$, respectively. That is, we get **p** by the solution of initial value problems for ODEs. In (4), the matrices $\mathcal{P}(E, z)$ and $\mathcal{S}_2(z)$ depend on the concrete pseudo-basis functions, T denotes the transposed of a matrix. With $q = li$ or $q = r$, $Y^q(z)$-s are the solutions of the initial value problems

$$\frac{dY^q}{dz} + [I_{2N} - Y^q(Y^{qT}Y^q)^{-1}Y^{qT}]\mathcal{P}Y^q = 0 \tag{5}$$

with initial values $Y^{li}(0)$ and $Y^r(\infty)$ and for both E_m and E_n, respectively. Next, for each eigenfunction, the normalization is taken into account by a splitting similar to (2), (3). The change consists in setting $m = n$, $i = j$ and replacing K_{mm}^p by H^q. The initial value problems for H^q are only slightly different from those for K_{mm}^p. Then

$$(\mathbf{c}^{li}(z_c), \mathbf{c}^r(z_c)) = (V_1^T H^{li}(z_c)V_1 - V_2^T H^r(z_c)V_2)^{-\frac{1}{2}} \cdot (V_1, V_2) \tag{6}$$

where V_1 and V_2 are nontrivial solutions of the system of linear algebraic equations $(Y^{liT}(z_c)Y^r(z_c)Y^{rT}(z_c)Y^{li}(z_c) - I)V_1 = 0$, $V_2 = Y^{rT}(z_c)Y^{li}(z_c)V_1$.

Together with the method of getting the pseudo-basis functions, the main point of the approach is that the truncated one-dimensional non-adjoint eigenvalue problem (5) replaces the original one. In fact, (5) realizes a transfer of boundary conditions due to Bakhvalov [1].

The advantages of both this and a similar treatment together with the proper handling of singularities has been discussed in our recent papers where the method was applied to and discussed for the diamagnetic Coulomb problem (hydrogen atom in strong magnetic field) [2] related to the model-spectra of magnetized white dwarfs and neutron stars.

When computing the quadratic functional (1) by this process, the most important gain is that we do not need the eigenfuntions Ψ to be computed! Thus, their non-uniform accuracy disappears, too. The computational errors can be kept under control. The qualitative theory of singular ODEs and that of singular boundary value problems allow us to reduce the problem to regular and stable initial value problems.

[1] Bakhvalov, N. S.: 1973, *Numerical Methods*, Nauka, Moscow, (in Russian)
[2] Balla K. and Benkő J. M.: 1996, *J. Phys. A: Math. Gen.* **29**, 6747 and 1997, *J. Phys. A: Math. Gen.* (to be submitted)

DIAGNOSTIC OF ASTROPHYSICAL PLASMA IN NEIGHBORHOOD OF NEUTRON STARS

M. MIJATOVIĆ
Institute of Physics, Faculty of Science, P.O. Box 162, Skopje, Macedonia
Institute for Nuclear Sciences "Vinča", Belgrade, Yugoslavia

AND

E.A. SOLOV'EV
Macedonian Academy of Sciences and Arts, P.O. Box 428, Skopje, Macedonia

If in the neighborhood of neutron stars exist clouds of hydrogen atoms, they are the natural astronomical object for realization of the model of hydrogen atom in a strong magnetic field $\sim 10^8$ T.

When the charged particle of velocity $\vec{v}$ moves in the homogeneous magnetic field $\vec{H}$ in the coordinate system fixed on the particle, there appears an electrical field $\vec{E} = \left(\vec{v} \times \vec{H}\right)/c$, where c is the velocity of light and $\hbar = m = e = 1$.

So, our problem of motion of the hydrogen atom in a strong magnetic field is equivalent of consideration of hydrogen atom in crossed electric $\vec{E}$ and magnetic field $\vec{H}$. The Hamiltonian $\mathcal{H}$ has the form

$$\mathcal{H} = \mathcal{H}_0 + V_1 + V_2,$$

where

$$\mathcal{H}_0 = -\frac{1}{2}\Delta - \frac{1}{r}, \quad V_1 = \vec{E} \cdot \vec{r} + \frac{1}{2c}\vec{H} \cdot \vec{L}, \quad V_2 = \frac{1}{8c^2}\left(\vec{H} \times \vec{r}\right)^2,$$

and $\vec{L} = \vec{r} \times \vec{p}$ is an angular momentum. The eigenvalue problem was first investigated in first order perturbation theory (neglecting the term V_2) (see Born, Pauli and Zimmerman et al.). The basic energy term is $\mathcal{E}_0 = -1/n^2$.

The first correction to the energy has the form $\mathcal{E}_1 = \omega q$, where $q = n' + n''$ and $n', n'' = -j, -j+1, \cdots, j$, $j = (n-1)/2$, while ω is the modulus of the vector $\vec{\omega} = \vec{H}/(2c) - 3n\vec{E}/2$.

The quadratic, in the field intensities, correction $\mathcal{E}_2$ to the energy is the sum of the second order correction from V_1 and the first order correction from V_2. The problem is solved through the separation of the variables in elliptic cylindrical coordinates on a sphere in four-dimensional space (Solov'ev and Braun). Introducing the operators $\vec{I}_1 = \left(\vec{L} + \vec{A}\right)/2$, $\vec{I}_2 = \left(\vec{L} - \vec{A}\right)/2$, where $\vec{A}$ is the Runge-Lenz vector $\vec{A} = \vec{p} \times \vec{L} - \vec{r}/r$, and $\gamma = 3ncE/H$. The second-order correction in energy has the form

$$\mathcal{E}_2 = \frac{n^4 E^2}{16}\left[3q^2 - 17n^2 - 19 - \frac{6}{1+\gamma^2}\left(n^2 - 3q^2 - 1\right)\right]$$

$$+ \frac{n^2 H^2}{16 c^2}\left(3n^2 + 1 - q^2 + \lambda\right).$$

Here λ is the eigenvalue of the operator $\Lambda = b(I_{1\alpha} - I_{2\alpha})^2 - 16 I_{1\beta} I_{2\beta}$, in which $b = \gamma^2 - 1 - 2/\left(1+\gamma^2\right)$, and $I_{i\alpha}$ is the component of $\vec{I}_i$ along the vector $\vec{\omega}_i$, while $I_{i\beta}$ is the component of $\vec{I}_i$ in the direction lying in the $(\vec{\omega}_1, \vec{\omega}_2)$ plane and orthogonal to the vector $\vec{\omega}_i$.

The eigenvalues λ cannot be computed analytically but the problem reduces to the solution of the difference equation

$$\left\{\left[(n-q)^2 - (k-1)^2\right]\left[(n+q)^2 - (k-1)^2\right]\right\}^{1/2} C_{k-2} + (bk^2 - \lambda) C_k$$

$$+ \left\{\left[(n-q)^2 - (k+1)^2\right]\left[(n+q)^2 - (k+1)^2\right]\right\}^{1/2} C_{k+2} = 0,$$

where $n' - n'' = k$ while C_k are the coefficients in the expansion of the correct zeroth-order functions in terms of the basics functions.

Contrarry to the first-order correction which is a linear function of principal quantum number n, the second-order correction in energy is proportional to $n^6 E^2$, and also to $n^4 H^2/c^2$. In the case of a strong magnetic field (neutron star) and big n (Rydberg atom) the last term is not negligible. The presence of the electrical field leads to a spectral modification, by interpreting which we can, in principle, determine the velocity v and, after that, the temperature of the plasma.

References

Born, M. (1925) *Vorlesungen über Atommechanik*, Springer, Berlin.
Pauli, W. 1926 *Z. Phys.*, **36**, 336.
Zimmerman, M.L., Litman, L.G., Kash, M.M. and Kleppner, D. (1979) *Phys. Rev.*, **A20**, no. 6, 2251.
Solov'ev, E.A. (1983) *Sov. Phys. JETP*, **56** (1), 63.
Braun, P.A. and Solov'ev, E.A. (1984) *Sov. Phys. JETP*, **59** (1), 38.

Session 1: Plasma and Fresh Nucleosynthesis Phenomena

1-3. Galaxies and Their Clusters

EVOLUTION OF MULTIPHASE HOT INTERSTELLAR MEDIUM IN ELLIPTICAL GALAXIES

Y. FUJITA
Department of Physics, Tokyo Metropolitan University, Minami-Ohsawa 1-1, Hachioji, Tokyo 192-03, Japan

J. FUKUMOTO
Nihon Silicon Graphics Cray K.K., Cuore Bldg., 9th Floor, 12-25,
Hiroshiba-cho, Suita-si, Osaka 564, Japan

AND

K. OKOSHI
Department of Earth and Space Science, Faculty of Science, Osaka University, Machikaneyama-cho, Toyonaka, Osaka 560, Japan

1. Introduction

Theoretical arguments indicate that the ISM is inhomogeneous; Mathews estimated that the $\sim 1 M_\odot$ of metal ejected by each supernova event into the ISM is trapped locally within the hot bubbles [1]. Since in elliptical galaxies, there is no overlapping of expanding supernova remnants after galactic wind period [2], it is expected that this inhomogeneity persists for a long time. The observations also suggests that the ISM of elliptical galaxies is inhomogeneous [3][4]. Based on these arguments, we studied the evolution of the multiphase (inhomogeneous) ISM.

2. Conclusions

The main results and conclusions can be summarized as follows [5]:

1. The model predicts that the supernovae are *not* effective as heating sources of the ISM in the *inner* region of galaxies after the galactic wind stops. In the inner region, supernova remnant can cool rapidly

because of their high density and/or metal abundance. Since the remnants initially have large thermal energy, the energy ejected by supernova explosions is radiated and supernovae do not heat up the ISM. Thus, cooling flow is established even if supernova rate is large. In the *outer* region of the galaxies, the cooling time of the remnants is long. Thus, most of energy ejected by supernova explosions is not radiated and it is transfered into the circumferential ISM. This results shows that *heating efficiency of supernovae depends on their environment.* Mixing of SNRs with ambient ISM makes this transfer more effective.

2. In a inner region of a galaxy, the present iron abundance of the hot ISM can be less than that of the mass-loss gas or stars if the supernova rate is small, because the phases with higher metal abundance generally cool faster and gas inflows from outer region where the metal abundance of the mass-loss gas is small. However, the spectral simulations show that predicted metal abundances are still larger than the ones observed by *ASCA* in the central region, if the present supernova rate is < 0.01 SNu. In the outer region where the selective cooling is ineffective, metal abundance of the ISM directly reflects that of the gas ejected from stars. Our model predicts that iron line emission by SNRs is prominent in the central region ; [Si/Fe] and [Mg/Fe] decrease towards the galactic center, when SNRs mix with ambient ISM of only their volumes.

References

1. Mathews, W. G. 1990, ApJ, 354, 468
2. Fujita, Y., Fukumoto, J., & Okoshi, K. 1996, ApJ, 470, 762
3. Thomas, Fabian, Arnaud, Forman, & Jones, 1986, MNRAS, 222, 665
4. Kim, D. -W., & Fabbiano, G. 1995, ApJ, 441, 182
5. Fujita, Y., Fukumoto, J., & Okoshi, K. 1997, ApJ, 488, 585

SOLVING THE MYSTERIES IN HOT ISM IN EARLY-TYPE GALAXIES

K. MATSUSHITA, T. OHASHI
Dept. of Phys., Tokyo Metropolitan Univ.
1-1 Minami-Ohsawa, Hachioji, Tokyo 192-03, Japan

AND

K. MAKISHIMA
Dept. of Phys., University of Tokyo
7-3-1, Hongo, Bunkyoku, Tokyo 113, Japan

We have analyzed ASCA data of about 30 early type galaxies, and studied their X-ray emitting ISM (InterStellar Medium) properties. Our study has been motivated by the apparently very low metallicity of the ISM, which cannot easily be reconciled with theoretical predictions. By carefully examining the abundance ratios and uncertainties in the Fe-L complex, we have concluded that the ISM abundances in X-ray luminous galaxies are in fact about 1 solar. Therefore, the severe discrepancy between the ISM and stellar abundance has been relaxed. The ISM metallicity of X-ray fainter galaxies are uncertain, but at least SNe Ia contribution to the ISM abundance is smaller than in the X-ray luminous ones.

We have also discovered that X-ray emissions from X-ray luminous galaxies are very extended, and expressed with two beta models of different angular scales. This means that the X-ray luminous ellipticals are central galaxies of some larger-scale potential structures. We show that presence/absence of such a larger-scale potential can consistently account for several unsolved problems with the ISM.

Reference

Matsushita, K., 1997, PhD thesis, University of Tokyo
Matsushita, K. et al. 1997, ApJL, 488, L125
Matsushita, K. et al. 1997, submitted to Nature

X-RAY OBERVATION OF THE NORMAL SPIRAL GALAXIES NGC2903 AND NGC628 WITH ASCA

T. MIZUNO, H. OHBAYASHI, N. IYOMOTO AND K. MAKISHIMA
Department of Physics, University of Tokyo
7-3-1 Hongo bunkyo-Ku, Tokyoo 113, Japan

1. Introduction

X-ray emission from spiral galaxies without activity is thought to consist of low-mass X-ray binaries (LMXBs), and the Einstein observations established the relation between X-ray (0.2–4 keV) and optical luminosities as $\log(\frac{L_X}{L_B}) \sim -4$ (Fabbiano, 1992). This relation has been used when discussing the activity other than LMXBs (Iyomoto, 1996). However, spectral information of Einstein Observatory was rather poor above 3 keV, where LMXBs would emit significant energy flux. Therefore we performed ASCA observations of two normal spirals, NGC2903 and NGC628, in order to better calibrate the $L_X - L_B$ relation.

2. Spectral Analysis

For both galaxies, the spectrum was extracted from a circular region centered on the source center and background spectrum was made from the source free region. The summary of the fit result are listed table 1 and 2.

3. Discussion

We compared L_X/L_B ratios of both garaxies with that of M31 (Makisnima, 1989). When comparing X-ray and optical luminosities, we calcurated X-ray luminosity of a hard component (L_X^{hard}) of energy band 2-10 keV, at this band LMXBs emit significant energy flux. Comparison are summarized in table 3. The L_X/L_B ratios of M31 and NGC628 are similar to the value obtained by Einstein ($\sim$ -4.4), but for NGC2903 ASCA exhibits smaller value than Einstein ($\sim$ -3.9) and the scattering of L_X/L_B ratios of these three galaxies becomes smaller. This is owing to the fact that ASCA has sensitivity up to 10 keV, and separated the LMXBs emission from the soft component. About these three sources, $\log(\frac{L_X}{L_B}) \sim -4.4$.

References

Fabbiano, G. et al. 1992 ApJS 80 531
Makishima, K. et al. 1989 PASJ 41 697
Iyomoto, N. et al. 1996 PASJ 48 231
Tully, R.B. Nearby galaxies catalog (Cambridge Univ. Press)

Parameter	bremss + Raymond-Smith	powerlaw + Raymond-Smith
kT (keV) of bremss or photon index	$8.2^{+3.7}_{-2.0}$	$1.62^{+0.07}_{-0.05}$
kT (keV) of Raymond-Smith	$0.44^{+0.05}_{-0.03}$	$0.47^{+0.06}_{-0.04}$
abundance	0.20 ± 0.03	$0.28^{+0.05}_{-0.04}$
flux of hard component (erg/s/cm^2, 2–10keV)	6.14×10^{-13}	6.86×10^{-13}

TABLE 1. best fit parameters and single-parameter 90% confidence limits of NGC2903.

Parameter	bremss	powerlaw
kT (keV) or photon index	$12.1^{+18.5}_{-6.1}$	$1.56^{+0.15}_{-0.12}$
flux (erg/s/cm^2, 2–10keV)	2.43×10^{-13}	2.49×10^{-13}

TABLE 2. best fit parameters and single-parameter 90% confidence limits of NGC628.

	M31	NGC2903	NGC628
L_X (erg/s, 2-10 keV)	3.9×10^{39}	2.9×10^{39}	2.7×10^{39}
L_B (erg/s)	1.1×10^{41}	5.2×10^{43}	8.0×10^{43}
$\log(\frac{L_X}{L_B})$	-4.5	-4.3	-4.5

TABLE 3. comparison of X-ray and optical luminosities from three normal spiral galaxies. X-ray luminosity is calculated from the 2–10 keV flux of thermal bremsstrahlung model. Optical luminosities are taken from Tully (1988). X-ray luminosity of M31 are calculated from the Ginga result (Makishima 1989).

HYDRODYNAMICAL MODEL OF X-RAY EMITTING GAS AROUND ELLIPTICAL GALAXIES

RYO SAITO AND TOSHIKAZU SHIGEYAMA
Department of Astronomy, University of Tokyo
7-3-1 Hongo, Bunkyo-ku, Tokyo 113, JAPAN

We have performed spherically symmetric, time-dependent hydrodynamical calculations for the X-ray emitting gas around elliptical galaxies, NGC4472, for the purpose to reproduce profiles of temperature and density. Stellar mass loss rate and type Ia supernova rate are assumed to be constant. Thermal conduction is introduced to stabilize against thermal instability due to radiative cooling.

If the gas should evolve in the intracluster medium with $n \sim 10^{-4}$, our model can reproduce the observed density profile (see below). The thermal conduction keeps the gas nearly isothermal for at least 4×10^9 years if the total stellar mass loss rate is less than $1 M_\odot$/year. Thus, we should consider the effect of resonance scattering to create cool component at the center (Shigeyama *et al.*, this volume).

When we assume that part of cooling gas sinks, the rest can reach steady state of central temperature less than $10^7 K$, although amount of cooled gas has not observed by CO and infrared.

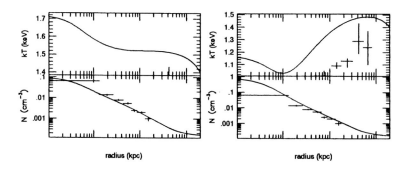

Figure 1. Distributions of temperature and density for the best-fit model at $t = 4 \times 10^9$ years. The right figure includes sink term. Crosses show ROSAT data (Irvin and Sarazin 1996,ApJ,471,683)

SIMULATED X-RAY EMISSION FROM STARBURST DRIVEN WINDS

D. K. STRICKLAND, I. R. STEVENS AND T. J. PONMAN

School of Physics & Astronomy, University of Birmingham, Edgbaston, Birmingham, B15 2TT, U.K.

Abstract. Winds from massive stars and supernovae in starburst galaxies drive global outflows of hot X-ray emitting plasma, as seen in M82 and NGC 253. These galactic winds are important for understanding galaxy evolution & formation, chemical enrichment of the IGM, and the starburst phenomenon itself.

X-ray observations provide the only direct probe of the hot gas in these winds. However, the limitations of current X-ray observatories and factors such as complex temperature structure, mass loading by ambient material and projection effects all make the link between the observed data and existing 1 & 2-D modeling and theory difficult to make.

We have therefore begun a program of numerical simulations of galactic winds, concentrating on predicting their *observable* X-ray properties. We present some initial results, comparing them to the archetypal starburst wind system M82.

1. Introduction

Starbursts create hot, 10^6–10^8 K, X-ray emitting bubbles of metal enriched gas from thermalised SN ejecta and massive star winds. These superbubbles sweep up and shock heat the ISM in the starburst galaxy, eventually blowing out of the galaxy as galactic winds (*c.f.* Heckman et al. 1990).

The presence of large scale starburst driven winds in many nearby edge-on galaxies is well established. However, little quantitatively is known about these winds, given the uncertainties in the state and filling factor of the hot gas, and many important questions remain unanswered.

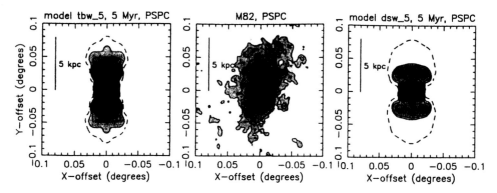

Figure 1. Simulated *ROSAT* PSPC images for two models at $t = 5$ Myr, shown at the same physical and intensity scale as the PSPC observation of M82. The lowest solid contour is roughly equivalent to a 3σ detection on a PSF scale in a 20 ksec exposure. The dashed line shows the true extent of the X-ray emitting volume, as would be seen by an instrument 10 times more sensitive than *ROSAT*.

2. Hydrodynamical modelling and selected results

Previous simulations (*c.f.* Suchkov et al. 1994), while illuminating in regard to some of the processes at work, are very difficult to compare to real observations, and have only studied a limited parameter space. It is impossible to predict the observable consequences of the complex, multi-temperature gas distributions of the simulations, without considering absorption, projection, instrument and spectral fitting effects. All of these must be taken into account before comparing the simulations with real observational data.

We simulate galactic winds using a high resolution 2-D hydrodynamics code, from which we create artificial X-ray observations (*e.g. ROSAT*, *XMM*). These are analysed in the same way a real observation (morphology, temperature structure etc.), allowing a direct comparison between observation and theory.

Fig. 1 compares *ROSAT* images of two wind models to the PSPC data for M82. The ISM distribution in model tbw_5 was criticized by Suchkov et al. (1994) for creating an artificially high level of collimation for the wind. However, model dsw_5, with the same energy and mass injection but a thinner and non-collimating disk never looks remotely like M82 in the PSPC, as it fades rapidly once the wind blows out of the disk. This demonstrates the need to consider the *observable* properties of the simulations.

References

Heckman T. M., Armus L., Miley G. K., 1990, ApJS, 74, 833
Suchkov A. A., Balsara D. S., Heckman T. M., Leitherer C., 1994, ApJ, 430, 511

ASCA X-RAY OBSERVATION OF THE LOBE DOMINANT RADIO GALAXY NGC 612

M. TASHIRO, H. KANEDA[1], K. MAKISHIMA
Department of Physics, University of Tokyo
Hongo 7-3-1, Tokyo, Japan 113-0033

The radio galaxy NGC 612 ($z = 0.0290$; Spinrad et al. 1985) exhibits bright (~ 11 Jy at 843 MHz) double lobes at a large scale of about 500 kpc $\times$ 130 kpc with a very faint core (Jones & McAdam 1992). We performed *ASCA* observations (AO-4; PI = Kaneda) of NGC 612 on July 12–14, 1996, in search for inverse-Compton (IC) X-rays from the radio lobes.

Figure 1 shows X-ray images of the central $\sim 10'$ field around NGC 612, obtained with the GISs in 0.7–3 keV and 3–10 keV bands separately. The soft-band brightness map reveals an anisotropical diffuse emission in comparison with that in hard-band. The emission extends in the direction of the radio structure up to about 200 kpc away from the core, exceeding typical spatial extent of diffuse X-ray emission observed from elliptical galaxies (Matsushita 1997). On the other hand, the point-like hard-band emission originates from the host galaxy. The X-ray spectrum obtained from the host galaxy region suggests heavily absorbed active nucleus.

We examined the X-ray spectrum of the extended emission and found it described with a power-law of photon index 1.8 ± 0.4. The agreement between the obtained spectral slope and that of the radio emission from the lobes strongly suggests that the X-rays are produced via the IC process. This is the third detection of the IC X-rays from radio lobes, following the cases of Fornax A (Kaneda et al. 1995; Feigelson et al. 1995) and Centaurus B (PKS 1343−601: Tashiro et al. 1997). The estimated flux from NGC 612 lobes is 2.7×10^{-13} erg s^{-1} cm^{-2} in the 0.5–10 keV energy band. We derived the physical quantities to summarize in Table 1, according to Harris and Grindlay (1979). We also show derived quantities from Fornax A and Centaurus B for comparison. We conclude that the obtained results

---

[1] present address: Institute of Space and Astronautical Science Yoshinodai 3-1-1, Sagamihara, Japan 299-8510

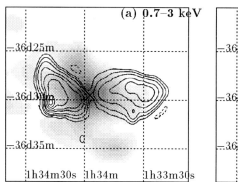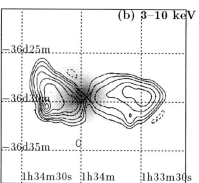

Figure 1. X-ray images of NGC 612 observed with the two GISs in (a) 0.7–3 keV and (b) 3–10 keV. The non-X-ray background and the cosmic X-ray background have been subtracted using the night earth and blank sky data, respectively. The images have been vignetting corrected and smoothed with a Gaussian kernel of $\sigma = 1'$. Radio (843 MHz) intensity contours from NGC 612 is superposed the X-ray images (Jones and McAdam 1992), and the cross is the optical core (Westerlund and Smith 1966).

from NGC 612 lobes are consistent with the energy equipartition hypothesis, although it prefers the particle dominant picture as Tashiro et al. (1997) showed in the case of Centaurus B.

TABLE 1. Evaluated Physical Quantities

source	Fronax A	Centaurus B	NGC 612
magnetic energy density $[10^{-13}\text{erg cm}^{-3}]$	3.6 ± 1.1	3.8 ± 1.8	1.0 ± 0.7
electron energy density $[10^{-13}\text{erg cm}^{-3}]$	3.0 ± 1.3	24 ± 9	2.6 ± 1.7

References

Ekers, Goss, Kotanyi, & Skellern, 1978, A&A 69, L2

Harris, & Grindlay, 1979, MNRAS 188, 25

Kaneda, et al. 1995, ApJ 453, L13

Jones, & McAdam, 1992, ApJ SS 80, 137

Matsushita, 1997, PhD thesis, University of Tokyo

Spinrad, Djorgovski, Marr, & Aguilar, 1985, Publ. Astron. Soc. Pac. 97. 932

Tashiro, et al. 1997, ApJ submitted

Westerlund, B. E. & Smith, L. F. 1966, Australian J. Phys. 19, 181

X-RAY OBSERVATIONS OF M32 WITH ASCA

T. TONERI, K. HAYASHIDA
Department of Earth and Space Science, Osaka University
1-1 Machikaneyama-cho, Toyonaka, Osaka 560, JAPAN

AND

M. LOEWENSTEIN
Laboratory for High Energy Astrophysics
Goddard Space Flight Center, Greenbelt, MD 20771, USA

1. Introduction

M32 is the nearest dwarf elliptical galaxy. Its center is known to have a mass concentration of 3×10^6 $M_\odot$, which is usually interpreted as an evidence of a super massive black hole. We observed M32 with *ASCA* two times in July and August of 1996. An X-ray source was detected at the center of M32 and its first broad-band X-ray spectra were obtained. *ASCA* observations of M32 limit the activity of the central black hole to be less than 10^{-6} times of the Eddington limit. We also found two other bright sources within 12 arcmin from the M32 center. One is the newly appeared X-ray source and the other is G144. In this paper, we summarize the results on the new source and G144. For M32, please refer to the publication (Loewenstein et al. 1997).

2. Analysis and Result

We discovered the new source at the north of M32 in the ASCA X-ray images taken in July and August of 1996. No bright sources were anticipated based on the previous observations at this region(figure 1). Although this region was observed with *ASCA* in 1993, the source was not apparent and the upper limit of the X-ray flux was smaller by one order of magnitude than the X-ray flux observed in 1996. The coordinate (J2000) of the new source is $(\alpha,\delta) \sim (00^h 42.6^m, 40°57')$, which coincides with a globular cluster in M31 observed in optical band(Battistini et al. 1980). The extracted X-ray spectrum is well-fitted by a power-law model ($\Gamma \sim 1.6$) or a thermal bremsstrahlung model($kT \sim 16\text{keV}$). The X-ray luminosity in the 0.7-10keV band was $\sim 1.8 \times 10^{38}$erg/s, assuming a distance to M31 (700kpc). The hard spectrum (the index of the power-law fit) and the high X-ray luminosity of the new source suggest a black hole binary as its likely origin. We cannot rule out a distant AGN as an alternative origin, though the X-ray flux increase by more than one order of magnitude for 3 years is not common for AGNs.

G144 is identified as a globular cluster in M31, too. This source was detected in the previous X-ray observations of this region. Its ASCA X-ray spectrum was also fitted with apower-law model as well as a thermal bremsstrahlung model. The X-ray luminosity in the 0.7-10keV band was $\sim 2.7 \times 10^{38}$erg/s at 700kpc. One of the candidates for the origin of this sources is a black hole binary, because of its hard X-ray spectrum and high X-ray luminosity. Note, however, we don't know an (active) black hole binary in globular clusters in our Galaxy.

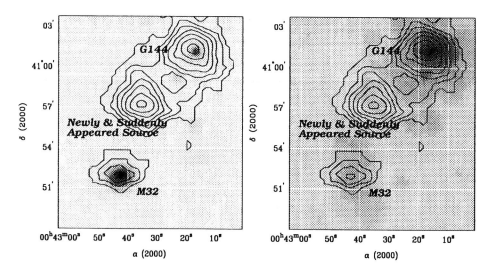

Figure 1. *ASCA* GIS contours [July 1996] superimposed on Left:*ROSAT* PSPC image [July 1991] ,Right:*ASCA* PV phase GIS image [July 1993]

References

Loewenstein, M., Hayashida, K., Toneri, T. and Davis, D.S. (1997) *ApJ*, submitted
Eskridge, P.B., White, R.E., III, and Davis, D.S. (1996) *ApJ*, 463, L59
van der Marel, R.P., de Zeeuw, P.T., Rix, H.W. & Quinaln, G.D. (1997) *Nature*, 385, 610
Battistini, P., Bònoli, F., Braccesi, A., Fusi Pecci, F., Malagnini, M.L. and Marano, B. (1980) *A&AS*, 42, 357

A CIRCULATION HYPOTHESIS OF SPIRAL GALAXIES

CHEN LINFEI

Yunnan Observatory, the Chinese Academy of Science, China

1. Circulation Hypothesis

A circulation hypothesis of spiral galaxies is proposed here to explain their spiral structure: the spiral arm material is going to the galactic center along a spiral orbit and then ejects from the galactic center in the form of two-way jets in opposite directions and finally, the ejected jets return to the galaxy in the form of spiral arms again, thereby forming a closed circulatory system.

2. Examination and Application

It could be inferred from the circulation hypothesis that a spiral galaxy should have the two-armed spiral pattern. Observations have shown that most spiral galaxies have the two-armed spiral pattern (the reason why a few spiral galaxies haven't the two-armed spiral pattern needs to be further discussed).

While the electing directions of the jets don't coincide with the galactic symmetric plane, there must be a warp in the spiral arms regarded as the backflows of the jets since the two-way jets locate in the opposite hemispheres respectively. Observations have shown that the warp of the gaseous disk of disk-shaped galaxies is a general phenomenon.

It is known from the circulation hypothesis that the velocity of the spiral arm material should have an inward radial component and the bar material of a barred spiral galaxy should have an inward flow along the bar. Researches on the Galaxy have shown that the velocity of the early type stars near the sun has a large-scale inward radial component v, $v = -25 \pm 6$ km s^{-1} (Zhao 1984); It has also been shown from observations that the bar material has indeed an inward flow along the bar (Carranza & Aguero 1989).

Now let's explain the flat rotation curve of the gases of the spiral arms in the prerequisite of the circulation hypothesis. It may be caused by the following three elements: (1) magnetic fields in the spiral arms, (2) the magnetic Reynolds number R_M of the gases of the spiral arms far greater than 1 and (3) the specific value k of the magnetic energy density of the magnetic field of the spiral arms to their thermal energy density equal to 3.3. So the magnetic field is certainly frozen in the spiral arm plasma and restricts the motion of spiral arm material, so that the magnetic field of the spiral arms and the spiral arm material controlled by the magnetic field could only be treated as a whole. it is obvious that the magnitude of the going velocity of the spiral arms should be the same everywhere, thereby inferring easily the flat rotation curve of the gases of spiral arms.

The bar structure in barred spiral galaxies may be explained by analyzing the influence of action upon the spiral arms also in the prerequisite of the circulation hypothesis. For convenience, a spiral arm is assumed as a logarithmic spiral line with a spiral angle α, $\rho = \rho_0 e^{a\theta}$ (where $a = tg\alpha$. The spiral arm material in with a galactic centric distance r is acted upon in the normal direction by two forces: One is the normal component of the gravitation towards the galactic center $F_1 = GmM(r)\cos\alpha/r^2$ (the formula is an approximate one, where $M(r)$ is the mass of the material within radius r) and the other is the outward normal centrifugal force $F_2 = mv^2 \cos\alpha/r$. The resultant force acting on the spiral arm in the outward normal direction is $F = F_2 - F_1$. There might be one position $r = r_0$ where $F_1 = F_2$. So F is greater than zero in the region $r > r_0$, and the direction of stress of one spiral arm is just opposite to that of the other spiral arm, i.e., the resultant forces of two spiral arms in the region $r > r_0$ just from a pair of shearing forces. Acted by the shearing forces, the bar structure could form in some spiral galaxies.

According to the circulation hypothesis, we may foretell that there should be the two-way jets at the center of every spiral galaxy, and there should be some link between the jet material and the spiral arms.

The author considers that the cooling flows going to the galactic center in some giant elliptical galaxies may also be the backflows of the observed two-way jets and that the disk-like material near young stars is going to the centric region along a spiral orbit, and the disk-like structure itself is also the backflow of the observed two-way jets.

References

Carranza, G.J., Aguero, E.L. (1989) *Ap & SS*, **152**, 279
Zhao, J.L. (1984), *Science of China, Series A*, **7**, 647

X–RAY PROPERTIES OF *ASCA* OBSERVED 43 CLUSTERS OF GALAXIES

F. AKIMOTO, M. WATANABE, A. FURUZAWA, Y. TAWARA,
Y. KUMAI, S. SATOH AND K. YAMASHITA
*Department of Astrophysics, Nagoya University, Furo-cho,
Chikusa-ku, Nagoya 464-01, Japan*
akimoto@satio.phys.nagoya-u.ac.jp

1. Introduction

Much attempts of statistical approach have been made to study the origin of heavy elements, distribution of dark matter and evolution of clusters of galaxies. Henry et al.(1991) reported a power–law relation; $L_X \propto kT^\gamma$, $\gamma \sim 2.7$. Edge and Stwert(1991) found significant scatter in the correlation using 45 clusters. David at al.(1993) reported $\gamma \sim 3.4$ using 104 clusters.

2. Results

We presented preliminary results of analyses for 43 clusters and groups of galaxies (z = 0.004–0.7). This sample is not selected under a certain criterion. For each group or cluster, we investigated the correlations of redshift and X–ray spectral parameters (using Raymond-Smith model); flux–weighted temperature kT[keV] and abundance(Fe) Ab[solar], luminosity of 2–10keV energy band L_X[erg s^{-1}].

Figure 1 shows that L_X and kT have a correlation as many previous works reported. We added the lines whose γ is 2.5, 3, 3.5 respectively.

The correlation has a significant scatter from this power–law relation. Then we ploted Ab and deviation of the relation in figure 2. At that time, we assumed γ is 3 and selected the targets whose 90% confidence error of abundance; $\delta[\log_{10} Fe]$ is smaller than 0.2. This correlation has a significant relation of 99.4% confidence level by Fisher's exact test. Scharf and

Mushotzky(1997) have already reported about this correlation. They concluded the correlation was explained by taking into account the range of cluster formation epochs expected within a hierarchiacl universe. Our result is consistent with their correlation and gives a stronger constraint.

According to their proposition, the x-coordinate should increase with earlier cluster formation epoch and the observed correlation indicated that earlier formed cluster of galaxies has higher abundance. Generaly in a hierarchical scenario, clusters have started as a single, dominant potential well, into which many smaller clumps have fallen, or have formed by the merger of intermediate sized clumps. And it evolved by merging and/or contraction and infall of small lump of gas. The subclumps involved in the formation of luminous clusters at low redshift are themselves typically low luminosity systems, such as groups, and the abundances measured in groups are low. These observational results are consistendt with this correlation. Then we made correlations of the dispersion of Lx–kT relation and R_c, β, gas fraction and found some have correlations.

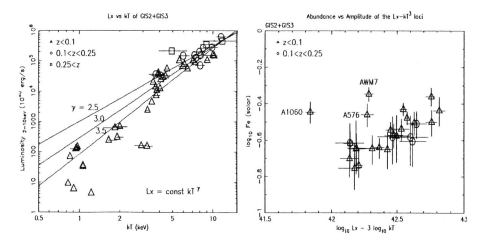

Figure 1. Correlation between L_X and kT Figure 2. Correlation between Abundance and Amplitute of the L_X-kT^3 loci

References

Edge and Stwert 1991, MNRAS, 252, 414
David, L. et al. 1993, ApJ, 412, 479
Henry, J. et al. 1991, ApJ, 372, 410
Scharf and Mushotzky 1997, ApJL

GENERATION AND IMPLICATIONS OF POST-MERGER TURBULENCE IN CLUSTERS OF GALAXIES

I. GOLDMAN

School of Physics and Astronomy, Tel Aviv University, Tel Aviv 69978, Israel; goldman@post.tau.ac.il

1. Introduction

Observations in X-ray and optical suggest that mergers of sub-clusters with galaxy clusters are quite common (for Coma see e.g., White et al. 1993; Colless & Dunn 1996; Ishizaka & Mineshige 1996). A merger leads to violent relaxation of the dissipationless dark matter resulting in a time-dependent gravitational potential. This in turn generates large-scale flows and shocks in the collisional baryonic intracluster gas (Takizawa & Mineshige 1997). Both the large scale flows and the shocks will excite turbulence in the gas. We focus here on turbulence generated by shocks, which is less dependent on the specifics of the merger. This paper is based on a more detailed work (Goldman 1997).

2. Intracluster Turbulence Generated by Shocks

For a shocked region of size ~ 0.8 Mpc and merging velocities of ~ 1000 km s^{-1}, we obtain for the largest turbulence scale and the corresponding turbulent velocity: $l_0 \sim 250$ Kpc and $v_0 = v_{turb}(l_0) \sim 250$ km s^{-1}, respectively. The turbulence timescale is thus $\tau_0 \sim \frac{l_0}{v_0} \sim 1$Gyr.

The energy input to the turbulence is mainly to the largest eddies. The non-linear eddy interactions result in an energy cascade to the small scales, down to a scale where it is dissipated. Over more than a decade in wavenumbers, the spectrum is a power-law $v(l) = v_0 \left(\frac{l}{l_0}\right)^m$ with the Kolmogorov value $m = 1/3$. In case of stable stratification the spectrum is quasi two dimensional with $m = 1$. In case that the local rotation rate is high, $m = 1/2$.

The heating rate per unit mass is $\epsilon \sim \frac{v_0^3}{l_0} \sim 0.02$ erg gr^{-1}s^{-1} and the resulting luminosity is $L_{turb} = \epsilon M_{mg} \sim 4 \times 10^{44} \left(\frac{M_{mg}}{10^{13} M_\odot}\right)$ $erg\ s^{-1}$ with M_{mg} denoting the merged gas mass. Thus, the turbulence is potentially an important heat source for the intracluster gas.

The turbulence timescale, on spatial scale l, is $\tau(l) \sim \frac{l}{v(l)}$. For $l = 10$ Kpc and Kolmogorov spectrum it is $\sim 10^8$ yrs. Given that the available time for the turbulence is $\tau_0 \sim 10^9$ yrs, amplification of magnetic fields by a large factor is possible. If the magnetic field is amplified to equipartition with $v(l)$, it can reach a level of $\sim 1\ \mu$G.

Turbulent heat conduction and turbulent mixing will be effective in smoothing out temperature and abundance gradients over a spatial scale $l < l_0$, if the available time exceeds $\tau(l)$. For truly stationary turbulence this could have occurred also on scales $> l_0$, provided time spans exceeding τ_0 were available. However, the lifetime of the turbulence is itself of the order of τ_0. Thus, one may expect that scales $\lesssim l_0 \sim 250$ Kpc will be affected.

3. Concluding Remarks

The observations suggest occurrence of multiple mergers, so a turbulent region could be shocked again. As a result, the largest turbulence scale will decrease by ~ 1.4 and the corresponding turbulent velocity will increase by a similar factor. This will increase the heating rate of the intracluster gas by a factor of ~ 4, will shorten the turbulent timescales by a factor of ~ 2 and thus could boost up magnetic field amplification.

Future X-ray spectrometers with high spectral and spatial resolutions could measure the underlying turbulence spectrum and thus decide on its nature. This could help to constrain the parameters of the merger, notably the amount of angular momentum in the merged gas, and stratification.

If a strong enough magnetic field is present, it can trap the cascaded turbulent energy before it reaches the viscous scales, and thus generate a magnetic turbulence. In such a case, a change in the turbulence spectrum is expected to be observed at a scale $\gtrsim 10$ Kpc.

This work was supported by the US-Israel BSF grant 94-314.

References

Colless, M. & Dunn, A.M. 1996, ApJ, 458, 435
Goldman, I. 1997, in preparation
Ishizaka, C. & Mineshige, S. 1996, PASJ, 48, L37
Takizawa, M. & Mineshige, S. 1997, astro-ph/9702047
White, S.D.M., Briel, U.G. & Henry, J.P. 1993, MNRAS, 261, L8

CLUSTERS OF GALAXIES IN A FLAT CHDM UNIVERSE

A.HABE, C.HANYU, AND S.YACHI
Hokkaido University, Sapporo, Japan

1. Introduction

Cold and hot dark matter (CHDM) model is one of viable models which can reproduce the large scale structure of the universe. HDM may affect structure of clusters of galaxies in CHDM universe. Bryan *et al.* (1994) gave numerical results of CHDM model that explain some statistical features of X-ray clusters of galaxies, e.g. X-ray luminosiry-temperature realtion, $L \propto\sim T^{3.5}$, without considering radiative processes. However their numerical resolution is insufficient to resolve the cores of X-ray clusters. So, we simulate the formation of clusters in CHDM universe more carefully.

2. Models and Numerical Results

We assume three cosmological models in a flat universe with different mass fraction of dark matter components ($\Omega_{\rm HDM} = 0.3, 0.2, 0$, $\Omega_{\rm baryon} = 0.05$). Hubble constant is $H_0 = 50$ km/s/Mpc and power spectra are normalized by $\sigma_{8h^{-1}} = 0.667$ for all models. Generating constrained density fluctuation for the initial conditions, we calculate by GRAPE + GRAPESPH code for dark matter motion and hydrodynamics.

From our results, we find that gas density profiles well agree with X-ray observation. We cannot find clear difference among three models on the X-ray luminosity-temperature relation, $L \propto\sim T^{2.0}$, at $z = 0$. This result is consistent with the recent results by Choi & Ryu (1997), and Bryan & Norman (1997).

References

Bryan, G.L., Klypin,A. Loken,C., Norman,M.L., and Burns, J.O. : *Ap.J.*, **347**, L5 (1994).
Bryan, G.L., and Norman,M.L. : *preprint*, astro-ph/9710107 (1997).
Choi, E., and Ryu, D. : *preprint*, astro-ph/9710078 (1997).

ASCA STUDY OF SHAPLAY SUPERCLUSTER

H.HANAMI
Physics Section, Iwate University
Morioka, 020 JAPAN

1. Introduction

We observed five clusters of galaxies in Shapley supercluster with ASCA; A3562, SC13290-313, SC1327-312, A3558 and A3556, which are candidates for interacting clusters since their separation and relative velocities are only 1Mpc and 700-1000km/s (e.g. Raychaudhury et al. '91, Breen et al. '94, Scaramella et al. '89). Main purpose of mapping observations is to make clear the formation and evolution process of the clusters of galaxies in a supercluster. ASCA gives essential informations of ICM like the temperature distribution related to the dynamics; the interactions due to the infalling and merging events. They are important tests for observational cosmology (e.g. Hanami '93).

2. ASCA Observations

We will present ASCA X-Ray observations of the Shapley Concentration in the following table.

objects	$L_X/10^{44}$ ergs s^{-1}	z_X	$k_B T$ keV	$E.M./10^{-16}$ cm^{-5}	Abandance (solar)	$N_H/10^{20}$ cm^{-2}
A3562	1.91	$0.033^{+0.011}_{-0.007}$	$4.55^{+0.06}_{-0.33}$	$2.01^{+0.11}_{-0.09}$	$0.39^{+0.12}_{-0.1}$	$8.93^{+3.49}_{-3.69}$
SC1329$_{in}$	0.268	< 0.0132	$3.64^{+0.44}_{-0.4}$	$4.70^{+0.46}_{-0.38}$	$0.30^{+0.11}_{-0.18}$	< 13.2
SC1329$_{out}$	0.382	< 0.0160	$6.45^{+3.95}_{-0.21}$	$4.30^{+3.90}_{-0.28}$	$0.21^{+0.2}_{-0.21}$	< 6.8
SC1327	0.762	$0.036^{+0.013}_{-0.026}$	$3.58^{+0.17}_{-0.31}$	$1.08^{+0.08}_{-0.05}$	$0.27^{+0.11}_{-0.1}$	< 5.67
A3558	5.99	$0.045^{+0.008}_{-0.006}$	$5.84^{+0.31}_{-0.26}$	$6.26^{+0.17}_{-0.17}$	$0.316^{+0.05}_{-0.05}$	$3.6^{+1.7}_{-1.8}$
A3556	0.199	$0.031^{+0.025}_{-0.031}$	$3.24^{+0.84}_{-0.62}$	$2.92^{+0.53}_{-0.54}$	$0.740^{+0.78}_{-0.43}$	$23.8^{+6.1}_{-15.7}$

The values of the rich ones (A3562 and A3558) are higher than that from ROSAT (Bardelli et al. '94). Our estimation, however, is consistent with GINGA (Day et al. '91) and the velocity dispersion (e.g. Breen et al '94).

We can see that the poor cluster SC1329-313 is very peculiar; the outer part is hotter than the inner part. The temperature of the inner compornent is consistent with that in hydrostatic equiribium like typical poor clusters. On the other hand, the hot envelope cannot be confined with its higher internal energy than the potential energy. In soft-photon (0.2-2keV) and hard-photon (>2keV) maps, we have also found the elongated cold core structure and the hot halo structure, which extends in the open angle $a \simeq 10$ min; $r_{hot} = 438 kpc (H_0/100 km s^{-1} Mpc^{-1})(a/10min)(z/0.05)$.

As shown in numerical simulations, the observed complex structure can be formed in a large merger event. A strong circular shock around a cluster originates with the merger at the center and moves outwards at speed of $v_s \simeq 300 km/s$, which is comparable to the sound velocity of the hot gas. This feature may be somewhat tentative. Because the hot envelop can be formed by the expanding shock with the age of $\simeq r_{hot}/v_s = 1.4 \times 10^6 yrs$ which is much smaller than the Hubble time. Then, SC1329-313 may be dynamically interacting or just merged. It is also supported from the fact that the x-ray redshift z_X determined with Raymond model < 0.016 is lower than 0.05 obtained from optical observations. It is suggests that the plasma is not in thermal equilibrium state in this region.

From the hard-photon map, we can see also a faint ridge strucutre between SC1327 and SC1329. This strucutre seems to be the filament-like or pancake-like shock which is formed with the interaction of the infalling plasma onto the supercluster plane from A3562 to A3556 in the Shapley concenration. This feature can be related to the existence of the supercluster structure.

This study was corrabrated with K.Shimasaku, Y. Ikebe, T. Tsuru, S. Yamauchi, and K. Koyama. The authors would like to thank all the ASCA Team members for their support in observations and in data analysis.

References

Bardelli, S. et al., 1994, MNRAS, 267, 665
Breen, J. et al., 1994, ApJ. 424, 59
Briel, U.G., Henry, J.P., & Bohringer, H.J., 1992, A&A, 424, 59
Day, C., S. et al., 1991, MNRAS, 252, 394
Hanami, H., 1993, ApJ, 415, 42
Raychaudhury, S. et al., 1991, MNRAS, 298, 101
Scaramella, R. et al., 1989, Nature, 338, 562

ASCA OBSERVATION OF "FAILED CLUSTER" OF GALAXIES CANDIDATE, 0806+20

K. HASHIMOTODANI AND K. HAYASHIDA
Dept. of Earth and Space Science, Graduate School of Science, Osaka University
Machikaneyama-cho 1-1, Toyonaka, Osaka, 560, Japan

AND

T. T. TAKEUCHI
Dept. of Astrophysics, Faculty of Science, Kyoto University
Kitashirakawa-Oiwake-cho, Sakyo, Kyoto, 606-01, Japan

The hypothetical object called "failed cluster" of galaxies is described by Tucker et al. (1995, ApJ,444,532) as a large cloud of X-ray emitting hot gas without any visible galaxies. They made extensive survey of this type of objects using Einstein IPC database and found only one candidate, 0806+20.

We successfully detected the X-ray emission from 0806+20 with ASCA. The ASCA X-ray image of 0806+20 shows no significant extension. The time variation of X-ray flux is not significant, either. Although we can not determine whether the spectra of 0806+20 are thermal or non-thermal, we find the emission line like feature around 4keV at 95% significance level(figure 2). If we identify it with an Fe-K line, which should contradict that 0806+20 is a "failed" cluster, the redshift value is implied to be around 0.6.

Moreover, we made additional deep R-band follow up observation using Kiso Observatory[1] 105cm Schmidt telescope. However, we could not find any clear counterparts of this X-ray source above R~24 magnitude(figure 1). We thus conclude that 0806+20 is a very distant, rich but no "failed" cluster of galaxies (and so galaxies could not be detected) or an optically faint quasar.

---

[1] Kiso Observatory is operated by Institute of Astronomy, Faculty of Science, University of Tokyo, Japan.

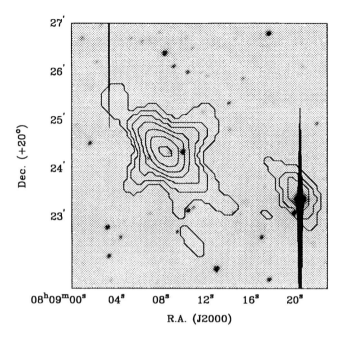

Figure 1. ASCA SIS X-ray contour superposed on the deep R-band image obtained at Kiso observatory. All the stellar images around the X-ray center are identified with Galactic stars.

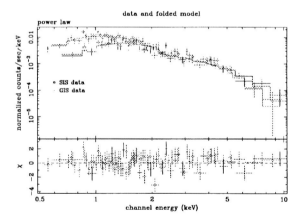

Figure 2. The power law fit to the X-ray spectra of 0806+20 obtained with ASCA, assuming the Galactic N_H. If we identify 4keV feature with a redshifted Fe-K line, the redshift of 0806+20 is proved to be around 0.6. The best fit photon index is $1.88^{+0.11}_{-0.09}$ when Fe-K line model is included with this redshift. Adopting the thermal plasma emission model, we obtain kT=$6.4^{+2.0}_{-1.2}$keV and metal abundance of $0.47^{+0.36}_{-0.30}$ times the solar value.

ASCA OBSERVATIONS OF THE ABELL 496 CLUSTER OF GALAXIES

I. HATSUKADE, J. ISHIZAKA, M. YAMAUCHI, K. TAKAGISHI
Faculty of Engineering, Miyazaki University
1-1 Gakuen-Kibanadai-Nishi, Miyazaki 889-21, Japan

1. Introduction

The metal in the intracluster medium (ICM) has been ejected or stripped from galaxies. Thus measurements of the metal distribution and the relative abundance of elements, in particular Si/Fe, are important to study the evolution of galaxies, as well as to study the chemical evolution of the ICM. We present the results from ASCA observations of Abell 496 cluster of galaxies. A496 is a nearby rich cluster with a central cD galaxy. At the redshift z=0.0327 of A496, 1 arcmin is 53kpc, where we assumed $H_0 = 50 kms^{-1} Mpc^{-1}$, $q_0 = 0.5$. A496 is known as a cooling flow cluster. Edge and Stewart (1991) obtained the mass flow rate of $\sim 100 M_\odot yr^{-1}$ and the cooling radius of 177 ± 52kpc.

2. Analysis and results

We made the GIS and SIS spectra accumulated for the ring-cut regions centered on the emission peak. The plasma temperature, the iron abundance, and the silicon abundance were obtained using Mewe–Kaastra model in XSPEC 9.00 (Mewe et al. 1985). The best fit parameters and the 90% confidence errors are shown in Figure 1. Abundances were presented by the ratio to solar values ($n(Si)/n(H) = 3.55 \times 10^{-5}$ and $n(Fe)/n(H) = 4.68 \times 10^{-5}$) given by Anders and Grevesse (1989). Temperature decrease and the abundance increase were seen in central region ($r < 200$kpc) whose size is nearly equal to the cooling radius.

3. Discussion

The relative abundance of silicon to iron is the key parameter to study the origin of the metal, because Type I SNe produce mainly iron, while

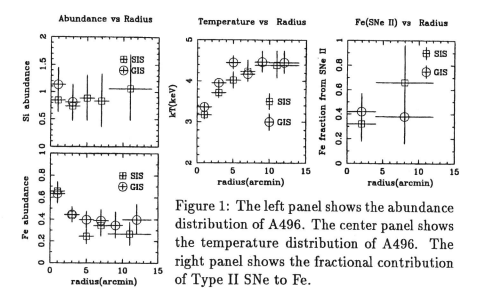

Figure 1: The left panel shows the abundance distribution of A496. The center panel shows the temperature distribution of A496. The right panel shows the fractional contribution of Type II SNe to Fe.

type II SNe produce mainly α–process elements. Observed number ratio of silicon to iron $(\frac{Si}{Fe})_{obs}$ is given by

$$\left(\frac{Si}{Fe}\right)_{obs} = \frac{Fe_{SNI} \cdot (\frac{Si}{Fe})_{SNI} + Fe_{SNII} \cdot (\frac{Si}{Fe})_{SNII}}{Fe_{SNI} + Fe_{SNII}} \quad (1)$$

where Fe_{SNI} and Fe_{SNII} are the number of iron atoms produced by Type I and Type II SNe, respectively, and $(\frac{Si}{Fe})_{SNI}$ and $(\frac{Si}{Fe})_{SNII}$ are the number ratio of silicon to iron of Type I and Type II SNe, respectively. Using theoretical values of $(\frac{Si}{Fe})_{SNI}$, $(\frac{Si}{Fe})_{SNII}$ (Tsujimoto et al. 1995) and $(\frac{Si}{Fe})_{obs}$, we can estimate the contribution of Type II SNe to the iron. The results are shown in Figure 1. About 40% of the iron is the product of Type II SNe.

The origin of the central metal excess is related to the presence of cD galaxies, because non-cD clusters do not show metal concentration. If the central metal is ejected from cD galaxies, $(\frac{Si}{Fe})_{obs}$ is expected to decrease toward the cD galaxy. However, the observed $(\frac{Si}{Fe})_{obs}$ is constant. Additional process is required to explain the metal concentration in the ICM.

References

Anders, E. and Grevesse, N. (1989) *Geochimica et Cosmochimica Acta*, Vol. 53., pp. 197
Edge, A. C. and Stewart, G. C. (1991) *MNRAS*, Vol. 252, pp. 414
Mewe, R. and Gronenschild, H.B.M. van den Oord, G.H.J. (1981) *A&AS*, Vol. 62, pp. 197
Tsujimoto, T. Nomoto, K., Yoshii, Y., Hashimoto, M., Yanagida, S., and Thielemann, F.K. (1995) *MNRAS*, Vol. 277, pp. 945

THE SUPPLY OF MAGNETIC FIELDS FROM A CD GALAXY TO INTRA-CLUSTER SPACE

H. HIRASHITA AND S. MINESHIGE
*Department of Astronomy, Kyoto University,
Sakyo-ku, Kyoto 606-01, Japan*

K. SHIBATA
*National Astronomical Observatory, 2-21-1 Osawa,
Mitaka, Tokyo 181, Japan*

AND

R. MATSUMOTO
*Department of Physics, Faculty of Science, Chiba University,
Inage-ku, Chiba 263, Japan*

1. Introduction

Intra-cluster spaces are filled with intra-cluster medium (ICM), whose typical temperature and density are $T_{\rm ICM} \sim 10^{7.5}$ K and $n_{\rm ICM} \sim 10^{-3}\,{\rm cm}^{-3}$, respectively (e.g., Sarazin 1988). Recent Faraday rotation measurements have revealed the existence of magnetic fields in ICM with $B_{\rm ICM} \sim$ a few $-10\,\mu{\rm G}$ (e.g., Ge & Owen 1993). In ICM, the plasma β (the ratio of gas pressure to magnetic pressure) is almost "equipartition" value as follows:

$$\beta \simeq 1 \times \left(\frac{B_{\rm ICM}}{10\,\mu{\rm G}}\right)^{-2} \left(\frac{n_{\rm ICM}}{10^{-3}\,{\rm cm}^{-3}}\right) \left(\frac{T_{\rm ICM}}{10^{7.5}\,{\rm K}}\right). \qquad (1)$$

However, the origin of the magnetic fields in ICM remains to be understood. We consider that fields may originate from central cD galaxies in clusters of galaxies. Here, we conjecture that the magnetic fields in cD galaxies can be lifted up to intra-cluster space by MHD instabilities. The most possible instability is the Parker instability (Parker 1966), which is a global MHD instability in a gravitational field. The growth time for the instability is estimated by the free-fall time.

2. Growth Rate of the Parker Instability

2.1. FIELD CONFIGURATION

To estimate the growth time for the Parker instability in a cD galaxy, we examined linear stability of the magnetized medium in cylindrical coordinate (r, θ, z), whose origin is located at the center of the cD galaxy. The gravitational field of the cD galaxy is radially inward and the magnetic field is assumed to be axially symmetric with curvature radius of r. Parker (1966) briefly discussed that the instability grows with the timescale of free-fall time in spite of the stabilizing effect of magnetic tension. Based on the formulation of Horiuchi et al. (1988), we confirmed Parker's result and derived the dispersion relation (Hirashita et al. 1998 in preparation).

2.2. GROWTH RATE

Using the dispersion relation we derived, we estimate typical growth time τ for the Parker instability in a cD galaxy:

$$\tau \sim 10^8 \left(\frac{H}{1\,\mathrm{kpc}}\right) \left(\frac{v_\mathrm{A}}{100\,\mathrm{km\,s^{-1}}}\right)^{-1} \mathrm{yr}, \tag{2}$$

where H and v_A are typical pressure scale height and Alfvén velocity in the cD galaxy, respectively. The growth time τ turns out to be shorter than a galaxy-evolution timescale [$\sim 10^9$ yr (Binney & Tremaine 1987)], which means that the Parker instability is an effective mechanism to supply magnetic fields to intra-cluster space.

3. Conclusion

Having examined properties of the Parker instability in a cD galaxy, we conclude that the Parker instability is an effective mechanism to supply magnetic fields to intra-cluster space.

The lifted-up magnetic fields may dissipate through magnetic reconnections, which may contribute effectively to the heating of ICM.

References

Binney, J. & Tremaine, S. (1987) *Galactic Dynamics*, Princeton University Press, Princeton, p. 552
Ge, J. P. & Owen, F. N. (1993) *AJ*, **105**, pp. 778–787
Horiuchi, T., Matsumoto, R., Hanawa, T., & Shibata, K. (1988), *PASJ*, **40**, pp. 147–169
Parker, E. N. (1966) *ApJ*, **145**, pp. 811–833
Sarazin, C. L. (1988) *X-Ray Emission from Clusters of Galaxies*, Cambridge University Press, Cambridge

TEMPERATURE STRUCTURE IN MERGING COMA CLUSTER OF GALAXIES

H. HONDA
The Institute of Space and Astronautical Science,
3-1-1, Yoshinodai, Sagamihara, Kanagawa, 229, Japan

M. HIRAYAMA, H. EZAWA
Department of Physics, University of Tokyo,
7-3-1 Hongo, Bunkyo-ku, Tokyo 113, Japan

K. KIKUCHI, T. OHASHI
Department of Physics, Tokyo Metropolitan University,
Hachioji, Tokyo 192-03, Japan

AND

M. WATANABE, H. KUNIEDA, K. YAMAHITA
Department of Physics, Nagoya University,
Chikusa-ku, Nagoya 464-01, Japan

1. Introduction

The Coma cluster has been recognized as an archetype of rich and relaxed clusters, until recent *ROSAT* observations reveal that the intracluster medium (ICM) has a complex distribution (Briel et al. 1992; White et al. 1993). The X-ray surface brightness distribution shows a secondary peak around the galaxy NGC 4839, at 40' SW from the cluster center.

Works by Hughes (Hughes et al. 1988a; Hughes et al. 1988b; Hughes 1989) show that the core of the cluster is nearly isothermal out to a radius of 1° and suggest that the temperature falls with radius beyond that. Hughes (1993), based on the recent analysis of the *Ginga* scanning observations, shows that the ICM is approximately isothermal in the core and becomes cooler beyond several core radii. Here, we report on the recent mapping observations of the Coma cluster with *ASCA* (see Honda et al. 1996 for details).

2. Analysis and Results

The observed data can be regarded as a weighted sum of the whole cluster emission, and one need to know the complete distribution of the surface brighness and temperature all over the cluster in order to derive the temperature in a given position. In principle, this is obtained by solving the whole temperature map in a self consistent way. Therefore, we are developing the new analysis system called TERRA (Kikuchi et al. in this volume).

Using the TERRA system, we obtain significant variation of the temperature. The temperature is lower than 4.5 keV in the west region and higher than 10 keV in the east one, respectively, both offset by 40' from the center. Systematic drop of the temperature in the outer region of the cluster has been observed in other systems.

3. Temperature structure

The azimuthal variation of temperature is explained that while central regions of these clusters have been heated up due to recent mergers, the disturbance is not reaching the outer parts of the clusters. In other words, these cool outer gas may be under a pre-merger equilibrium state.

When a merger occurs, dark matter particles take a longer time than the gas to settle into a sigle body. This is seen in numerical simulations (see e.g. Evrard 1990). In a cluster experiencing a merger, diffence in spatial distribution between gas and dark matter would be larger than that in relaxed clusters. Since dark matter is the dominant source of gravitational potential, part of the gas may be pulled by the extended dark matter field. Such an effect may cause temperature variation in the outer part of the cluster. Detailed calculation or simulation would be necessary to look into these possibilities. Mapping observation with ASCA of other rich clusters would be useful to obtain more knowledge on mergers and cluster evolutions.

References

1. Briel, U. G., Henry, J. P., and Boehringer, H. 1992, A&A, 259, L31
2. Evrard, A.E. 1990, Clusters of Galaxies, ed. W.R. Oegerle (Cambridge: Cambridge Univ. Press), 287
3. Honda, H., Hirayama, M., et al. 1996, ApJ, 473, L71
4. Hughes, J.P., Butcher, J. A., Stewart, G. C., and Tanaka, Y. 1993, ApJ, 404, 611
5. Hughes, J.P. 1989, ApJ, 337, 221
6. Hughes, J.P., Gorenstein, P., and Fabricant, D. 1988a, ApJ, 329, 82
7. Hughes, J.P., et al. 1988b, ApJ, 327, 615

THE FATE OF INTRA-GROUP MEDIUM

YUHRI ISHIMARU
National Astronomical Observatory,
2-21-1, Mitaka, Osawa, Tokyo 181, Japan.
email: ishimaru@th.nao.ac.jp

Abstract. Combining X-ray data for clusters of galaxies, groups, and elliptical galaxies, we have obtained the evidence for that groups eject parts of the intra-group medium (IGM) via supernovae-driven 'group winds' like ellipticals. This scenario is confirmed by 1D hydrodynamical simulations.

1. Elliptical-Group-Cluster sequence

Figure 1 shows the relation between gravitational mass $M_{\rm grav}$ and $M_{\rm gas}/L_B$. While the gas mass to galaxy luminosity ratio is almost constant for rich clusters, it decreases with a reduction in gravitational mass for groups smaller than $M_{\rm grav} \sim 5\ 10^{13} M_\odot$. One can clearly identify the sequence from elliptical to clusters (hereafter EGC sequence). The EGC sequence is also established in the relation between $M_{\rm grav}$ and FeML with the same critical mass. If this correlation is due to accretions of baryons, FeMLs must take the universal value. The decreases of gas and iron suggests that smaller groups lose more intra-group medium (IGM). Especially, the EGC sequences are smoothly connected to ellipticals. Thus, this correlation can be interpreted as follows: While clusters have sufficient binding energy to keep the intra-cluster medium throughout their evolution, groups eject parts of the IGM via supernovae-driven 'group winds' like ellipticals. The efficiency of group winds should be negatively correlated with the binding energy, and therefore, the EGC sequence would emerge.

2. Group wind scenario

We examine that group winds can indeed reproduce the EGC sequence under the assumption of the universal initial baryon fraction, using the 1D

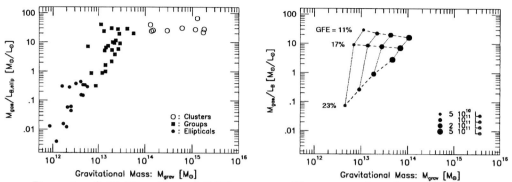

Figure 1. (left) The observed EGC sequence: The open circles, closed squares, and closed diamonds represent clusters, groups, and ellipticals, respectively.

Figure 2. (right) The calculated final gas-to-luminosity ratio as a function of gravitational mass: The total luminosities of each models are $5\ 10^{10}$, 10^{11}, $2\ 10^{11}$, $5\ 10^{11} L_\odot$, which are given by smaller to larger closed circles. The dotted and dashed lines connect models with the identical luminosities and galaxy formation efficiencies, respectively. The values of GFE are put at the side of corresponding lines.

hydrodynamical model. We apply the 1D hydrodynamical scheme for the gas flows in ellipticals (e.g., David et al. 1990; Ciotti et al. 1991) to the groups. The evolution of IGM is examined for groups with $L_B \sim 5\ 10^{10} - 5\ 10^{11} L_\odot$. The initial galaxy fraction is determined by the galaxy formation efficiency GFE= $M_{\rm galaxy}/M_{\rm baryon}$. The observed value of M/L_V; $\sim 70 - 150 M_\odot/L_\odot$ determines the possible range of GFE as $17 \pm 6\%$. Figure 2 shows the relation between the final $M_{\rm gas}/L_B$ ratio and total mass. As shown in this figure, a small dispersion in GFE causes a large dispersion in the efficiency of gas ejection. Because the GFE represent the energy source in a unit gas mass, groups with higher GFE eject more IGM. In addition, smaller groups is more sensitive to the GFE. Therefore, a possible dispersion in GFE easily generate the observed trend and dispersion.

If rich clusters are formed through hierarchical clusterings, the EGC sequence requires that *all* group wind ejecta come back successfully at merging of groups. On the other hand, if each galaxy system has evolved isolately, the EGC sequence can be understood simply by group winds. Moreover, unnatural baryon enhancement in rich clusters is not required.

References

Ciotti, L., D'Ercole, A., Pellegrini, S., & Renzini, A. 1991, ApJ, 376, 380
David, L. P., Forman, W., & Jones, C. 1990, ApJ, 359, 29

DEVELOPMENT OF NEW ANALYSIS METHOD FOR MAPPING OBSERVATIONS OF CLUSTERS OF GALAXIES

K. KIKUCHI, T. OHASHI
Department of Physics, Tokyo Metropolitan University

H. EZAWA, M. HIRAYAMA
Department of Physics, University of Tokyo

AND

H. HONDA, R. SHIBATA
Institute of Space and Astronautical Science

1. Introduction

Mapping observations of nearby large-extended clusters of galaxies (Coma, Perseus, Virgo, etc.) are being performed with ASCA. Such clusters allow us to map physical parameters of hot gas in the clusters, such as temperature, metal abundance, and X-ray surface brightness. To determine such parameters at each part of a cluster, one should take careful care of X-ray contamination from outside of a pointed field, which is mainly due to "stray-light" X-rays (Honda et al. 1997). For this reason, the only way to obtain the distribution of hot gas parameter is to process the whole cluster data in a self-consistent way. For this purpose, we are developing the new analysis system called TERRA.

2. The TERRA System

The new analysis system TERRA (TEchnique of Reproducing the Response for ASCA) is characterized by following three points.

- Response calculation by Monte Calro simulation
- Simultaneous fit to spectra from multi-pointing observations
- Database to store the response simulated

The stray light contamination is evaluated correctly utilizing Monte Carlo simulation with a ray-tracing code. In addition, the system is adapted to multi-pointing observations and performing a simultaneous fit to multiple spectra. Furthermore, the simulated ASCA response is accumulated in the form of a database, so that users do not have to run the same ray-tracing code for each analysis. This brings us a considerable save of time compared with the existing analysis method in which the response has to be calculated by the ray-tracing simulation for every analysis of each observation.

To verify performance of the TERRA system, we fit simulated pulse-height spectra of a cluster placed at various offset angles referred to a pointing direction. Pointing directions were taken to be similar to those of the Coma cluster observations with ASCA (Honda et al. 1997). We fit these simulated spectra jointly with the TERRA system, with kT and normalization factor set free and others fixed at each region. The best-fit values are obtained with $\chi^2 = 1478$ for 1344 degrees of freedom and consistent with the simulated ones. This suggest that the TERRA is able to derive meaningful parameters even when the stray light creates severe contaminations. Also, the plots for large offset simulations indicate that the system predicts the stray light contamination very well.

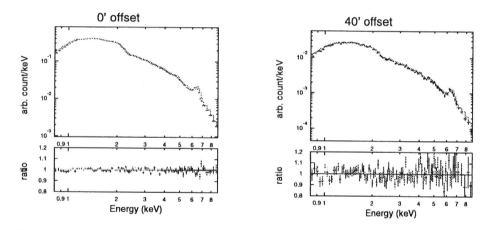

Figure 1. Fitting result by TERRA. Lower panel shows the ratio of simulated data and fitted model.

References

Ezawa, H., Fukazawa, Y., Makishima, K., Ohashi, T., Takahara, F., Xu, H., Yamasaki, Y.Y., (1997) *ApJ Letters*, Nov. 20 in press.

Honda, H., Hirayama, T., Watanabe, M., Kunieda, H., Tawara, Y, Yamashita, K., Ohashi, T., Hughes, J.P., Henry, J.P., (1997) *ApJ*, **473**, L71.

Honda, H., Ezawa, H., Hirayama, M., Kikuchi, K., Ohashi, T., Watanabe, M., Kunieda, H., Tawara, Y, Yamashita, K., (1997) *Proc. ASCA/ROSAT WORKSHOP ON CLUSTERS OF GALAXIES*, p. 153.

CONSTRAINTS ON THE MATTER FLUCTUATION SPECTRUM FROM X-RAY CLUSTER NUMBER COUNTS

TETSU KITAYAMA AND YASUSHI SUTO
Department of Physics, The University of Tokyo
e-mail(TK): kitayama@utaphp2.phys.s.u-tokyo.ac.jp

We find that the observed $\log N$–$\log S$ relation of X-ray clusters (Ebeling et al. 1997; Rosati et al. 1997) can be reproduced remarkably well with a certain range of values for the fluctuation amplitude σ_8 and the cosmological density parameter Ω_0 in cold dark matter (CDM) universes (Kitayama & Suto 1997). The 1σ confidence limits on σ_8 in the CDM models with $n = 1$ and $h = 0.7$ are expressed as $(0.54 \pm 0.02)\Omega_0^{-0.35-0.82\Omega_0+0.55\Omega_0^2}$ ($\lambda_0 = 1 - \Omega_0$) and $(0.54 \pm 0.02)\Omega_0^{-0.28-0.91\Omega_0+0.68\Omega_0^2}$ ($\lambda_0 = 0$), where n is the primordial spectral index, and h and λ_0 are the dimensionless Hubble and cosmological constants. The errors quoted above indicate the statistical ones from the observed $\log N$–$\log S$ only, and the systematic uncertainty from our theoretical modelling of X-ray flux in the best-fit value of σ_8 is about 15%. In the case of $n = 1$, we find that the CDM models with $(\Omega_0, \lambda_0, h, \sigma_8) \simeq (0.3, 0.7, 0.7, 1)$ and $(0.45, 0, 0.7, 0.8)$ simultaneously account for the cluster $\log N$–$\log S$, X-ray temperature functions, and the normalization from the *COBE* 4 year data. The derived values assume the observations are without systematic errors, and we discuss in details other theoretical uncertainties which may change the limits on Ω_0 and σ_8 from the $\log N$–$\log S$ relation. We have shown the power of this new approach which will become a strong tool as the observations attain more precision.

References

Ebeling H., et al. (1997) *MNRAS* submitted
Kitayama, T. and Suto, Y. (1997) *ApJ*, 490, in press
Rosati, P., Della Ceca, R., Norman, C. and Giacconi, R. (1997) *ApJL* submitted

MAGNETIC FIELD AMPLIFICATION AND INTERGALACTIC PLASMA HEATING THROUGH MAGNETIC TWIST INJECTION FROM ROTATING GALAXIES

R. MATSUMOTO
Department of Physics, Faculty of Science, Chiba University, 1-33 Yayoi-Cho, Inage-Ku, Chiba 263, Japan

A. VALINIA
Laboratory for High Energy Astrophysics, NASA/Goddard Space Flight Center, Greenbelt, MD 20771, USA

T. TAJIMA
Institute for Fusion Studies, the University of Texas at Austin, Austin, TX 78712, USA

S. MINESHIGE
Department of Astronomy, Kyoto University, Sakyo-ku, Kyoto 606-01, Japan

AND

K. SHIBATA
National Astronomical Observatory, Mitaka, Tokyo 181, Japan

We propose a mechanism of amplification of magnetic fields and plasma heating in clusters of galaxies. Recent observations indicate the existence of $\sim \mu G$ magnetic fields in clusters of galaxies (e.g., Kronberg 1994). There should be some mechanism which locally amplify magnetic fields. In clusters of galaxies, individual motions of galaxies may create locally strong field region by stretching and tangling the magnetic fields threading the galaxies. Magnetic reconnection taking place in the tangled magnetic fields may convert the kinetic energy of the galaxy motion into the inter-galactic plasma heating (Makishima 1996).

Here we present the results of three-dimensional magnetohydrodynamic simulations of large-scale magnetic fields threading the rotating galaxies. The initial state consists of two, constant angular momentum tori with polytropic index $n = 3$ rotating in a static isothermal halo. The gravita-

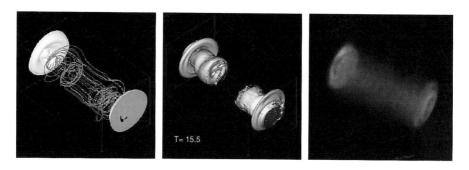

Figure 1. Numerical results for a typical model with $\beta = 100$ at $t = 15.5 r_0/V_{K0}$. The left panel shows the density isosurface and magnetic field lines. The middle panel shows the isosurface of magnetic field strength $B/B_0 = 10$. The right panel is the pseudo X-ray image rendered by using $\rho^2 T^{1/2}$.

tional field is assumed to be given by two point masses at $(r, z) = (0, 0)$ and $(0, z_{max})$ in a cylindrical coordinate. The initial magnetic field is assumed to be uniform and axial ($\mathbf{B} = B_0 \hat{z}$). Symmetric boundary conditions are imposed at $z = 0$ and $z = z_{max}$. We solved the magnetohydrodynamic (MHD) equations by using the modified Lax-Wendroff method with artificial viscosity. Typical number of grid points is $(N_r, N_\varphi, N_z) = (122, 32, 240)$. Figure 1 shows the snap shot of magnetic field lines (left), magnetic field strength (middle), and pseudo X-ray image (right) at $t = 15.5 r_0/V_{K0}$ where r_0 is the reference radius and V_{K0} is the Keplerian rotation speed at $r = r_0$. The initial plasma β ($= P_{gas}/P_{mag}$) at $(r, z) = (0, r_0)$ is $\beta = 100$.

Since torsional Alfvén waves generated by the rotation of galaxies extract angular momentum, the rotating disk infall toward the galactic center. The infalling gas further twists the magnetic fields and bunch them into a twisted filament along the rotational axis. Numerical results show that the magnetic pressure of this filament is comparable to the thermal pressure of the intergalactic medium. When the magnetic twist accumulates, the flux tube deforms itself into a helical structure due to the kink instability. Magnetic reconnection taking place in such kink-unstable flux tube may further heat the intergalactic plasma.

We thank K. Makishima for discussion. Numerical computations were carried out by using VPP300/16R at the Astronomical Data Analysis Center of the National Astronomical Observatory, Japan.

References

Kronberg, P.P. (1994), Rep. Prog. Phys., 57, 325.
Makishima, K. (1996), in Imaging and Spectroscopy of Cosmic Hot Plasmas, eds. F. Makino and H. Inoue, Universal Academy Press, Tokyo, p.137.

MAPPING THE VIRGO CLUSTER OF GALAXIES WITH ASCA

T. OHASHI, K. KIKUCHI, K. MATSUSHITA, N. Y. YAMASAKI,
A. KUSHINO AND ASCA VIRGO PROJECT TEAM
Department of Physics, Tokyo Metropolitan University
1-1 Minami-Ohsawa, Hachioji, Tokyo 192-03, Japan

1. Introduction

In the Virgo cluster, we can perform a close study of the gas injection mechanism from galaxies into the cluster space and the interaction between the injected gas and the sorrouding cluster medium. In 1996 to 1997, we carried out mapping observations of a $2.°5 \times 2.°5$ area in the north-west region of the cluster. There are 16 pointings in total in this region, and the observed results are briefly reported here.

2. Observed Results

The mosaic map of the GIS image (Kikuchi et al. 1997) shows enhanced X-ray emission from bright galaxies. To look into a large-scale distribution of the ICM properties, we examined GIS spectra in each pointed region with $40'$ diameter. Since the cluster emission is only $20-30\%$ brighter than the diffuse X-ray background in the outermost regions ($> 3°$ offset from M87), selection of the background affects the results significantly. Because of the long-term increase of the GIS non X-ray background by about 5% yr^{-1} (Ishisaki 1997), we used black sky data taken in 1997 for the background.

The pulse-height spectra corrected for the background are fitted with thermal models by Raymond and Smith. Figure 1(a) and (b) show distribution of the ICM temprature and metal abundance as a function of the distance from M87. The error bars indicate 90% limits allowing for a $\pm 10\%$ fluctuation of the diffuse background intensity. As seen clearly, the temprature shows a systematic drop with radius and becomes $50-60\%$ of the central level at $r \sim 1$ Mpc. The results are not yet corrected for the stray light which tends to smear the temperature gradient.

The metal abundance shows a narrow peak centered at M87 (Matsumoto et al. 1996) and then shows a gradual drop in large scale. It is $0.2 - 0.3$ solar within 400 kpc from M87 and less than 0.2 solar at $r > 600$ kpc. Note that the abundance at this temperature is mainly detemined by Fe-L lines, for which discrepancy exists among theoretical models.

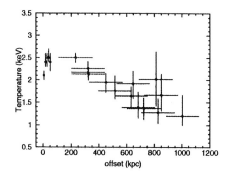

Figure 1. ICM Temprature as a function of distance from M87

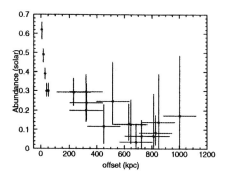

Figure 2. Distribution of metal abundance

3. Discussion

The temperature drop in the outer region of clusters is predicted from numerical simulations (e.g. Eke et al. 1997), and recently observed as an average properties of about 30 clusters by Markevitch (1997). If temerature gradient is steeper in clusters which are in the early stage of gravitational collapse (i.e. outer region is not heated enough), this would be an additional indication that the Virgo cluster is a young system (Böhringer et al. 1994). It is intersting that other systems such as AWM7 and Perseus cluster show more uniform temperature distribution than the Virgo cluster.

The abundance gradient is roughly consistent with the galaxy distribution as already seen in AWM7 (Ezawa et al. 1997). This suggests that metals (Fe) are mostly injected from the present population of galaxies and no strong mixing has occured in the ICM.

References

Böhringer, H. et al. (1994) *Nature*, **368**, 828
Eke et al. (1997) submitted to *ApJ*, (astro-ph/9708070)
Ezawa et al. (1997) *ApJ Letters*, Nov. 20 in press
Ishisaki, Y. (1997) Ph. D. Thesis, University of Tokyo (ISAS RN-613)
Kikuchi, K. et al. (1997) *ASCA/ROSAT Workshop on Clusters of Galaxies*, p. 221
Markevitch, M. (1997) in preparation
Matsumoto, H. et al. (1996) *PASJ*, **48**, 201

ASCA OBSERVATIONS OF THREE GRAVITATIONAL LENSING CLUSTERS OF GALAXIES; CL0500-24, CL2244-02, AND A370

NAOMI OTA, KAZUHISA MITSUDA

Institute of Space and Astronautical Science

AND

YASUSHI FUKAZAWA

Department of Physics, University of Tokyo

Abstract. We determined the X-ray temperatures of three gravitational lensing clusters, CL0500-24, CL2244-02, and A370, and obtained significant constraints on the surface brightness profile assuming the β-model and the King model profiles. The mass of the cluster estimated from these X-ray data is by a factor of two to three smaller than the mass estimated from lens models for two of the clusters.

It is suggested by several authors that the X-ray measurements gives systematically smaller masses than the lens masses (e.g. [1]). In order to estimate mass from X-ray observations, both the X-ray temperature and the X-ray surface-brightness profile are necessary. However, these quantities have been directly measured only for a limited number of lensing clusters. With ASCA we have determined the X-ray temperatures and constrained the surface-brightness profile of three lensing clusters. The results of spectral fits, spatial fits and mass estimates are summarized in Tables 1, 2, and 3. For the two of the clusters, CL0500-24 and A370, the X-ray masses were found to be significantly smaller than the mass estimated from lensing models. The discrepancy is statistically significant and by factor of two to three. On the other hand consistent X-ray and lensing masses were obtained for CL2244-02. We suggest that the major cause of mass discrepancy is in the substructures of the two clusters. The results are presented in [2] in more detail.

References

1. Wu, Z-P. & Fang, L-Z. 1997, ApJ, **483**, 62.
2. Ota, N., Mitsuda, K., & Fukazawa, Y. 1998, to appear in ApJ, March 1998 issue.
3. Wambsganss, J., Giraud, E., Schneider, P., & Weiss, A. 1989, ApJ, **337**, L73.
4. Kovner, I 1989, ApJ, **337**, 621.
5. Kneib, J.-P., Mellier, Y., Fort, B., & Mathez, G. 1993, A&A, **273**, 367.

TABLE 1. Results of Spectral Fits

Cluster	$N_H [\text{cm}^{-2}]$[†]	$kT[\text{keV}]$	$L_X(2-10)[\text{erg/s}]$	Abundance	χ^2 (dof)
CL0500-24	2.6×10^{20}	7.2(5.4–10.9)	1.7×10^{44}	< 1.5	24.7 (26)
CL2244-02	4.8×10^{20}	6.5(5.2–8.3)	1.3×10^{44}	< 0.2	72.2 (58)
A370	3.1×10^{20}	6.6(5.7–7.7)	8.3×10^{44}	0.3 ± 0.2	118.4 (124)

[†] N_H is fixed at the galactic value. The quoted errors correspond to a single parameter error at 90% confidence.

TABLE 2. Results of β model fits

Cluster	β-model profile			King-model profile	
	β	$r_c[\text{Mpc}]$[†]	χ^2 (dof)	$r_1[\text{Mpc}]$[†]	χ^2 (dof)
CL0500-24	0.9 (0.6–1.4)	0.41 (0.17–0.62)	61.9 (43)	0.44 (0.34–0.58)	63.9 (44)
CL2244-02	0.30 ± 0.05	< 0.10	72.1 (45)	1.3 (0.9–1.5)	90.0 (46)
A370	0.95 (0.6–1.7)	0.48 (0.26–0.86)	42.9 (45)	0.52 (0.41–0.60)	44.2 (46)

[†]1 arc minute corresponds to 0.342, 0.348, and 0.373 Mpc for CL0500-24, CL2244-02, and A370, respectively for the Cosmological parameters of $\Omega_0 = 1, \Lambda = 0$, and $H_0 = 50$ km/sec/Mpc.

TABLE 3. Comparison of Projected Mass of Clusters inside the arc radius

Cluster	$\theta_{\text{arc}}/r_{\text{arc}}$ ["/Mpc]	X-ray Mass [$\times 10^{13} M_\odot$]		Lens Mass [$\times 10^{13} M_\odot$]	
		β-model	King-model	Spherical	Detailed [ref]
CL0500-24	22/0.13	4.0 (2.6–8.4)	3.8 (2.3–7.5)	13	13 [3]
CL2244-02	9.9/0.057	1.6 (0.9–2.3)	0.27 (0.18–0.46)	2.1	1.5 [4]
A370	25/0.16	5.0 (3.9–7.2)	4.7 (3.6–6.80)	26	17 [5]

EVOLUTION OF X-RAY CLUSTERS OF GALAXIES WITH ACTIVE PROTOGALAXIES

H. SAGA, S. YACHI AND A. HABE
Graduate School of Science, Hokkaido University, Sapporo, JAPAN

1. Introduction

We consider heating due to proto galaxies in the formation process of clusters of galaxies, since much metal is observed in the intracluster gas which must be ejected from protogalaxies vir strong galactic winds and the metal abundance in the intracluster gas correlates with the fraction of early type galaxies in clusters (Arnaud 1993). We also consider radiative cooling. From the difference between the observed $L_X - T_X$ relation of X-ray cluster (Hatsukade 1989), $L_X \propto T_X^{2.7-3.3}$, and the prediction from the self-similar model (Kaiser 1986), $L_X \propto T_X^2$, it is pointed out that physical processes which are not taken into account in the self similar model, e.g., effect of radiative cooling, and/or effect of proto galaxy heating, play a important role in the formation process (Evrard and Henry 1991, Kaiser 1991).

2. Model and Numerical Results

We assume that the galaxy formation begins from $z = 3$ for $10^{8.5}$yr at density maxima selected by the friend-friend algorithm at z=3. During the galaxy formation, supernovae explode and add metal and heat energy to gas in these regions. Our numerical calculation code is N-body code by using the GRAPE3A for dark matter and SPH code for gas motion. Numbers of both particles are 50000.

From our numerical results of the non-heating models, power index of the $L_X - T_X$ relation is 2.3, which is larger than the self-similar model prediction. This power index value is well agree with the numerical results by Kang et al. (1994). The power index value of the non-heating models is larger than the power index value of the self-similar model. From the heating models of $\epsilon = 0.4$ the power index of $L_X - T_X$ is 2.6, which is larger than the non-heating model results, and are close to the observed power index of the $L_X - T_X$ relation, where $\epsilon = 0.4$ correspond to heating due to about half enegy released by supernova in protogalaxies if the initial mass function is the Salpeter' one. Considering radiative cooling in the heating model, the power index of $L_X - T_X$ is 2.2,

3. Discussion

If we assign metal abundance of intracluster gas, we can estimate the formation efficiency of galaxies and the heating efficiency from our numerical results.

From our numerical results, we suggest that (1) high z X-ray luminous clusters are metal rich, and (2) small clusters are metal abundant. These properties are consistent with GINGA data (Arnaud 1990), although ASCA data is different from it (Ohashi 1995).

Radiative cooling reduces the effect of heating by protogalaxies, especially for small clusters, since small clusters have low temperature of gas. As a result, the $L_x - T_x$ relation becomes similar to the non-heating case. Additional heating, such as heating due to type I supernovae after initial star burst, can make a steep $L_x - T_x$ relation.

references

Arnaud,M, 1994, in Cosmological Aspects of X-ray clusters of galaxies, ed. by W,C,Seitter, Kluwer Academic Publishers, p 197.
Evrard,A.E. and Henry J.P., 1991, Ap.J. 383, 95.
Hatsukade I, 1989, Phd. thesis in ISAS research note 435.
Kaiser,N. 1986, M.N., 222, 323.
Kaiser,N. 1991, Ap.J., 383, 104.
Kang, H. et al., 1994, Ap.J., 430, 83.
Ohashi, T., 1995, private communication

ASCA OBSERVATION OF A1674

Detection of Metal-Free (Primordial) Hot Gas ?

T. TAKAI, K. HAYASHIDA AND K. HASHIMOTODANI
Dept. of Earth & Space Science, Faculty of Science, Osaka University
1-1 Machikaneyama-cho Toyonaka-shi Osaka Japan

AND

W. KAWASAKI
Dept. of Astronomy, School of Science, University of Tokyo
7-3-1 Hongo Bunkyo-ku Tokyo Japan

1. Introduction

A cluster of galaxies Abell 1674 is a nearby cluster($z=0.106$) and an unique sample among Briel & Henry's(1993, A&A 278,379) catalogue. Although it has the largest number of galaxies within the Abell radius, 165, its X-ray luminosity measured in the ROSAT all-sky survey is 5×10^{43}erg/s (in 0.5-2.5keV), about one order of magnitude lower than the brightest one.

2. X-ray Images

A1674 was observed by ASCA in the AO-4 phase with 60ksec observation time. X-ray emitting gas is extending for diameter of 6arcmin, corresponding to 1Mpc. X-ray luminosity is 5.9×10^{43}erg/s(0.5-2.5keV), 1.25×10^{44}erg/s(0.5-10.0keV), which is consistent with ROSAT observation.

3. X-ray Spectrum

We fitted the X-ray spectrum with Raymond-Smith model (with absorption of our galaxy, hydrogen column density of 1.9×10^{20}cm^{-2} fixed) and found the gas temperature of $3.19^{+0.62}_{-0.50}$keV and the metal abundance of $0^{+0.20}$ solar. This abundance is very lower than typical metal abundance of about 0.4 solar for other clusters. Although we have tried 2-temperature model or introduced an excess absorption, the best fit parameter for the metal abundance did not change.

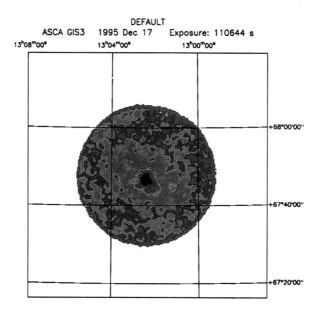

Figure 1. ASCA GIS image of A1674

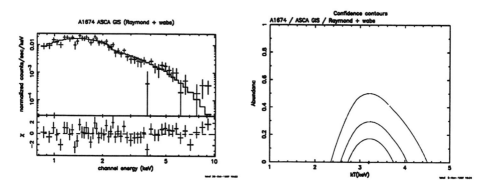

Figure 2. ASCA GIS spectrum of A1674

Figure 3. Confidence Contour

UNCERTAINTY IN MASS DETERMINATION OF GALAXY CLUSTERS DUE TO THE BULK MOTION OF INTRACLUSTER MEDIUM

M. TAKIZAWA AND S. MINESHIGE
*Department of Astronomy, Faculty of Science, Kyoto University,
Sakyo-ku, Kyoto 606-01,JAPAN*

1. Introduction

Determination of the mass of cluster of galaxies (CG) is very important mainly because it relates to determination of cosmological density parameter (Ω_0).

However, mass obtained from gravitational lensing tends to be larger than that from X-ray observation with the assumption of hydrostatic equilibrium (HSE) in some CGs (Miralda-Escudeé & Babul 1995, Schindler et al. 1997). We suggest one of the reason for this discrepancy is that the assumption of HSE is inaccurate because bulk motion of ICM is left. We, therefore, performed the simulation of spherical CG consisting of dark matter and gas.

2. Methods of Analysis

We use the model cluster in Takizawa & Mineshige (1997). Cosmological model is Einestain de Sitter ($\Omega_0 = 1, \Lambda_0 = 0$). Spherically symmetric is assumed.

We derive two kind of the estimated mass from the model cluster. One is the mass derived from the density profile and the temperature profile, which we call $M_{\mathrm{est},1}$. The other is the that from the density profile and the emissivity weighted temperature, which we call $M_{\mathrm{est},2}$. Therefore,

$$M(r)_{\mathrm{est},1} = -\frac{kT(r)r}{G\mu m_{\mathrm{p}}}\left[\frac{d\ln n}{d\ln r} + \frac{d\ln T}{d\ln r}\right], \qquad (1)$$

$$M(r)_{\text{est},2} = -\frac{kr}{G\mu m_{\text{p}}} T_{\text{ew}} \frac{d\ln n}{d\ln r}. \qquad (2)$$

We will compare $M_{\text{est},1}$ and $M_{\text{est},2}$ to the true mass, M.

3. Results and Discussions

Figure 1 shows the radial profiles at $z = 0$ of $M_{\text{est},1}/M$ (solid line) and $M_{\text{est},2}/M$ (dotted line), respectively.

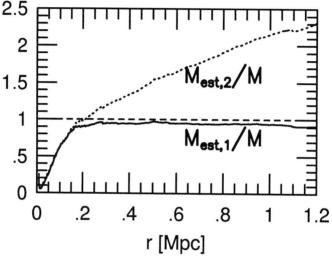

Figure 1. Radial profiles of $M_{\text{est},1}/M$ (solid line) and $M_{\text{est},2}/M$ (dotted line), respectively.

$M_{\text{est},1}/M$ is about 0.95 except in the central region. Therefore the effect of the deviation from HSE is not important. On the other hand, the behavior of $M_{\text{est},2}/M$ is rather different, which is monotonically increasing outwards. This is mainly due to the temperature gradient. Therefore, in the case of spherical symmetry, the effect of bulk motion of ICM is not very importance. Note that non isothermality of ICM can have great influence.

According to our results, the observational discrepancy cannot be explained solely due to the effect of bulk motion of ICM. The deviation from spherical symmetry (substructure, projection effect) and uncertainty of modeling of gravitational lensing should be considered in more detail.

References

Miralda-Escudeé, J. & Babul, A. 1995, ApJ, 449, 18
Schindler, S., Hattori, M., Neumann, D. M. & Böhringer, H. 1997, A&Ap, 317, 646
Takizawa, M. & Mineshige, S., 1998 accepted for publication in ApJ, astro-ph/9702047

ASCA OBSERVATION OF GROUPS OF GALAXIES

Y. TAWARA, S. SATO, A. FURUZAWA AND K. YAMASHITA
Dept. of Physics, Nagoya University, Furo-cho, Chikusa, Nagoya 464-01, Japan

K. ISOBE
Toshiba Corporation, Shibaura 1-1-1, Minato, Tokyo 105-01, Japan

AND

Y. KUMAI
Kumamoto-gakuen University, Ohe 2-5-1, kumamoto,862, Japan

1. X-ray Properties of Group of Galaxies

Based on the ASCA observations, the X-ray features like spatial extentions and spectral properties of their component were analyzed for nearby four compact groups of galaxies; HCG62, NGC2300 group, HCG42 and HCG48. We found wide variety in their X-ray features. One of the brightest source of HCG62 shows the presence of cool component and enhanced abundance at its center, while HCG42 shows less extention and low abundance ratio of α-element to iron. For NGC 2300 group, we also found interesting feature which may be related to galaxy-intragroup medium interaction.

2. Abundance and Group Environment

There is wide variety of the hot gas property in the present group samples. To consider this point, we first derived stellar mass, gas mass and total gravitating mass for each group under the spherically symmetric distribution of gas. The results are summarized in the table 1, together with values for typical cluster of galaxies. We also derive the indicator of the strength of galactic wind relative to gravitational binding force. If we normalize this value by that for HCG 62, then the same value for NGC 2300, 10 times larger value for HCG 42 and 10 times lower value for typical cluster are obtained. Then very low gas mass or X-ray luminosity or peculiar abundance can be interpreted by strong galactic wind, which can blow out their gas out side the group potential region.

E_{wind} $L_B(E,So)$: galactic wind energy
E_{bind} $M_{total} M_{gas} / R$: binding energy for gas

Sysytem	L_B ($10^{11} L_o$)	M_{galaxy} ($10^{12} M_o$)	R (kpc)	M_{gas} ($10^{12} M_o$)	M_{total} ($10^{12} M_o$)	E_{wind} / E_{bind} (HCG62--->1.0)	kT (keV)	Ab(Fe) (Solar)	Ref
Cluster	130	65	3000	300	1170	0.1	2-10	0.3	a
NGC3923	0.49	0.39	16	0.0001	1.2	3600.	0.67	0.12	e
NGC4636	0.29	0.27	350	0.6	10	0.9	0.75	0.93	d
HCG62	1	0.8	300	1.4	12	1.0	0.86	0.26	a
N2300 Gr.	0.5	0.6	250	0.6	10	1.2	0.87	0.16	a
HCG51	1.9	1.5	370	1.4	20	1.4	1.1	0.25	b
WP23	2.3	1.6	350	1.5	15	2.0	0.85	0.29	b
HCG42	3	2	120	0.2	10	10.0	0.36	-	a
HCG92	1	0.5	80	0.02	6	37.0	0.76	0.08	c

a) this work, b) Fukazawa et al. '96, c) Awaki et al. '96, d) Matsushita et al. '97, e) Sato et al. '97

Table 1. Relative Strength of Galactic Wind

3. References

Fukazawa, Y. et al. 1996, PASJ, 48, 395
Awaki, H. et al. 1997, PASJ, 49, 445.
Matsushita, K. et al., 1997, Ap. J., 488, L125.
Sato, S. et al., 1997, Ap. J. sumitted.

MATTER DISTRIBUTION IN THE GALAXY CLUSTERS A 539 AND A 2319

D. TRÈVESE, G. CIRIMELE AND M. DE SIMONE
Istituto Astronomico, Università di Roma "La Sapienza"
Via G.M. Lancisi 29, Roma I-00161

We performed a combined X-ray and optical analysis of the two clusters A539 and A2319, based on ROSAT PSPC 0.4-2.4 keV images of the public archive and F band photometry from microdensitometric scans of Palomar 48 inch plates (Trèvese et al. 1992, A&AS, 94, 327). Assuming spherical symmetry and following the methods adopted in Cirimele, Nesci, and Trèvese (1997, ApJ, 475, 11 (CNT97)) we derived the radial distribution of gas and galaxy densities ρ_{gas} and ρ_{gal} and we have computed the morphological parameter $\beta_{xo} \equiv d\ln\rho_{gas}(r)/d\ln\rho_{gal}(r)$, introduced in CNT97. This allows to check the validity of the hydrostatic equilibrium condition, which reads $\beta_{spec}(r) \equiv \frac{\mu m_p \sigma_r^2}{kT(r)} = \beta_{xo}(r) + d\ln T/d\ln\rho_{gal}$, for an isotropic and uniform velocity distribution of r.m.s. dispersion σ_r. In the case of A539, adopting σ_r=629 km s^{-1} from Fadda et al. (1996, ApJ, 473, 670) and T=1.57 keV David et al. (1996, ApJ, 473, 692), we obtained marginally consistent values of β_{spec}=1.54±0.50 and β_{xo}=1.08±0.11. In the case of A2319 we took into account the presence of the secondary component A2319B (Oegerle et al. 1995, AJ, 110, 32) and the temperature gradient (Markevitch M. 1996, ApJ, 465, L1). The resulting radial increase of β_{spec} is consistent with that of $(\beta_{xo}(r) + d\ln T(r)/d\ln\rho_{gal})$, suggesting that the hydrostatic equilibrium holds also in the presence of a temperature gradient. The radial distribution of the total binding mass, the mass in galaxies and intergalactic gas show that in both clusters the gas mass profile is steeper than galaxies and total masses consistently with our previous results (CNT97). Adopting a constant gas temperature, the relevant baryon fractions are larger than 20 %, adding new evidence to the "baryon catastrophe". Taking into account the radial decrease of gas temperature, the baryon fraction is further increased. This implies that either $\Omega_o < 0.25$, or that large halos of dark matter surround galaxy clusters, as suggested by White & Fabian (1995, MNRAS, 273, 72).

SUNYAEV-ZEL'DOVICH EFFECT OBSERVATION PROJECT WITH THE NOBEYAMA 45-M TELESCOPE

M. TSUBOI, T. OHNO AND A. MIYAZAKI
Ibaraki University, Mito, Ibaraki, 310, Japan

T. KASUGA
Hosei Univeristy, Koganei, Tokyo, 184, Japan

AND

A. SAKAMOTO AND T. NOGUCHI
Nobeyama Radio Observatory, Nagano, 384-13, Japan

1. Introduction

The combination of X-ray and Sunyaev-Zel'dovich (S-Z) effect observations toward cluster of galaxies gives the independent estimation of Hubble constant (Sunyaev and Zel'dovich 1970). The measurement of S-Z effect is one of the most difficult observations in radioastronomy because of the weakness of the effect, $\Delta T = 0.1 - 1$ mK (e.g. Rephaeli 1995). Because the field of view of the exist interferometers is smaller than the extended distribution of S-Z effect of low redshifted clusters, single-dish telescopes gain an advantage over interferometers. In addition, to reduce the contaminations from Galaxy and galaxies in the cluster, the mm-wave observation is preferable. Thus, we have started the project of S-Z effect observation with the Nobeyama 45-m telescope, which is the largest mm-wave telescope in the world. Our scientific goal is reliable measurement of S-Z effect of many clusters. To realize this we have made a multi-feed PCTJ-SIS mixer receiver at 40 GHz as a sophisticated tool for the observation of S-Z effect (Noguchi et al. 1995, Kasuga et al. 1995, Tsuboi et al. 1997).

2. Receiver System

The receiver system consists of 3 dual feeds differential radiometers in order to reject weather and ground effects. Scanning observation can depict

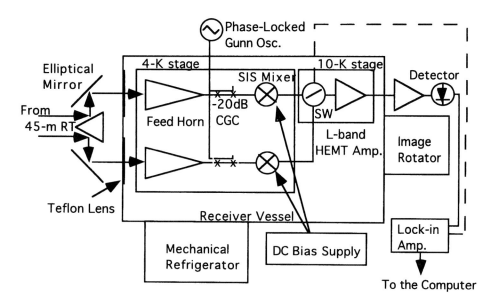

Figure 1. The block diagram of the receiver system

the spatial distribution of S-Z effect. Figure 1 shows the block diagram of the receiver system. The SIS mixers and the horn antennas are located in the Dewar and cooled to 4 K by a mechanical refrigerator. The low-noise amplifiers are on 15-K stage in the Dewar. The receiver noise temperature is below 60 K and the instantaneous bandwidth is 700 MHz. The beam size is 34". The neighboring incident beams are divided by the image dicing mirror on the focal plane. The beams are located at 2×3 grid with 90" interval on the sky. The system was installed on the focal plane platform of the 45-m telescope in May 1997. Now, only one beam is available. The beam efficiency and the aperture efficiency are 0.5 and 0.8, respectively. We are focalizing other beams of the receiver system for the test observation in the winter of 1997-8.

References

Kasuga, T., Tsuboi, M., Miyazaki, A., Noguchi, T., and Sakamoto, A. (1995) *URSI General Assembly*
Noguchi, T., Shi, S. -C., and Inatani, J. (1995) *IEEE Trans. Appl. Superconductivity*, **Vol.5**,, pp 2228-2231
Rephaeli,Y. (1995) *Ann Rev. Ann. Astrophys.*, **Vol.33**, p541
Sunyaev, R. A. and Zel'dovich, Y. B. (1970) *Astrophys. and Space Sci.*, **Vol.7**, p3
Tsuboi,M., Miyazaki,A, Kasuga,T.,Sakamoto, A, and Noguchi, T. (1997) *TWAA'96*, pp99-103

Session 2: Future Space Programs

DEVELOPMENT OF STJ AS A NEW X-RAY DETECTOR

N. Y. YAMASAKI, T. OHASHI, K. KIKUCHI, H. MIYAZAKI,
E. ROKUTANDA AND A. KUSHINO
Dept. of Phys., Tokyo Metropolitan Univ.
1-1 Minami-Ohsawa, Hachioji, Tokyo 192-03, Japan

AND

M. KURAKADO
Advanced Technology Research Laboratories,
Nippon Steel Corporation
1618 Ida, Nakahara-ku, Kawasaki, Kanagawa 211, Japan

1. Introduction

STJs are promising X-ray detectors as high energy resolution spectrometers due to the small excitation energy to break the Cooper pairs to product detectable electrons. The expected energy resolution is about 5 eV for a 6 keV incident X-rays (see review by Kraus et al. and Esposito et al.). We have developed a large area ($178 \times 178 \mu m^2$) Nb/Al/AlO$_X$/Al/Nb STJs (Kurakado et al. 1993) and series-connected STJs with a position resolution of $35 \mu m$ for α particles (Kurakado 1997) at Nippon Steel Corporation. As a focal plane detector in future X-ray missions, we are developing STJs whose targert characteristics are; an energy resolution of 20 eV at 6keV, an effective area of 1 cm^2, and position resolution of $100 \mu m$.

2. Experiment at TMU

A measurement system for STJs with an X-ray generator attached to a ^{3}He cryostat to be cooled to 0.35 K is fabricated at Tokyo Metropolitan University. Using this system, we succeed to detect X-ray signals from both single and series-connected STJs.

From a single STJ detector of $178 \times 178 \mu m^2$, an energy resolution of 112 eV at 5.9 keV including 95 eV electrical noise is obtained (see Fig.1). Utilizing the X-ray generator, we measure the X-ray energy vs. pulse height relation for 6 incident X-ray energy at the same condition. We found that the energy scale is well represented by a parabolic function(see Fig.2), because the recombination rate of quasi-particles in the superconductor is proportional to the power of the density of the pseudo particles, i.e. the square of the incident energy.

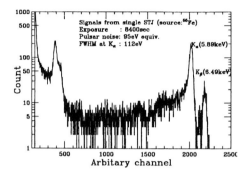

Figure 1. The energy spectrum obtained by a STJ

Figure 2. The X-ray line energy vs pulse height

3. Future developments

To obtain the good energy resolution, it is necessary to reduce the readout noise. Frank et al. obtained a good energy resolution (29 eV at 5.9keV) with a SQUID readout. Recombination loss of the pseudo particles in the superconducting layer drops the signal current. Multi-trap structure (for example, Nb/Ta/al/AlO$_X$/Al/Ta/Nb) to gain the efficient tunneling of the quasi-particles is proposed (Kurakado 1997). And we are starting the position sensing of X-rays with series-connected STJs analyzing the rise-time and pulse height of each events.

References

Esposito, E., Frunzio, L., Parlato, L., Barone, A. (1996) Superconductive tunnel junction detector: ten years ago, ten years from now, *NIM*, **A370**, 26

Frank, M., Mears, C.A., Labov, S.E., Azgui, F., Lindeman, M.A., Hiller, L.J., Netel, H., Barfknecht A., (1996) High-resolution X-ray detectors with high-speed SQUID readout of superconducting tunnel junctions, *NIM*, **A370**, 41

Kraus, H., Jochum J., Kemmather, B., Gutsche, M., (1993) Progress on Detectors with Superconducting Tunnel junctions, *SPIE* **2006**, 211

Kurakado, M., Takahashi, T., Matsumura, A., (1993) High Resolution Detection of X-rays with a Nb-based STJ Detector, *Journal of Low Temperature Physics*, **93**, 567

Kurakado, M. (1997) Further developments of series connected superconducting tunnel junction to radiation detection, *Rev. Sci. Instr*, in press

MULTILAYER X-RAY OPTICS FOR FUTURE MISSIONS

K. YAMASHITA, H. KUNIEDA, Y. TAWARA, AND K. TAMURA
Department of Physics, Nagoya University
Furo-cho, Chikusa-ku, Nagoya 464-01, Japan

1. Introduction

Multilayers have a great potentiality to improve the image quality, spectral resolution and energy coverage of x-ray optical systems. The angular resolution of a normal incidence telescope aims at approaching the diffraction limit in the soft x-ray region. Multilayer supermirror makes it possible to fabricate a grazing incidence telescope with high sensitivity in hard x-ray region. Multilayer coated gratings are also useful dispersive elements with high efficiency and spectral resolution in the 2-10keV region. The application of multilayers is expected to open up a new field in astronomical imaging and spectroscopic observations which are not accessible by present telescopes.

2. Normal incidence telescope

Mo/Si multilayered spherical mirror with 20cm in diameter and 30cm in focal length was fabricated, which was on board a sounding rocket for the observation of a hot white dwarf(HZ43) and hot interstellar medium in the direction of north galactic pole within the field of view of 4 deg.[1]. The wavelength bands were tuned at 130-140A and 170-180A. The peak reflectivity was obtained to be 60%. For future missions we have to develop a multilayered normal incidence telescope with reflectivity of more than 30% in 50-80A band.

3. Grazing incidence telescope

In order to extend the energy region to hard x-rays, Pt/C supermirrors(depth-graded multilayer) were successfully deposited on float glass and replica foil mirrors used for Astro-E[2]. The reflectivity was obtained to be 30% in 24-37keV band at the incidence angle of 0.3 deg., as shown in figure 1. The

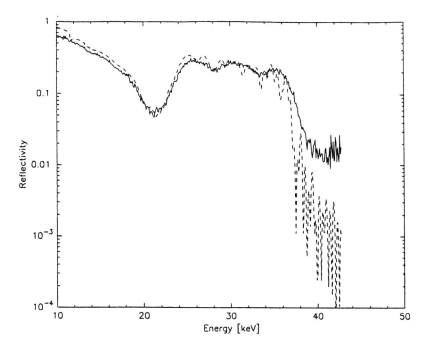

Figure 1. X-ray reflectivity of Pt/C supermirror at the incidence angle of 0.3deg.

supermirror consists of five different multilayers with combinations of periodic length and number of layer pairs of (106, 1), (51-49, 4), (43, 8), (38, 13), (35, 18) and (33, 25) from outermost layer to substrate. This fact promises that supermirror is a real break-through to construct a hard x-ray telescope.

4. Spectrometer

Pt/C(2d=100Å, N=10) multilayer coated laminar gratings(500 grooves/mm, groove depth 400Å) were fabricated and characterized with Cu-K_α (8.04keV) and monochromatized synchrotron radiation in 1-3keV[3]. The first order peak reflectivity and resolution were obtained to be 20% and more than 100 for Cu-K_α, respectively. We are aiming at making a spectrometer with $E/\Delta E=1000$ in 2-10keV region and even with imaging capability.

References

[1] Yamazaki, T. et al., (1996), *J. Elect. Spec. Rel. Phenom.*, **80**, 299
[2] Tamura, K. et al., (1997), *Proc. SPIE*, **3113**, 285
[3] Ishiguro, E., et al., (1995), *Rev. Sci. Instrum.*, **66**. 2112

DIRECT MEASUREMENT OF THE CCD WITH A SUBPIXEL RESOLUTION BY USING A NEW TECHNIQUE

K. YOSHITA, S. KITAMOTO, E. MIYATA AND H. TSUNEMI
Graduate School of Science, Osaka University
1-1 Machikaneyama-cho, Toyonaka, Osaka 560, Japan

AND

K. C. GENDREAU
Goddard Space Flight Center

1. Introduction

The response function of the front-illuminated CCD, like SIS on ASCA, can be devided into three parts: the gate structure transmission, the absorption efficiency in the depletion region and the charge spreading after the photo-absorption. These effect depend on both the incident X-ray energy and the landing posion of the X-ray inside the CCD pixel. Then, the measurement of the X-ray efficiency within the pixel is important for the response function of the CCD. By using a new technique, we performed an experiment using the SIS CCD chip in the GSFC in order to measure how the various types of events are formed.

2. Experimental Setup

We placed a copper mesh in front of the CCD to restrict the landing position of the X-ray inside the CCD pixel. The CCD we used has 420×420 pixels of 27μm square. The mesh has 925×925 holes at 27μm intervals. The holes are circular and about 4μm in diameter. The X-ray beam line we used is a 40m long tube with an X-ray generator in the GSFC.

The alignment between the mesh and the CCD is slightly titled, which creates a moire pattern. Refering to the moire pattern, we can identify the mutual alignment between the mesh and the CCD. Then, we can determine the input position of the X-rays within the pixel.

3. Data Analysis

If we know the input position of the X-rays within the pixels, we can reconstruct the pixel image. We reconstructed the pixel image for single pixel events which are formed when the X-ray is photo-absorbed well inside the pixel. We drew the image of 3×3 pixels as shown in the Figure 1 to see the pixel boundary clearly. The dark regions in it show the regions where there are very few single pixel events. These correspond to the pixel boundaries.

We also reconstructed for vertically split events, horizontally split events, L/square events which are formed when the incident X-ray is photo-absorbed near the pixel boundary. Figure 1, 2, 3 and 4 is the reconstructed image of Y-L X-ray(1.9keV) data. We can clearly see how the various types of events are formed by the landing position inside the pixel.

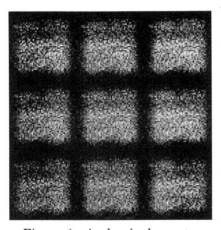

Figure 1. single pixel events

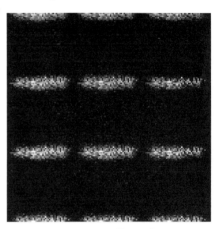

Figure 2. vertically split events

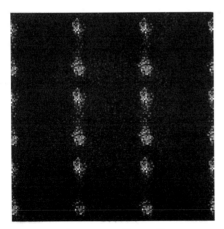

Figure 3. horizontally split events

Figure 4. L/square events

FINAL PERFORMANCES OF THE X–RAY MIRRORS OF THE JET–X TELESCOPE

E. PORETTI, S. CAMPANA, O. CITTERIO, P. CONCONI, M. GHIG
F. MAZZOLENI AND G. TAGLIAFERRI
Osservatorio Astronomico di Brera
Via E. Bianchi, 46 – 23807 Merate, Italy

G. CUSUMANO AND G. LA ROSA
Istituto di Fisica Cosmica CNR
Via U. La Malfa, 153 – 90146 Palermo, Italy

W. BRÄUNINGER AND W. BURKERT
Max Planck Institut
Giessenbachstrasse – 85748 Garching, Germany

AND

C.M. CASTELLI AND R. WILLINGALE
Dept. Physics and Astronomy, Leicester University
University Road – Leicester, United Kingdom

The Joint European X–ray Telescope (JET–X) is one of the core scientific instruments of the SPECTRUM RONTGEN-γ astrophysics mission. The project is a collaboration of British, Italian and Russian consortia, with the participation of the Max Planck Institut (Germany). JET–X was designed to study the emission from X–ray sources in the band of 0.3–10 keV. Citterio et al. (1996 and references therein) describe its structure, composed by two identical and coaligned Wolter I telescopes. Focal plane imaging is provided by cooled X–ray sensitive CCD detectors which combine high spatial resolution with good spectral resolution, including coverage of the iron line complex around 7 keV at a resolution of $\Delta E/E \sim 2\%$.

To measure the effective area of the telescopes, the detectors (PSPC or CCD) were either directly exposed to the incoming X–ray beam or put in the focal plane collecting the photons reflected by the telescopes. The ratio between the two exposures gave the effective area values. The mean values obtained by using the two flight models were 161 cm^2 at 1.5 keV and 69 cm^2 at 8 keV, respectively. These values match very well the theoretical

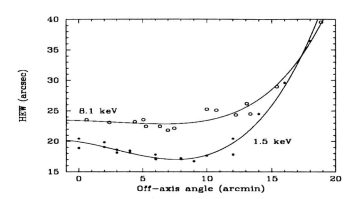

Figure 1. The results of the end–to–end tests at the Panter facility confirm that the required HEW is achieved also at large off–axis angles.

ones (see Figure 3 in Citterio et al. 1996). Another test, concerning the evaluation of the angular resolution, was carried out using a CCD detector in the focal plane; after half the exposure time the detector was moved by a distance equivalent to 20". The two sources are well resolved on the resulting CCD image, confirming the excellent angular resolution achieved by JET–X (see Figure 2 in Citterio et al. 1996).

Crowded field regions and extended X–ray emitting sources represent key targets for the JET–X telescopes. Thus, it is required that the Half Energy Width (HEW) remains good also at large off–axis angles. To illustrate the JET–X capabilities, we plot in Fig. 1 the HEW of one full telescope as a function of the off–axis angle, as measured during the end-to-end test at the Panter facility in München. As it can be seen, up to 12' at 1.5 keV the HEW is below ~ 20" (filled circles) and at 8.1 keV it is almost constant at a level of ~ 23" (open circles). Beyond this off-axis angle, the mirror figuring errors dominate, producing almost the same HEW at all energies. The best HEW at 1.5 keV is obtained at $\sim 8'$ off-axis, due to a small displacement in the JET–X focus, introduced to improve the off-axis response. As a general result, the end–to–end calibration tests provided a successful confirmation of all the JET–X performances.

References

Citterio O., et al. (1996), Characteristics of the flight model optics for the JET–X telescope on board the SPECTRUM X-Gamma satellite, Proc. SPIE 2805, 56

Session 3: Diagnostics of High Gravity Objects with X- and Gamma Rays

3-1. White Dwarfs and Neutron Stars

SPIN-DOWN OF NEUTRON STARS AND COMPOSITIONAL TRANSITIONS IN THE COLD CRUSTAL MATTER

KEI IIDA[1] AND KATSUHIKO SATO[1,2]
[1] *Department of Physics, University of Tokyo*
7-3-1 Hongo, Bunkyo, Tokyo 113, Japan
[2] *Research Center for the Early Universe, University of Tokyo*
7-3-1 Hongo, Bunkyo, Tokyo 113, Japan

Transitions of nuclear compositions in the crust of a neutron star induced by stellar spin-down are evaluated at zero temperature. We construct a compressible liquid-drop model for the energy of nuclei immersed in a neutron gas, including pairing and shell correction terms, in reference to the known properties of the ground state of matter above neutron drip density, 4.3×10^{11} g cm^{-3}. Recent experimental values and extrapolations of nuclear masses are used for a description of matter at densities below neutron drip. Changes in the pressure of matter in the crust due to the stellar spin-down are calculated by taking into account the structure of the crust of a slowly and uniformly rotating relativistic neutron star. If the initial rotation period is of the order of milliseconds, these changes cause nuclei, initially being in the ground-state matter above a mass density of about 3×10^{13} g cm^{-3}, to absorb neutrons in the equatorial region where the matter undergoes compression, and to emit them in the vicinity of the rotation axis where the matter undergoes decompression. Heat generation by these processes is found to have significant effects on the thermal evolution of old neutron stars with low magnetic fields; the surface emission predicted from this heating is compared with the *ROSAT* observations of X-ray emission from millisecond pulsars and is shown to be insufficient to explain the observed X-ray luminosities (Iida and Sato, 1997).

References

Iida, K. and Sato, K. (1997), *Astrophys. J.* **Vol. 477**, pp. 294–312

THE EQUATIONS OF MOTION FOR BINARY SYSTEMS WITH RELATIVISTIC QUADRUPOLE-QUADRUPOLE MOMENTS INTERACTION

C. XU AND X. WU
Dept. of Phys., Nanjing Normal Univ.,
Nanjing 210097, P.R.China

The study of binary systems is one of the most important problems in astronomy. Especially recently, gravitational wave detection made possible by Laser Interferometer Gravitational wave Observatory - LIGO and VIRGO, LISA opens up a completely new window for the observation of our universe, it becomes one of the most important and forward area in the modern general relativistic astrophysics. Coalescing binary neutron star (NS) systems are believed to be the most important source emitted high-frequency gravitational wave. Therefore the study of NS coalescence is regarded as a major challenge in modern relativistic astrophysics. Indeed, if the two-body problem could be solved with a sufficient accuracy, the wealth of information might be extracted from the waveforms of coalescing binaries. Many early works to derive and investigate the gravitational two-body system with spin and quadrupole moment interaction have been done already. A detail appraisal of their works has been made by Xu, Wu and Schäfer[1] where they derived the first post-Newtonian equations of motion for binary systems with monopole, spin and quadrupole interaction by making use of the scheme developed by Damour, Soffel and Xu (DSX)[2,3] As we know, in the last stages of coalescence in binary system, the distance between two stars is closer, the tidal force is stronger, so the nonspherical size l is larger, l^2/r^2 can reach the level of v^2/c^2, where r is the distance between two bodies. In this case, the relativistic qudrupole-quadrupole term is of 3-PN order, so one can not neglect the 1-PN contribution of the q-q terms when 3-PN equations of motion (for mass-monopole) are considered. In order to fit the requirement of more accurate solution for binary system, the relativistic q-q terms in the post Newtonian equation of motion have been calculated in this paper. Our work is the first to obtain explicit 1-PN equations of motion for binary systems with relativistic quadrupole-quadrupole interaction in terms of only collective coordinates and B-D moments.

Because of complicated calculation, we use Maple Computer Algebra System as an auxiliary medium to save time and guarantee the correctness. According to Eqs. 5.26, 5.27, A7 and A8 of DSX paper II[3], we first calculated $W^{B/A}$ and $W_a^{B/A}$, which are the local external potential of body A produced by body B. Substitute the calculated external potential $W^{B/A}$ and $W_a^{B/A}$ into Eqs. A14, A15 of DSX paper II, we obtain the tidal moments. Substitute them into Eq. 6.17c of DSX paper II, we finally obtain the equations of motion for binary systems truncated to relaticistic quadrupole-quadrupole moments level in the local coordinate system of body A. From the physical meaning we can decompose them into nine coupling parts of different type (monopole-spin-quadrupole moments of body A coupled with monopole-spin-quadrupole moments of body B):

$$\begin{aligned} M^A A_a^A &= F_a^A(M^A \times M^B) + F_a^A(M^A \times S^B) + F_a^A(M^A \times Q^B) \\ &+ F_a^A(S^A \times M^B) + F_a^A(S^A \times S^B) + F_a^A(S^A \times Q^B) \\ &+ F_a^A(Q^A \times M^B) + F_a^A(Q^A \times S^B) + F_a^A(Q^A \times Q^B). \end{aligned} \quad (1)$$

Compare our results to the paper by Xu et al.[1], the 1st, 2nd, and 4, 5, 6, 8rd terms are the same, the another three terms enclosed more relativistic q-q terms. Each q-q term is composed of four parts: the first part is G/r_{AB}^6 (r_{AB} is the distance between body A and B) with a constant coefficient sometimes which include in a ratio of mass M^A (body A) to M^B (body B) or M^B to M^A; the second part is quadrupole(Q) multiplied by quadrupole(Q) which could be $Q^A \times Q^B$ or $Q^A \times Q^A$ or $Q^B \times Q^B$; the third part is a post Newtonian coefficient v^2/c^2 or u/c^2 where v might be v_A (velocity of body A), v_B (velocity of body B) or v_{AB} (relative velocity from B to A) and u might be the potential of M^A or M^B; the fouth part is a spatial position tensor $n_{AB}^{abc\cdots}$ and/or a spatial Levi Civita symbol ε_{abc}, the indices of which would be summation with the indices of quadrupoles to just make the index at left side (the index of the acceleration). Since the results are very long, we do not have enough room to write down here detail. We will publish in another place.

If we neglect all the relativistic quadrupole-quadrupole terms, our results are total the same as the paper by Xu et al.[1]. Furthermore, the Eq. 1 for binary systems can be easily extended to n-body equations of motion.

References

1. C. Xu, X. Wu and G. Schäfer, *Phys. Rev.* D **55**, 528 (1997).
2. T. Damour, M. Soffel and C. Xu, *Phys. Rev.* D **43**, 3273 (1991).
3. T. Damour, M. Soffel and C. Xu, *Phys. Rev.* D **45**, 1017 (1992).

RADIATIVE WINDS FROM ACCRETION DISKS IN CVS

M. HACHIYA, Y. TAJIMA, J. FUKUE
*Astronomical Institute, Osaka Kyoiku University,
Asahigaoka, Kashiwara, Osaka 582, Japan*

P Cyg profiles observed in CVs strongly suggest the existence of accretion disk winds in CVs. Recently, X-ray observations revealed bipolar outflows in supersoft X-ray sources. The wind velocity measured in these observations is about 3000–5000 km s^{-1}, which is of the order of the escape velocity of white dwarfs. Hence, it is supposed that the winds would originate from the inner disk and/or the boundary layers between the disk and white dwarf. We examine radiatively-acclerated accretion-disk winds, and found that the terminal speed of the wind emanated from the inner region becomes $\sim 0.6\sqrt{GM/r_{\rm in}}$, where M is the white dwarf mass and $r_{\rm in}$ the inner radius of the disk.

1. Radiation Fields of the Disk and the Boundary Layer

For the present purpose, we quantitatively calculate radiation fields produced by the standard disk and the boudary layer. Examples of radiation fields are shown below, where the isoflux contours are displayed.

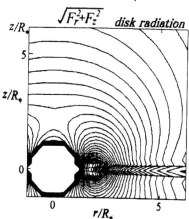
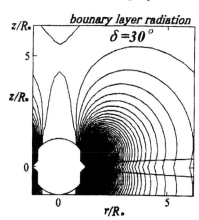

2. Particle Winds

Using the radiation fields obtained numerically, we calculate motions of the winds, which is assumed to consist of normal plasmas.

Trajectories of particle winds ejected from the disk surface are shown below (left panel). The abscissa is the distance from the central object and the ordinate is the height from the disk surface measured by a white dwarf radius R_*.

Terminal velocities of radiatively-accelerated particle winds are also shown for various disk luminosities (right panel), where the normalized disk luminosity Γ_d is set to be equal to the normalized star luminosity Γ_*. The abscissa is the initial radii of the winds in units of R_*, and the ordinate is the terminal velocity in units of the Keplerian rotation speed at R_*, GM/R_*. The normalized luminosities are attached on each curve.

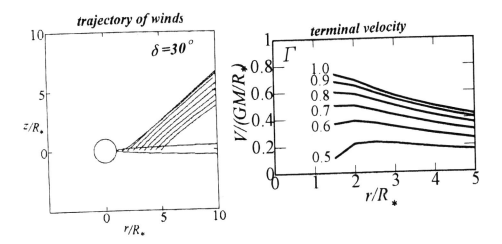

3. Concluding Remarks

We calculate the radiation fields produced by a standard accretion disk and a boundary layer, and the radiatively-accelerated accretion-disk winds, numerically. We found that the terminal speed of the wind emanated from the inner region becomes $\sim 0.6\sqrt{GM/r_{\rm in}}$, that is consistent with observations.

Reference

Tajima Y., Fukue J. 1996, PASJ 48, 529
Tajima Y., Fukue J. 1997, in this volume

SUPERSOFT X-RAY SOURCE RX J0019.8+2156

KATSURA MATSUMOTO
Department of Astronomy, Kyoto University, Japan
E-mail: katsura@kusastro.kyoto-u.ac.jp

AND

MASAMITSU OKUGAMI AND JUN FUKUE
Astronomical Institute, Osaka Kyoiku University, Japan

1. Irradiation by White Dwarf

RX J0019.8+2156 (RX J0019) is one of supersoft X-ray sources (SSXSs) which were rarely found in the Galaxy, and the brightest one upto now (Beuermann et al. 1995). Although the binary systems of SSXSs are still not perfectly revealed, if the model with nuclear burning white dwarf (van den Heuvel et al. 1992) is adopted, it is supposed that most of radiative energy are generated on the surface of the white dwarf. The radiation strongly irradiates the accretion disk and companion, and the reprocessing has an influence upon an observed flux of SSXSs in lower frequencies (Popham, DiStefano 1996).

2. Modeling of Irradiation Effect and Light Curve

We assumed the nuclear burning on the white dwarf to be isotoropic, and the accretion disk obeys the standard model. Then, the heating rate on the disk by irradiation (e.g., Fukue 1992) can be estimated for RX J0019. The main parameters are mass and inclination angle; where the mass of the white dwarf is set to be 1.2 $M_\odot$, and $L_{wd} = 4 \times 10^{36}$ erg s^{-1}. The distance to the object is fixed to 2 kpc for this calculation. In order to compare the theoretical energy distributions with observed fluxes, we used the UV (Gänsicke et al. 1996) and optical results (Matsumoto 1996).

With the standard accretion disk model, the optical and UV fluxes of expected luminosities are less than the observed value by factor of about 5–15. We next considered the accretion disk and companion with the irradi-

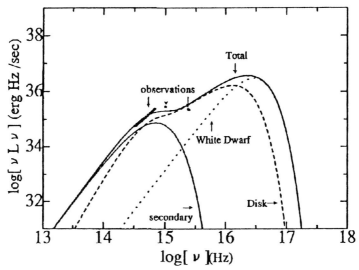

Figure 1. The expected energy distribution of RX J0019 with irradiation effects.

ation effect (figure 1). The energy distribution exhibits clearly increasing of UV and optical fluxes by factor of about 3-5 compared with the standard disk model. The expected energy distributions for the parameters reproduce close concord with the observed results for relatively lower inclination angle. The flux of the companion is accordingly smaller, and the importance of the irradiated accretion disk enlarges especially in the lower frequency. As a result, observations in UV or optical for the object provide reprocessed radiation of X-ray generated on the white dwarf.

Then we calculated optical light curves using this model. Inclination angles of $i = 70$–$80°$ can reproduce the shape well like the observations around the photometric minimum. It is supposed that the lower inclination system may be rejected, and the higher inclination system is favorite for observations in B and V band. While, a parameter of $i \sim 50°$ has relatively good reproducing for the amplitude. As for the variations of color index, the case of $i = 70$–$80°$ seems to be too high to reproduce the observations.

References

Beuermann, K. et al. 1995, *Astron. Astrophys.*, 294, L1
Fukue, J. 1992, *Publ. Astron. Soc. Japan*, 44, 669
Gänsicke, B. T., Beuermann, K. and deMartino, D. 1996, *Supersoft X-Ray Sources*, Springer-Verlag, pp 107
Matsumoto, K. 1996, *Publ. Astron. Soc. Japan*, 48, 827
Popham, R. and DiStefano, R. 1996, *Supersoft X-Ray Sources*, Springer-Verlag, pp 65
van den Heuvel, E.P.J., Bhattacharya, D., Nomoto, K. and Rappaport, S. A. 1992, *Astron. Astrophys.*, 262, 97

ASCA OBSERVATIONS OF THE SGR A REGION

Y. MAEDA, K. KOYAMA, H. MURAKAMI AND M. SAKANO
Department of Physics, Graduate School of Science, Kyoto University, Sakyo-ku, Kyoto, 606-01, Japan

K. EBISAWA AND T. TAKESHIMA
Laboratory for High Energy Astrophysics, NASA/Goddard Space Flight Center, Greenbelt, MD 20771, USA

AND

S. YAMAUCHI
Faculty of Humanities and Social Sciences, Iwate University, 3-18-34, Ueda, Morioka, Iwate, 020, Japan

1. Introduction & Observations

The complex radio source Sgr A is embedded in a region near our Galactic Center. The dynamical center of our Galaxy is considered to be Sgr A*, the compact non-thermal radio source. Dynamical mass within ~0.1 pc from Sgr A* has been estimated to be $\sim 3\times10^6$ M$_\odot$. This places Sgr A* to be a candidate of a massive blackhole (Eckart and Genzel, 1997 and reference therein).

In spite of extensive observations, direct evidence for high energy activities from Sgr A*, as those found in active galactic nuclei, is still lacking. *ASCA* observations near to Sgr A* were made on three occasions:'93 Autumn, '94 Autumn & '97 Spring. The results of the former two observations have been already published in Koyama et al.,(1996) and Maeda et al.,(1996). In the SIS/*ASCA* hard-band image (2-10 keV) during '93 Autumn observations, we found two bright spots within the inner 2 arcmin from Sgr A*. One named as the "hard source" is located 1'.3 away from Sgr A*. In the '94 observations, we discovered an X-ray burst and 8.4 hr period dips from the hard source, establishing that the hard source is a new eclipsing low-mass X-ray binary (Maeda et al.,1996, Kennea and Skinner,1996). The peak of the other spot named as the "soft source" corresponds to the position of Sgr A*. The soft source was dominated by diffuse X-rays and exhibited no time variability. in the former two observations. Thus, in the former two observations, no direct evidence of the high-energy activity from Sgr A* was found. However, we found a hint of intermittent and past activities from Sgr A* (Koyama et al.,1996). Accordingly, we performed the third Sgr A observation on '97 Spring in order to search the suggestive activities of Sgr A* in more detail. We report the results of this third observations, together with the previous two observations.

2. Results & Discussion

We detected significant X-ray emissions from the two sources of the Sgr A region in the '97 Spring observations. The fluxes of the two sources, together with those obtained with the previous two observations are given in figure 1. We found no direct evidence for the X-ray activity of Sgr A*; neither flux variability nor spectral change of the soft source was found from the three observations. Thus we infer that Sgr A* is currently quiescent, or at least, during the past several years. The absorption collected luminosity has been $L_{x,SgrA*} < 10^{36}$ erg s^{-1} at the distance of 8.5 kpc.

The flux of the hard source in Spring '97 was the same as that in Autumn '93, but was about 1/5 of that in Autumn '94. We confirmed the periodic dips of $\sim$ 8.4 hrs, and thus conclude that the time variability in the close vicinity of Sgr A* is due to the hard source located 1'.3 away from the Galactic Center. We note that the previous non-imaging observations of Sgr A*, if not all, might have been contaminated by the variable hard source.

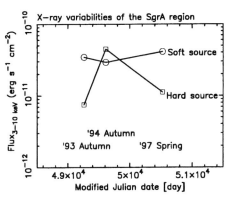

Figure 1. Plot of the 3–10 keV flux [ergs s^{-1} cm^{-2}] for the soft source(Circle) and for the hard source(Square), respectively.

References

Eckart,A. and Genzel,R. 1997, *MNRAS*, 284, 576
Kennea J.A. and Skinner G.K. 1996, *PASJ*, 48, L117
Koyama K., Maeda Y., Sonobe T., Takeshima T., Tanaka Y., and Yamauchi S. 1996, *PASJ*, 48, 249
Maeda, Y., Koyama, K., Sakano, M., Takeshima, Y., and Yamauchi, S. 1996, *PASJ*, 48, 417

ENERGY SPECTRA OF X 1636–536 OBSERVED WITH ASCA

K. ASAI, T. DOTANI, K. MITSUDA AND H. INOUE
Institute of Space and Astronautical Science,
3-1-1 Yoshinodai, Sagamihara, Kanagawa 229, Japan

Y. TANAKA
SRON Laboratory for Space Research,
Sorbonnelaan 2, 3584 CA Utrecht, The Netherlands

AND

W. H. G. LEWIN
Massachusetts Institute of Technology, Center for Space Research,
37-627, Cambridge, MA 02139, USA

1. Introduction

Absorption line features were detected at 4.1 keV from X 1636–536 with the Tenma satellite in the spectra of X-ray bursts (Waki *et al.*, 1984). Similar features were also detected from X 1608–52 and EXO 1747–214 during bursts (Nakamura *et al.*, 1988; Magnier *et al.*, 1989). These features at 4.1 keV may be interpreted as the redshifted Kα absorption line of helium-like iron atoms. However, such interpretation requires extremely soft equation of state for the nuclear matter, and confirmation with high resolution detectors is urged (Lewin *et al.*, 1993). To investigate the line features, we observed X 1636–536 with ASCA for $\sim$ 240 ksec.

2. Analysis and Results

X 1636–536 was observed at seven different epochs from 1993 to 1995. Total of 12 bursts were detected, six of which were observed when the telemetry rate was high. We used only the high bit rate data of SIS and GIS, because the telemetry was saturated for the medium bit rate data.

During the observations, X 1636–536 stayed in the banana state. We could not find any correlation of color/intensity when the bursts occured. We performed model fitting to the spectrum of the persistent emission of

TABLE 1. Equivalent width of narrow line of burst

Name	A	B
Line Energy [keV]	$4.3^{+0.1}_{-0.7}$	$4.0^{+0.2}_{-0.1}$
Line EW [eV]	42^{+41}_{-36}	41^{+26}_{-25}

each set of data. We adopted two component model consisting of a blackbody and a multi-color disk blackbody (Mitsuda et al., 1984). The fit was acceptable for all cases, but the best-fit disk temperatures were slightly lower than those obtained by Tenma. This may be due to the lower energy range covered by ASCA (Mitsuda et al., 1989). We also investigated the emission line at 6.7 keV. The equivalent width for a narrow line obtained with GIS (EW= 16 ± 4 eV) is larger than the upper limit with SIS (EW < 5 eV). This may be interpreted that the 6.7 keV emission line is broad.

No burst among the six showed photospheric expansion. We searched for line-like features in the energy spectra of six bursts. Two bursts (A and B) showed a hint of absorption feature in the decay phase, but only upper limits were obtained for other bursts (EW < 40 eV). Significance of the features are, when assumed as an absorption line, 1.96 σ and 2.575 σ for burst A and B, respectively. Best-fit parameters are listed in Table 1. Even if the presence of an absorption features was real, their EW is much smaller than those detected by Tenma, and they could have different origin.

3. Conclusion

Although we detected marginal line-like structure in the spectra of these bursts, the equivalent widths were much smaller than those detected by Tenma. Thus we could not obtain clear reconfirmation of the Tenma results.

References

Lewin, W. H. G. et al. 1993 Space Sci. Rev., **62**, 223
Magnier, E. et al. 1989 MNRAS, **237**, 729
Mitsuda, K. et al. 1984 PASJ, **36**, 741
Mitsuda, K. et al. 1989 PASJ, **41**, 97
Nakamura, N. et al. 1988, PASJ, **40**, 209
Waki, I. et al. 1984, PASJ, **36**, 819

THE CIRCUMSTELLAR MATTER OF CYGNUS X-3

S. KITAMOTO AND K. KAWASHIMA
Department of Earth and Space Science, Graduate School of Science, Osaka University,
1-1, Machikaneyama-cho, Toyonaka, Osaka, 560, JAPAN

1. Introduction

X-ray source in Cyg X-3 is embedded in the photo-ionized circumstellar gas. Recombination edges and emission lines were detected by ASCA observation (Kawashima, Kitamoto 1996; Liedahl, Paerels 1996). The strength of the $K\alpha$ line and the recombination edge are simply determined by the atomic process. Therefore, the comparison of observed these intensities is a good tool to study the Cyg X-3 system.

2. Analysis and Results

The emission measure can be defined as $\int n_e n_z^{i+1} dV \, \text{cm}^{-3}$. The line intensity is a function of a product of the emission measure and of the emissivity, which is a function of the electron temperature. Therefore observed line intensity makes a constraint on a map of the emission measure and electron temperature. The recombination edge intensity also gives a constraint on the map. Since the energy of the emission line is smaller than that of the recombination edge, absorption by the circum- and inter-stellar matter should be taken into account. However, the absorption measure has not well established. Here, we applied the absorption measure of 1.5×10^{22} H atoms cm^{-2}; this value is rather small value comparing previously reported values (e,g, Kitamoto et al. 1987). Figure 1 shows the derived constraint map for the H-like S. There is no overlapped region.

3. Discussion

Before we conclude the above result, we have to consider the effect of the absorption edge. The absorption edge has the same energy as the recom-

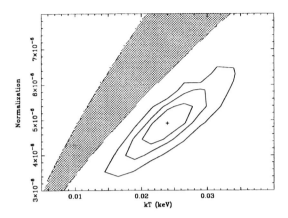

Figure 1. Comparison between the line intensity and the recombination edge intensity of H-like S. Contours are 68 %, 90% and 99% confidence region derived from the recombination edge on the (Electron Temperature v.s. Emission Measure) space. Normalization means the emission measure defined in text, in an unit of $\frac{10^{-14}}{4\pi D^2}$. Gray region is a 99% confidence region derived from the line intensity.

bination edge has. Therefore, the absorption edge feature could affect the intensity of the recombination edge. If we assume the spherically symmetric and ionization equilibrium plasma, the absorption optical depth can be estimated from the column density of the ions, which we can derive from the recombination edge. The obtained absorption optical depth at the threshold energy is less than 0.01 and the resultant effect on the intensity of the recombination edge is smaller than the statistical uncertainty. Consequently, the effect of the photoelectric absorption edge structure can be neglected in our case.

The most plausible reason of the observed strong line is an anisotropic configuration. Since the main lines that we are considering are resonance lines, which have a large scattering cross section. If geometrical obscuration occurs, for example, the line of sight to the continuum source is blocked, the ratio of the scattered photons to the recombination edge will become large.

4. References

Kawashima K., Kitamoto S. 1996, PASJ, 48, L113
Liedahl D. A., Paerels F. 1996, ApJ, 468, L33
Kitamoto S, Miyamoto S., Matsui W., Inoue H., 1987, PASJ, 39, 259

THE ASCA OBSERVATION CAMPAIGN OF SS433

T. KOTANI, N. KAWAI AND M. MATSUOKA
RIKEN
2-1 Hirosawa, Wako, Saitama 351-01, Japan

AND

W. BRINKMANN
MPIE
Giessenbachstrasse, D-85740 Garching, Germany

Here we present a very short review of the ASCA observation campaign of the enigmatic galactic jet system SS433. The campaign started in 1994 just after the launch, and ended in 1996. Various phases of the 162.5-day precession and 13-day orbital motion were sampled. With ASCA, the Doppler-shifted pairs from various ion species from Si to Ni were resolved for the first time (Kotani et al. 1994). The Doppler-shift parameters were determined with an accuracy comparable to optical spectroscopy (Kawai 1995). No velocity gradient was found between the X-ray emission region of the jet and the optical. The distance between them was constrained to be less than 10^{15} cm. Line intensity ratios of Fe XXVI/Fe XXV give the base temperature of the jet to be 20 keV (Kotani et al. 1996). The variation of the apparent base temperature of the jet can be explained in terms of the partial occultation of the jet by a precessing accretion disk (Kotani et al. 1997a). From the variation, the disk radius and the disk height in unit of the X-ray jet length were estimated to be 0.23 ± 0.10 and 0.0232 ± 0.0049, respectively. (These are an improved version of the values in Kotani et al. (1997a).) SS433 is also known as an eclipsing binary. Because the emission from each jet with ASCA, it is possible to know how much of which jet is occulted by the companion star during an eclipse. Relative size of the companion star gives Roche lobe size and thus mass ratio $M_X/M_C = 0.22^{+0.09}_{-0.16}$ (Kotani 1997b). With the help of Doppler modulation, compact star mass is constrained. However, the values of Doppler modulation reported from optical observations largely scatters. D'Odorico et al. (1991) reported 112 km s^{-1} and this gives $M_X = 0.68^{+0.43}_{-0.53}$ M$_\odot$, i.e., a white dwarf, while Fabrika and Bychkova (1990) reported 175 km s^{-1}, which gives $2.6^{+1.6}_{-2.0}$ M$_\odot$. (This error includes systematic errors of the X-ray data, and will be reduced in future analysis.) On the other hand, the absolute size of the system were determined with a satisfactory precision. For example, the X-ray jet length was determined to be 2×10^{13} cm, ten times larger than previous estimations (Kotani et al. 1997c). Other physical parameters of the jet can be derived from the X-ray jet length. Mass outflow rate and the kinetic luminosity of both jet were determined to be 8×10^{-6} $M_\odot$ yr^{-1} and 1.6×10^{40} erg s^{-1} (Kotani et al. 1997d), implicating a highly super critical accretion. Most of these values are first precise measurements and/or "radical" revisions of previous estimations. The new picture of SS433 drawn here is far stormy and highly energetic.

References

D'Odorico S., Oosterloo T., Zwitter T., Calvani M. (1991), *Nature* **353**, 329

Fabrika, S.N., Bychkova, L.V. (1990), *A&A* **240**, L5

Kawai, N. (1995), in *Multifrequency Behavior of High Energy Cosmic Sources*, ed. Giovannelli, F., Sabau-Graziati, L., (Italian Astronomical Society, Firenze), p.381

Kotani, T., Kawai, N., Aoki, T., Doty, L., Matsuoka, M., Mitsuda, K., Nagase, F., Ricker, G. White, N.E. (1994), *PASJ*, **46**, L147

Kotani, T., Kawai, N., Matsuoka, M., Brinkmann, W. (1996), *PASJ* **48**, 619

Kotani, T., Kawai, N., Matsuoka, M., Brinkmann, W. (1997a), in *X-Ray Imaging and Spectroscopy of Cosmic Hot Plasmas*, ed. Makino, F., Mitsuda, K., (Universal Academy Press, Tokyo), p. 443

Kotani T. (1997b), Doctroral Thesis, University of Tokyo

Kotani, T., Kawai, N., Matsuoka, M., Brinkmann, W. (1997c), in *Accretion Phenomena and Related Outflows, IAU Colloq. 163*, ed. Wickramasinghe, D.T., Bicknell, G.V., and Ferrario, L., (Astronomical Society of the Pacific, San Francisco)

Kotani, T., Kawai, N., Matsuoka, M., Dotani, T., Inoue, H., Nagase, F., Tanaka, Y., Yamaoka, K., Ueda, Y. et al. (1997d), in *The Proc. of the Fourth Compton Symp.*, ed. Dermer, C.D., Strickman, M.S., and Kurfess, J.D., (American Institute of Physics, New York), in press

THE ASCA RESULTS OF GRO J1744-28

MAMIKO NISHIUCHI, YOSHITOMO MAEDA, KATSUJI KOYAMA AND JUN YOKOGAWA
Department of Physics, Kyoto University, Sakyo, Kyoto, Japan
TADAYASU DOTANI, KAZUMI ASAI AND YOSHIHIRO UEDA
KAZUHISA MITSUDA, HAJIME INOUE AND FUMIAKI NAGASE
ISAS
AND
CHERISSA KOUVELIOTOU
Universities Space Research Association

1. Introduction

We present the ASCA/GIS results of the transient source GRO J1744-28, the bursting X-ray pulsar, for the two times observations about one year apart. (the first and the second observations are on Feb.27th 1996 and Mar.16th 1997 respectively.) Since the discovery of GRO J1744−28 on Dec. 2 1995 with the BATSE observatory (Fishman et al. 1995; Kouveliotou et al.), thousands of Type II X-ray bursts, and sinusoidal pulsations have been observed in the X-ray band. No other source has ever been found to exhibit such unusual characteristics. Following each burst the flux decreases below, and recovers to the pre-burst persistent level (here after the dip) within a few seconds to a few minutes, depending on the total flux of the burst. We examined the spectrum of GRO J1744−28 during each persistent, dip, burst phases at the two observations. This is the first detail study of the spectra (1–10keV band) of GRO J1744−28.

2. Results and Discussion

The persistent spectra of both observations are quite similar and the acceptable model for them is an absorbed power-law with a broad line whose center energy is around 6.7keV and with a soft black body component. Since the value of hydrogen column density is comparable to that of our

Galactic center region, here and after we assume that GRO J1744−28 is near the Galactic center(d=8kpc). At the first observation, GRO J1744−28 was as bright as the Eddington limit of a neutron star, but at the second observation, this source was about ∼ 1/5 times bright as that of the first observation. The normalization of the black body component remain constant between these two observations. The size of this black body emission region is about the same as that of the surface of a neutron star. This suggests that the black body component is from the surface of the a neutron star. The gaussian center energy is constant from the first and second observations, but the gaussian sigma is broader at the first observation than at second one. The equivalent width of the Fe line at the first and second observations were ∼ 260eV and ∼150eV respectively. It is reasonable to suppose that the broad gaussian is caused by the Doppler broadening of an iron line and that the difference of the widths is explained as the difference of the Alfven radius between two observations. As for the spectra among three phases, persistent, dip and burst, the photon indices are slightly different at the first observation ;the photon index in the persistent is larger (steeper) than that in the burst but smaller (flatter) than that found in the dip. However, at the second observation, no clear difference among spectra during three phases was found. We also investigated the spectra separately accumulated according to the sinusoidal pulse phase during persistent phase. The photon index in the upper phase of the sinusoidal curve is smaller than that in the lower phase at both observations. Because the X-ray pulsation would be due to the appearing and hiding of the part of the polar caps to our direction, the emission from the polar cap region has harder spectrum than that from the other region.

References

Fishman G.J. *et al.* (1995), *IAUC*, **6272**,
Kouveiotou C.*et al.*(1996), *NATURE*, **379** ,

STRUCTURE AND TIME-DEPENDENT BEHAVIOR OF BE STAR DISKS IN BE/X-RAY BINARIES

A.T. OKAZAKI
*College of General Education, Hokkai-Gakuen University,
Toyohira-ku, Sapporo 062, Japan*

We consider the structure and time dependent behavior of the outflow in disks of Be stars in Be/X-ray binaries, based on the viscous decretion disk scenario (Lee et al. 1991). In this scenario, the matter ejected from the star with the Keplerian velocity at the equatorial surface of the star drifts outward because of the effects of viscosity, and forms the disk.

In the present study, we adopt the Shakura-Sunyaev's α-viscosity prescription, and assume the disk to be isothermal. The disk radius is fixed to be $3^{-2/3}$ times the mean binary separation, because in a Be/X-ray binary the disk of the Be star is likely truncated at the radius where the tidally-induced eccentric instability occurs.

Figure 1(a) illustrates a typical structure of viscous decretion disks around Be stars in Be/X-ray binaries: The outflow velocity increases as r, the surface density decreases as r^{-2}, and the angular velocity of the disk decreases as $r^{-1/2}$.

In general, the viscous decretion discs are overstable for $m = 1$ perturbations. The growth rate is of the order $\alpha(H/r)^2\Omega$, where H is the scale-height of the disk and Ω is the angular frequency of disk rotation. Figure 1(b) shows the $m = 1$ overstable mode in the decretion disk shown in panel (a). We note that the perturbation pattern is of the leading, one-armed spiral. We also note also that the characteristics of the one-armed spiral modes are in agreement with the periodicity and the profile variability of the V/R variations of Balmer lines observed for some Be/X-ray binaries.

Finally, we discuss the orbital modulations of X-ray lightcurves observed for Be/X-ray binaries A0535+26 and 4U0115+63. Type I outbursts from A0535+26 occur close to the periastron passages of the neutron star, while those from 4U0115+63 are seen close to orbital phase of ~ 0.3 (Negueruella et al. 1997). The present model is consistent with these features. For small amplitude $m = 1$ perturbations, the eccentricity of the disk is negligible.

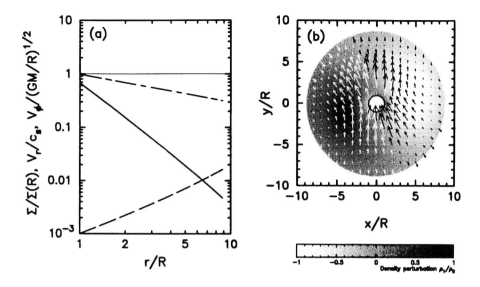

Figure 1. (a) Structure of an unperturbed, viscous decretion disk for a Be/X-ray binary A0535+26. Solid, dashed, and dash-dotted lines denote V_r/c_s, $V_\phi/(GM/R)^{1/2}$, and $\Sigma/\Sigma(R)$, respectively, where M and R are the mass and radius of the Be star, respectively, V_r and V_ϕ are the radial and azimuthal components of the vertically averaged velocity, respectively, c_s is the isothermal sound speed, and Σ is the surface density. We adopt $V_r/c_s(R) = 10^{-3}$ and $\alpha = 0.1$. The radiative force in the form of $\eta(r/R)^\epsilon \times GM/r^2$ with $(\eta, \epsilon) = (0.2, 0.1)$ is included. (b) Linear, one-armed fundamental mode in the disk shown in panel (a). The period of this mode is 1.9 yr and the growth time is 8.8 yr. The disk rotates counterclockwise. A gray-scale representation denotes the density perturbation, while arrows denote the perturbed velocity vectors.

Then, the Type I outbursts will occur close to the periastron passages of the neutron star. However, when the amplitude of the $m = 1$ perturbation is large enough to make the disk highly eccentric, the phase of the Type I outburst will also depend on the phase of the $m = 1$ mode which does not necessarily coincide with the periastron pasage of the neutron star.

In conclusion, the viscous decretion disk model for Be stars in Be/X-ray binaries agrees well with long-term V/R variations of Balmer lines and is consistent with the orbital modulations observed for some Be/X-ray binaries.

References

Lee, U., Saio, H., Osaki, Y., 1991, MNRAS 250, 432
Negueruela, I., Reig, P., Coe, M.J., Fabregat, J., 1997, submitted to A&A

VRI LIGHT CURVE ANALYSIS OF SS 433

MASAMITSU OKUGAMI, YOSHIMI OBANA, JUN FUKUE
Astronomical Institute, Osaka Kyoiku University,
Asahigaoka, Kashiwara, Osaka 582, Japan

1. Introduction

In the light curves of SS 433, in addition to a binary modulation with a period of 13.1 d, there is a precessional one with a period of 162.5 d (Kemp et al. 1986). In the current studies, however, the attention was focused mainly on the light curves during *eclipse* (see, however, e.g., Antokhina, Cherepashchuk 1987). Thus, in this paper we examine light curves during *precession* and compare them with observational ones (Fukue et al. 1997b).

2. Model

The expected light curves were calculated in a similar way as in the previous papers (Fukue et al. 1997a).

We suppose that SS 433 consists of a compact star with mass M_x and an early type "normal star" with mass M_v.

We show the assumptions briefly as follows: i) The compact star is surrounded by a *geometrically thick torus*. ii) The specific angular momentum l of the torus gas can be expressed by a power law of radius r as $l \propto r^{2-q}$, where the torus configuration index q is a constant. iii) The companion star fills its Roche lobe and the surface temperature of the companion is around 17000 K. iv) The SS 433 system has the orbital period (ϕ) is 13.082 d and the precessional period (Φ) is 162.5 d. v) The inclination angle is $i = 78°\!.8$ and the precession angle is $19°\!.8$. vi) When the precession phase Φ is 0.5, the jet/torus is mostly inclined to the observer. Assuming that $i = 78°\!.8$, we obtain the relation between M_x and M_v as $M_v/(1 + M_x/M_v)^2 = 2.12 M_\odot$.

The model parameters are thus the binary mass ratio Q ($= M_v/M_x$), the torus configuration index q (from thin($q = 1.51$) to thick($q = 1.8$)), and the surface temperature T_c of the companion.

3. Results

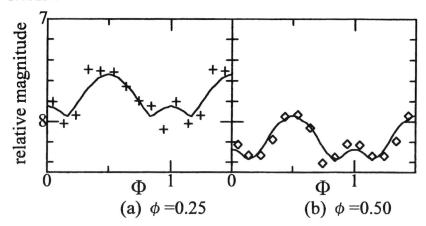

Figure 1. Typical examples of model light curves (best fit). The abscissa is the precession phase Φ, whereas the ordinate is the expected V magnitude. The parameters are: the binary mass ratio Q is 2, the torus configuration index q is 1.52 (the torus is geometrically thin), and the surface temperature T_c of the companion is 17000 K. The orbital phase ϕ is (a) 0.25 and (b) 0.5.

We show the light curves of the best fit parameter. The model for this parameters are well reproduced the observations.

4. Conclusion

We have calculated the precessional light curves under the picture in which SS 433 consists of a geometrically thick torus around a compact star and a companion star filling the Roche lobe.

The extremely geometrically-thick tori are rejected, since the calculated light curves are too flat to explain the observational ones. If this is the case, we should reconsider the model for SS433 jets; instead of the funnel jets we may consider the jet (wind) from a geometrically thin disk (e.g., Tajima, Fukue 1996).

Furthermore, the favarite combination of other parameters is that the binary mass ratio is about 2 (a black hole case) and the surface temperature of the companion is around 17000 K.

Reference

Antokhina É.A., Cherepashchuk A.M. 1987, SvA 31, 295
Fukue J. et al. 1997a, PASJ 49, 93
Fukue J. et al. 1997b, submitted to PASJ
Kemp J.C. et al. 1986, ApJ 305, 805
Tajima Y., Fukue J. 1996, PASJ 48, 529

BRIGHT X-RAY STARS NEAR THE GALACTIC CENTER

MASAAKI SAKANO, MAMIKO NISHIUCHI, YOSHITOMO MAEDA,
KATSUJI KOYAMA AND JUN YOKOGAWA
Department of Physics, Kyoto University, Sakyo, Kyoto, Japan

Abstract.
This paper report the *ASCA* observations of the three brightest persistent X-ray stars near the Galactic Center: an X-ray burster A1742−294, black hole candidate 1E 1740.7−2942, and unclassified source 1E 1743.1−2843. Emission mechanism is briefly discussed based on the new ASCA results.

1. Introduction

The Galactic Center region draws much attention in every wave band, in conjunction with a possible massive black hole, starburst activity and so on. In order to overview the high energy phenomena, we have performed repeated *ASCA* observations near the Galactic Center, consisting of a few tens of pointings with total exposure of about 1000 ksec. Several new facts are already published (e.g. Koyama *et al.* 1996). This paper is focused on the persistent X-ray stars locating near the Galactic Center. Systematic works on the whole X-ray stars in the Galactic Center region are presented by Sakano *et al.* (1997).

2. Results & Discussion

X-RAY BURSTER: A1742−294

A1742−294 is the brightest persistent X-ray binary near the Galactic Center. Several X-ray bursts have been reported by Pavlinsky *et al.* (1994). We also detected 10 bursts in the total observation time of 115 ksec, hence confirmed A1742−294 to be an X-ray burster. All these bursts showed black body spectra with spectral softening in the decay phase, hence are established to be type-I bursts. At the Galactic Center distance (8.5 kpc), the peak intensity of all the bursts is estimated to be below the Eddington limit of a neutron star. A1742−294 also exhibited a long-term variability by a factor of 3 in 3.5 years.

1E 1740.7−2942 ("THE GREAT ANNIHILATOR")

1E 1740.7−2942 is one of the most interesting sources in this region. It exhibits double radio jets (Mirabel et al., 1992), strong hard X-ray emission extending up to 100keV and intermittent 511keV bursts (Bouchet et al., 1991). The X-ray position coincides to the peak of a molecular cloud (Bally & Leventhal, 1991). All together, 1E 1740.7−2942 is referred to be a micro quasar, powered by direct gas accretion from the molecular cloud.

We detected long-term flux variability by a factor of 3, with no spectral change. The spectrum is found to be highly absorbed ($N_H \sim 1.5 \times 10^{23} H\ cm^{-2}$), although the overall spectral shape with 1–10keV is not sufficiently well represented with a single-power law. No significant fluorescent line from neutral iron is found, suggesting that 1E 1740.7−2942 is not in the dense cloud but would be behind it. Thus we infer that 1E 1740.7−2942 is a binary black hole rather than an isolated black hole in the dense cloud.

UNCLASSIFIED OBJECT: 1E 1743.1−2843

1E 1743.1−2843 has been observed many times in the past, but none is clear for the nature of this source. Although long-term variability was found with *ASCA*, neither short-term nor periodic variation was found. The spectrum is fitted by an absorbed power-law model with a photon index $\Gamma \sim 2$, and hydrogen column of $N_H \sim 1.9 \times 10^{23} H\ cm^{-2}$.

The large absorption may suggest that 1E 1743.1−2843 is a background object behind our galaxy, such as an extragalactic AGN. No significant flux above the 35keV band with past observations, which indicates spectral turn-off near the energy 10–50 keV, may rule out the AGN possibility. We hence prefer galactic origin. From the spectral shape, a low mass X-ray binary (LMXB) near the Galactic Center is the most probable candidate. X-ray luminosity is estimated to be $2 \times 10^{36} erg\ sec^{-1}$, a luminosity in which type-I bursts are expected. However, no burst has been detected through long term observations not only with *ASCA* but also with other previous instruments, with perhaps more than 1200ksec exposure in total.

References

Bally, J. & Leventhal, M. (1991), *Nature*, **353**, 234
Bouchet, L. et al. (1991), *ApJL*, **383**, L45
Koyama, K. et al. (1992), *PASJ*, **48**, 249
Mirabel, I.F. et al. (1992), *Nature*, **358**, 215
Pavlinsky, M.N. et al. (1994), *ApJ*, **425**, 110
Sakano, M. et al. (1997), Proc of IAU Symp. 184, ed. Y.Sofue

DISCOVERY OF QPO FROM X PERSEI WITH RXTE

TOSHIAKI TAKESHIMA
NASA Goddard Space Flight Center,
Universities Space Research Aassociation
takeshim@ginpo.gsfc.nasa.gov

X Persei is a Be X-ray binary pulsar with ~835-sec pulsation period[1]. Different from most other Be X-ray binaries, X Persei does not exhibit X-ray outbursts. A binary period of ~580 days has been suggested[2], but is not confirmed the by follow-up observations.

The Proportional Counter Array (PCA) onboard Rossi X-ray Timing Explorer (RXTE) observed X Persei twice on March 19 and April 1, 1996. Since a couple of flares in the timescale of ~100-sec were observed whose pulse phases are not constant, we removed such flares from data for the analysis of the coherent pulsation. Applying the folding technique to these data, the barycentric corrected pulse period are obtained as 840±1.8 and 837.6pm2.5 seconds for March and April data, respectively. The most plausible averaged pulse period for all data is 838.91pm0.04 seconds for the epoch of MJD=50617.9311, while the possibility for the period of 838.15pm0.04 or 839.41pm0.04 seconds cannot be rejected completely.

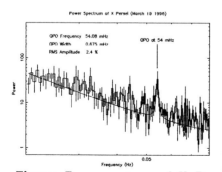

Figure: Power spectrum of X Persei around QPO frequency range (20–80 mHz). Histogram with error bars is the data and the solid line is the best fit model.

By calculating the power spectra for two observations separately, we found a broad peak structure at around 54 mHz from March 19 data. The figure shows the power spectrum between 20–80 mHz with the best fit model. The employed model was the sum of a power law component (for continuum), a Lorentzian (for QPO), and a constant (for Poisson noise). The addition of a QPO component (three parameters) reduces the χ^2 value by 14.4 (from 175.5 to 161.1) for the d.o.f. of 120. From the *F-test*, the addition of a QPO component is significant at the 99.5% level. The centroid frequency, the width (FWHM), and the root-mean-square (RMS) amplitude of QPO are $54.08^{+0.23}_{-0.28}$ mHz, $0.67^{+0.23}_{-0.25}$ mHz, and 2.4 ± 1.0 % (90% confidence), respectively.

QPO from X-ray binary pulsars have been reported from seven sources to date. These pulsars have relatively short to moderate pulse periods of 0.7-103 seconds and are believed to be disk-fed pulsars. The QPO frequencies concentrate to 10-200 mHz. Since the QPO observed from X Persei have typical frequency, width and amplitude of QPO from X-ray binary pulsars, it is natural to think that QPO from X Persei have the same origin as those observed from other X-ray binary pulsars.

The origins of QPO not only from X-ray binary pulsars but from low-mass X-ray binaries and black hole binaries are still unclear. The beat frequency model (BFM)[3] is generally regarded as the most plausible model for QPO from X-ray binary pulsars. In the BFM, QPO originate from the inhomogeneity of the accreting matter at the inner edge of the accretion disk. In other words, if the BFM works, the detection of QPO is evidence of the existence of an accretion disk. On the other hand, the facts that X Persei is a low luminosity Be X-ray binary system, the Roche lobe is not filled, and that the random-walk like behavior of pulse period change in relatively short time scale similar to Vela X-1 is known, strongly suggest that X Persei is a wind-fed system. If so, this is the first detection of QPO from a wind-fed pulsar.

Assuming the BFM and the accretion disk model for binary pulsars by Ghosh and Lamb[4, 5], and using the observed QPO frequency (54.1 mHz), pulsation frequency (1.2 mHz), and X-ray luminosity (1.1×10^{34} erg/sec), the strength of the surface magnetic field is estimated as 1.7×10^{11} Gauss. This value is relatively small for a binary X-ray pulsar. E-cut energy is thought to be roughly proportional to the strength of the magnetic field[6]. From this relation, the estimated strength of the surface magnetic field is 0.64-1.15×10^{12} Gauss, which is by 3.8-6.8 times larger than that derived by assuming the BFM and Ghosh & Lamb's model. The moderate pulse modulation of X Persei ($\sim$50%) does not support the very weak magnetic field, either.

References
[1] White et al. 1976, *Monthly Notices Roy. Astron. Soc.*, **176**, 201.
[2] Hutchings et al. 1974, *Astrophys. J. Letts.*, **191**, L101.
[3] Alper & Shaham 1985, *Nature*, **316**, 239.
[4] Ghosh and Lamb 1979a, *Astrophys. J.*, **232**, 259.
[5] Ghosh and Lamb 1979b, *Astrophys. J.*, **234**, 296.
[6] Makishima & Mihara 1992, Frontiers of X-ray Astronomy, Proc. of the Yamada Conf. p23.

NON-DETECTION OF KHZ QPOS IN GX 9+1 AND GX 9+9

RUDY WIJNANDS AND MICHIEL VAN DER KLIS
Astronomical Institute, University of Amsterdam

AND

JAN VAN PARADIJS
*Astronomical Insitute, University of Amsterdam, and
Department of Physics, University of Alabama at Huntsville*

1. Introduction

In numerous low-mass X-ray binaries quasi-periodic oscillations (QPOs) between 300 and 1200 Hz have been discovered (the kHz QPOs; see van der Klis 1997 for a recent review on kHz QPOs). Here we present the search for kHz QPOs in the atoll sources GX 9+1 and GX 9+9.

2. Observations and analysis

We observed GX 9+1 on 1996 Feb 29, Apr 21, May 29, and 1997 Feb 10 and Mar 9, and GX 9+9 on 1996 Aug 12, Oct 16, and Oct 30 with the RXTE satellite. We obtained a total of 23.3 ksec (GX 9+1) and 15.2 ksec (GX 9+9) of data. The X-ray hardness-intensity diagrams (HIDs) were made using the *Standard 2* data. Due to gain changes the GX 9+1 HID for the 1996 Feb 29 observation can not be directly compared with those of the other observations. The power density spectra were made using the $250\mu s$ time resolution data. We calculated rms amplitude upper limits (95% confidence) on QPOs with a FWHM of 150 Hz in the frequency range 100–1500 Hz.

3. Results

The HIDs for GX 9+1 and GX 9+9 are shown in Figure 1. According to the HID (Fig. 1a) and the high-frequency noise in the power spectrum GX 9+1 was on the lower banana during the 1996 Feb 29 observation. During the other observations GX 9+1 moved along the banana branch (Fig. 1b). The power spectrum and the HID of GX 9+9 suggest that this source was on the

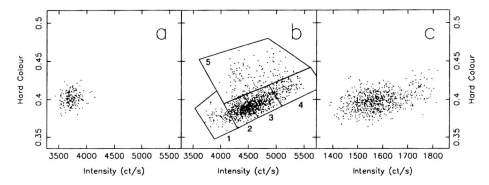

Figure 1. The HIDs of GX 9+1 (*a* and *b*) and GX 9+9 (*c*). The data of 1996 Feb 29 of GX 9+1 (*a*) were taken with a different PCA gain compared to the data of the other observations (*b* and *c*). The intensity is the count rate in the photon energy range 2.0–15.9 keV (*a*) or 2.1–16.0 keV (*b* and *c*); the hard colour is the count rate ratio between 9.7–15.9 kev and 6.5–9.7 keV in *a*, and between 9.7–16.0 keV and 6.4–9.7 keV in *b* and *c*. All points are 16s averages. The count rates are background subtracted, but not dead-time corrected. The regions in *b* have been used to calculate the upper limits.

banana branch during the observations. We find for GX 9+1 upper limits of 1.6% (the 1996 Feb 29 observation; energy range 2–60 keV) and 1.3% (all other observations combined; energy range 2–18.2 keV), and for GX 9+9 an upper limit of 1.8% (all data; energy range 2–18.2 keV). We divided the GX 9+1 banana in Figure 1b into different regions. For each region we calculated the rms upper limit on kHz QPOs. We find rms amplitude upper limits of 3.2%, 1.3%, 1.9%, 2.7%, and 3.4% (energy range 2–18.2 keV), for region 1, 2, 3, 4, and 5, respectively.

4. Discussion

The non-detection of kHz QPOs in GX 9+1 and GX 9+9 is consistent with the predictions of the sonic-point model proposed to explain the kHz QPOs (Miller et al. 1997). It is known from other atoll sources (e.g. 4U 1636−53: Wijnands et al. 1997; 4U 1820-30: Smale et al. 1997) that when they are in the upper banana branch the kHz QPOs are not detected. Thus, it remains possible that when GX 9+1 and GX 9+9 are observed longer on the lower banana, or even in the island state, kHz QPOs are detected in these sources.

References

Miller, C., Lamb, F. K., & Psaltis, D. 1997, ApJ, submitted (astro-ph/9609157)
Smale, A. P., Zhang, W., White, N., ApJ, 483L, 119
Van der Klis, M. 1997, to appear in Proceedings of the NATO Advanced Study Institute "The many faces of neutron stars", Lipari, Italy, 1996 (astro-ph/9710016)
Wijnands, R. A. D. et al. 1997, ApJ, 479L, 141

SPECTRAL EVOLUTION DURING DIPPING OF THE LOW MASS X-RAY BINARY XBT 0748-676

M. J. CHURCH AND M. BAŁUCIŃSKA-CHURCH
University of Birmingham, Birmingham B15 2TT, UK

AND

K. MITSUDA, T. DOTANI AND K. ASAI
ISAS, Yoshinodai 3-1-1, Sagamihara, Kanagawa 229, Japan

1. Introduction

XBT 0748-676 is a dipping LMXB source, with dips in X-ray intensity occurring at the orbital period of ~ 3.8 hrs. It is a member of the sub-group of dipping sources also including XB 1916-053 and XB 1254-690 in which the spectral evolution in dipping has previously been modelled by the "absorbed plus unabsorbed" approach, in which the dip spectra are modelled by two terms, each having the same form as used for non-dip emission, one of which is strongly absorbed but one which is not absorbed, which has a normalisation decreasing strongly in dipping. This energy-independent decrease has sometimes been taken to imply electron scattering in the absorber.

Previously we have proposed a physical model for the LMXB dipping sources (Church & Bałucińska-Church, 1995) consisting of two components: point-source blackbody emission from the neutron star, plus extended Comptonised emission from the accretion disk corona. This model has been shown to fit well all of the dipping sources to which it has so far been applied (X 1755-338, X 1624-490 and XB 1916-053 (see Church et al. 1997a). In the case of XB 1916-053, dipping often reaches a depth of 100% showing that the extended Comptonising emission region is totally covered by the absorbing region, and in spectral modelling we allowed progressive covering of this term. XBT 0748-676 is similar in that dipping also often reached 100%. In the present work we show that the two-component model with progressive covering of the extended emission also gives a very good explanation of dipping in XBT 0748-676.

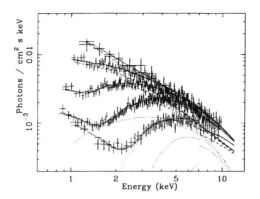

Figure 1. Fits to non-dip and dip data: solid lines - total model, dots - blackbody, dashes - power law.

2. Results

XBT 0748-676 was observed with ASCA on 1993, May 7th for 23 hours. Dipping reached a depth of 100% in the band 1 - 3 keV, and a depth of 80% in the band 3 - 10 keV. Spectral fitting results for GIS2 + GIS3 data are shown in Fig. 1 consisting of simultaneous fitting to non-dip data and 4 levels of dipping selected in intensity bands. The good fits shown were obtained with the 2-component model with $kT_{bb} = 1.99\pm0.16$ keV and photon index $\Gamma = 1.70\pm0.16$. As dipping develops, the partial covering fraction of the Comptonised emission increases from 0 to $\sim$1, N_H rising to $\sim 1.2 \cdot 10^{23}$ H atom cm^{-2}, while the blackbody N_H rises to $8 \cdot 10^{23}$ H atom cm^{-2}.

3. Conclusions

As we found previously in the case of XB 1916-053, dipping can be explained in XBT 0748-676 using our 2-component model in which the unabsorbed peak in dip spectra is the uncovered extended Comptonised emission. Dipping is due to photoelectric absorption alone without needing electron scattering in the absorber to play an important role in the band 1 - 10 keV. Detailed discussion can be found in Church et al. (1997b)

References

Church M. J. and Bałucińska-Church M., 1995, A&A **300**, 441
Church M. J., Dotani T., Bałucińska-Church M., Mitsuda K., Takahashi T., Inoue H. and Yoshida K., 1997a, ApJ **491**, Dec. 10th.
Church M. J., Bałucińska-Church M., Dotani T. and Asai K., 1997b, ApJ *Submitted*

TWO-DIMENSIONAL ACCRETION DISK MODELS OF A NEUTRON STAR

M. FUJITA
*Department of Phys., Hokkaido University
Kita-ku, Sapporo 060, Japan*

AND

T. OKUDA
*Hakodate College, Hokkaido Univ. of Education
1-2 Hachiman-cho, Hakodate 040, Japan*

We investigate the accretion disks around compact objects with high mass accretion rates near the Eddington's critical value $\dot{M}_{\rm E}$, where radiation pressure and electron scattering are dominant. This raises next problems : (a) whether stable disks could exist in relation to the theory of thermal instabilities of the disk and (b) what characteristic features the disks have if the stable disks exist. A non-rotating neutron star with the mass $M = 1.4 M_\odot$, radius $R_* = 10^7$cm and the accretion rate $\dot{M}_{\rm ac} = 2.0$ and $0.5 \dot{M}_{\rm E}$ (models 1 and 2) is considered as the compact object. We assume the α-model for the viscosity and solve the set of two-dimensional time-dependent hydrodynamic equations coupled with radiation transport. The numerical method used is basically the same as one described by Kley and Hensler (1987) and Kley (1989) but we include some improvements in solving the difference equations (Okuda et al. 1997). The initial configuration consists of a cold, dense, and optically thick disk which is given by the standard α-model (Shakura and Sunyaev 1973) and a rarefied optically thin atmosphere around the disk.

Numerical results are summarized as follows.

1. After several hundreds rotational periods at the inner edge of the disk, the disk attains to a nearly steady state. The disks are optically thick, geometrically thick, and thermally stable. The disk thickness (H/r) is $\sim$ 0.5 and 0.2 for models 1 and 2, respectively. The disk has approximately homogeneous vertical structures for the density and temperature (Figs. 1 and 2). The angular velocity exterior to the boundary layer ($r \leq 0.1 R_*$)

is nearly Keplerian in the whole disk region ($R_* \leq r \leq 10R_*$) considered here. The stable disks belong to a category of the advection dominated disks (*slim accretion disk*) proposed first by Abramowicz et al. (1988). However the disks are unstable against convection and many convective cells are formed in the disk region.

2. Due to the dominant radiation pressure force in the inner disk, the powerful winds with mass loss rates comparable to the input accretion rate $\dot{M}_{ac}$ are driven and a high velocity jet with $\sim 9 \times 10^4$ km s^{-1} is formed in the polar region. These velocity fields are rather variable. Resultantly the total luminosity emitted through the disk surface becomes lower by a factor of 5 than the theoretical one corresponding to the input accretion rate $\dot{M}_{ac}$.

3. The disks are surrounded by the optically thin, rarefied, and hot atmosphere.

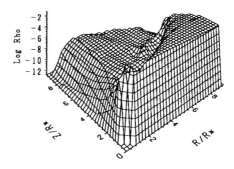

Fig. 1
The bird's-eye view of the density on the meridional plane for model 1 at $t = 10^3 P_d$, where P_d is the Keplerian orbital period at the inner edge of the disk. The densities are expressed by the logarithm of ρ (g cm^{-3}).

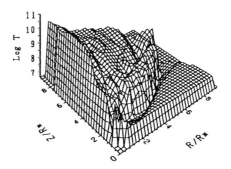

Fig. 2
The temperature distribution for model 1 at the same time as Fig. 1. The temperatures are expressed by the logarithm of T (K).

References

Abramowicz M.A., Czerny B., Lasota J. P., Szuszkiewicz E., 1988, ApJ 332, 646
Kley W., Hensler G. 1987, A&A 172, 124
Kley W. 1989, A&A 208, 98
Okuda T., Fujita M., Sakashita S. 1997, PASJ, in press
Shakura, N. I., Sunyaev R. A. 1973, A&A 24, 337

Session 3: Diagnostics of High Gravity Objects with X- and Gamma Rays

3-2. Black Hole Binaries

A MODEL FOR A THIN MAGNETISED DISC IN LMC X-3

P. HOYNG
SRON Laboratory for Space Research,
Utrecht, The Netherlands
(p.hoyng@sron.ruu.nl)

1. Introduction and summary

A model for the stationary radial distribution of the magnetic energy-stress tensor $<\boldsymbol{BB}>$ in a standard thin disc is presented, with allowance for magnetic torques and dissipation, see Fig. 1. The model is an extension of earlier work by Schramkowski et al. (1996). For LMC X-3, $<B^2>^{1/2}$ reaches $\sim 4 \times 10^5$ G near the inner edge of the disc, and the magnetic pressure is much smaller than the sum of gas and radiation pressure.

2. Equations for a magnetised thin disc

A description of $\boldsymbol{B}$ in terms of $<\boldsymbol{BB}>$ permits to allow for: (1) small-scale fields, (2) the influence of magnetic stresses on accretion, and (3) magnetic heating of a disc corona. The relevant equations are (Hoyng 1998):

$$(\partial_t - \beta\nabla^2)\,\varepsilon = (2\gamma - \zeta\Omega)\,\varepsilon - 3\Omega\,m \tag{1}$$

$$(\partial_t - \beta\nabla^2)\,m = (-\tfrac{2}{5}\gamma - \zeta\Omega)\,m - \tfrac{1}{2}\Omega\,\varepsilon, \tag{2}$$

where
$$\varepsilon \equiv <B^2>/8\pi \;;\quad m \equiv <B_r B_\theta>/8\pi, \tag{3}$$

with $\beta = \tfrac{1}{3}<v^2>\tau_c$ and $\gamma = \tfrac{1}{3}<|\nabla \times \boldsymbol{v}|^2>\tau_c$ ($\boldsymbol{v}$ is the turbulent flow); $\zeta = \nu/\beta$ is the turbulent Prandtl number and ν is the kinematic viscosity. The terms in (1) and (2) describe turbulent transport (β), amplification due to random field line stretching (γ), resistive dissipation ($\zeta\Omega$), and amplification by the keplerian shear flow Ω. No α-effect has been included.

The disc is modelled with the usual thin disc equations; those for energy and angular momentum have extra terms accounting for heating due to resistive dissipation and for magnetic stresses ($f = 1 - (r_0/r)^{1/2}$):

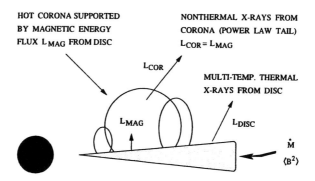

Figure 1. An optically thick magnetised disc in LMC X-3 radiates a multi-temperature blackbody X-ray spectrum. Like in the case of the solar corona, a magnetic flux density $-\beta \partial \varepsilon / \partial z$ supports a hot, optically thin corona emitting nonthermal Bremsstrahlung (the hard component of the X-ray spectrum).

$$\frac{4\sigma T^4}{3\tau} = \tfrac{1}{2}\nu\Sigma \left(r\frac{\partial \Omega}{\partial r}\right)^2 + \tfrac{1}{2}\zeta\Omega \int_{-H}^{H} \varepsilon \, dz \qquad (4)$$

$$\frac{\dot{M}\Omega f}{2\pi} = -\nu\Sigma r\frac{\partial \Omega}{\partial r} - 2\int_{-H}^{H} m \, dz \qquad (5)$$

By separating the vertical co-ordinate z in (1) and (2), two PDE's for ε and m in the central plane $z = 0$ are obtained, which are coupled to the algebraic thin disc equations that also refer to $z = 0$. Nonlinear effects are accounted for indirectly, by tuning (1) β so that the growth rate of the fundamental mode is zero, and (2) the constant determining the magnitude of $\varepsilon(r)$ so that the sum of the disc luminosity L_disc and magnetic luminosity L_mag equals the gravitational energy liberated during accretion.

3. Application to LMC X-3

The stationary solution applicable to LMC X-3 features: (1) a maximum r.m.s. field strength $<B^2>^{1/2}$ of 4×10^5 G at $r = 2r_0 = 6r_s$; (2) a $<B_r B_\theta>$ that is always negative, while $\varepsilon/\rho c_s^2 \lesssim 0.09$; (3) $L_\text{mag}/L_\text{disc} \simeq 0.02$, implying that the nonthermal X-ray flux should also be about 0.02 of the thermal X-ray flux, as is observed (Treves et al. 1990; Ebisawa et al. 1993).

References

Ebisawa, K., et al.: 1993, *Ap. J.* 403, 684.
Hoyng, P.: 1998, *A. & A.*, in preparation.
Schramkowski, G.P., Van Niekerk, E.C.M., Hoyng, P. and Achterberg, A.: 1996, *A. & A.* 315, 638.
Treves, A., et al.: 1990, *Ap. J.* 364, 266.

DISK OSCILLATIONS AS THE ORIGIN OF QUASI-PERIODIC OSCILLATIONS IN BLACK HOLE CANDIDATES

T. YAMASAKI
Hida Observatory, Kyoto University
Kamitakara, Gifu 506-13

AND

S. MINESHIGE AND S. KATO
Department of Astronomy, Faculty of Science, Kyoto University
Sakyo-ku, Kyoto 606-01

Quasi-periodic oscillations (QPOs) of a few Hz are observed in the very high state of some black hole candidates (GX 339-4 and GS 1124-68). This is the Kepler frequency at the radius of a few hundred Schwarzschild radii. As a possible mechanism of the QPOs in these objects, the trapped oscillations in the accretion disks are considered. The trapped oscillations of the disks were investigated by several authors. They studied the trapped oscillations in the standard radiative cooling-dominated disks. Recently, the advection-dominated accretion flow is considered, as a possible model to explain the hard X-ray spectra of the black hole candidates or the active galactic nuclei. In particular, in the very high state of some black hole candidates, the spectrum can be explained by the disk-corona model which comprises the cold standard accretion disk and the advection-dominated corona above the cold disk. We thus investigated the trapped axi-symmetric oscillations in the advection-dominated corona by the global linear analysis.

The results show that, i)the periods of the trapped oscillations are $0.1\Omega_{out} \sim 2\Omega_{out}$. Here Ω_{out} is the Kepler frequency in the outer boundary. In the case of the black hole of $10M_\odot$, the Kepler frequency is about $7s^{-1}$ at the radius of 100 Schwarzschild radii. ii)As the advective cooling becomes more efficient, the frequencies become large. iii)Owing to the radial inflow in the unperturbed flow, the phase of the oscillation differs with radius. Because of this, the shape of the light curve deviate from the sinusoidal one.

RAPID X-RAY VARIABILITY OF CYG X-1

Observed with the Indian X-ray Astronomy Experiment (IXAE)

P. C. AGRAWAL, B. PAUL, A. R. RAO, M. N. VAHIA AND
J. S. YADAV
Tata Institute of Fundamental Research
Homi Bhaba road, Mumbai (Bombay), 400 005, India

AND

T. M. K. MARAR, S. SEETHA AND K. KASTURIRANGAN
ISRO Satellite Centre, Bangalore, 560 017, India

Abstract: We have made observations of the black hole binary Cyg X-1 with the Indian X-ray Astronomy Experiment (IXAE). Observations made with time resolution ranging from 0.4 ms to 1 s showed variations and flaring activity on sub-sec and longer time scales. Results on time variability on different time scales and flaring characteristics in the two states of Cyg X-1 are presented.

1. Introduction

The bright X-ray binary Cyg X-1 considered to be a strong candidate for a black hole, has a bimodal behaviour with distinct 'soft' and 'hard' states and makes transition from one state to the other at irregular intervals. It exhibits rapid and chaotic intensity variations over time scales ranging from milliseconds to hours and days. Quasi-periodic oscillations have also been detected from it on several occasions. It has been found that changes in the PDS are associated with the spectral transition of the source. We report observations of Cyg X-1 made with the Indian X-ray Astronomy Experiment (IXAE) onboard the Indian Satellite IRS-P3 in both the hard and the soft states.

2. Observations

The observations were made with the three proportional counters in the pointed mode (PPCs) which form a part of the IXAE onboard the Indian Satellite IRS-P3 which was launched from India on 1996, March 21. Cyg X-1 was first observed during 1996, April 30 to May 11 period in a low-hard state and again during 1996, July 5–8 interval in a soft-bright state.

3. Results

The two plots in the left column of Figure 1 show the background subtracted light curves of the PPCs in the hard and the soft states respectively. Random and

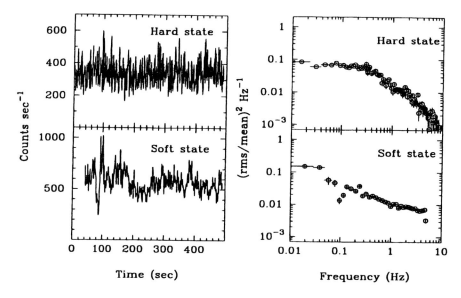

Figure 1. The light curves (left column) and power density spectra (right column) of Cyg X-1 in the two intensity states observed with the PPCs.

chaotic intensity variations on time scale of 1 s and longer are clearly visible in the figure. The short time scale variations are found to be more pronounced in the hard state compared to the soft one. Power density spectra (PDS) for the two states were obtained from individual data segments and then added together. The PDS obtained from our observations of Cyg X-1 in the two intensity states are shown in the two plots in the right column of Figure 1. The hard state PDS is based on data obtained with 1 s and 0.4 ms time resolution while the soft state PDS is based on data of 1 s and 0.1 s resolution. The high state PDS can be fitted by a power law with index of -1.1 between 0.3 to 10 Hz. Below 0.3 Hz the PDS is quite flat right upto .01 Hz with a break in the spectral slope at about 0.3 Hz. The PDS for the soft state can also be fitted with a power law of index -0.39 above a frequency of 0.03 Hz indicating that the soft state PDS is flatter compared to the hard state. There is an indication of flattening below 0.03 Hz similar to that seen in the hard state but no clear indication of the break in the slope.

The present observations confirm that the PDS characteristics are dependent on the intensity state of the source and change smoothly as the source makes transition from one state to another.

OPTICAL FLARES OF CYGNUS X-1

N.G. BOCHKAREV, E.A. KARITSKAYA AND V.M. LYUTY
Sternberg Astronomical Institute, Moscow, 119899, Russia
Institute of Astronomy, Pyatnitskaya st., 48, Moscow, Russia

Photoelectric observations of Cyg X-1 in 1996-97 fulfilled in Tien-Shan, Kazakhstan (WBVR), Mt. Maidanak, Uzbekistan (UBVR) and in Crimea, Ukraine (UBV) have detected unusual optical flare activity. During May-August 1996 Cyg X-1 was in high soft X-ray state and after August 9, 1996 it returned to its ordinary state but with some evidences of X-ray activity.

Karitskaya et al. (1997) have registered two large flares (June 2, 1996 = JD 2450267 and October 9, 1996 = JD 2450366) in U, B, V and R bands with the amplitudes about 0.04 magnitude and durations $\geq$ 2 days and $\geq$ 1 day. The amplitudes of the flares were approximately of the same magnitude in all the bands. According to X-ray AMS/RXTE data (http://space.mit.edu/XTE) the flares coinsided in time with 20 % X-ray 3-12 keV dips (Fig.1, upper dot set). The power of the flares is approximately 10^{36} erg/s (about $3*10^{35}$ erg/s per band).

On May 14, 1997 V.Lyuty and G.Zaitseva detected a very powerful single flare. Similar flares were not previously observed during more than 25 years of Cyg X-1 optical observations. During 2.5 hours long observations Cyg X-1 was 0.12 mag brighter than expected from the regular orbital light curve in bands U and B and only 0.02 mag brighter in V band. Preceding and following nights there was no noticeable brightness deviations. The power of the flare was about 10^{36} erg/s in each U and B bands.

The colors of the flare of May 14, 1997 correspond to optically thin hydrogen gas region with the temperature $T = (1 - 3) * 10^4$ K. Here the hydrogen lines dominate in R and B bands. The He II 4686 line is in B band. There are no strong permitted lines in V band and the luminosity of Pashen continuum can be small. This gas can be ejected by a quasipoint explosion with a velocity v = 1500 km/s. The gas parameters are as follows: $T = (10 - 30) * 10^4$ K; volume emission measure $MV = n_e * V \simeq 10^{61} cm^{-3}$; radius $R \sim 3 * 10^{12} cm$; electron number density $n_e \sim 3 * 10^{11} cm^{-3}$; mass $M \sim 3 * 10^{-8}$ solar mass; kinetic energy $E \sim 10^{40}$ erg.

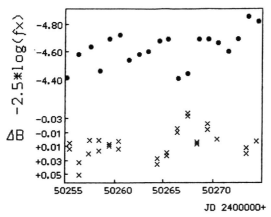

Figure 1. Optical flare observed June 2, 1996 in Mt.Maidanak observatory (Karitskaya et al., 1997). Comparison of AMS/RXTE 3–12 keV X-ray data (upper part of the figure) with the optical light curve in V band (lower part of the figure). X-ray data are given in the logarithmic scale of magnitudes. For V band deviation from orbital double wave curve is shown. Residuals of the daily averaged X-ray data are less than circle dimentions.

Hydrodynamic time of the outflow is only 5–6 hours. Therefore this outflow was to occur several hours before the observations were started and disappeared during the next 0.5 – 1 day.

The June 2, 1996 and October 9, 1996 flares were smaller and emitted in all the bands. In these cases the gas is assumed to be of higher density. It can be ff-emission of gas with MV $\simeq 10^{61} cm^{-3}$. It is probably the emission from optically thick gas balk on the outer part of the accretion disk. This "hot spot" can be originated as a result of interaction of the disk matter with the gas inflowing into the disk. The existense of similar gas balk on the outer part of Cyg X-1 accretion disk has been suggested by Kemp (1980) for interpretation of optical linear polarization orbital variations. Flare luminosity can be explained by re-radiation of X-ray emission of the central part of the accretion disk by the gas balk with $R \simeq 1/3$ of the disk radius. The X-ray dips arise as a result of eclipsing of X-ray source by the outer part of the gas balk, where the column density of gas $N \simeq 10^{23} cm^{-2}$ is lower then N following from MV for main part of the "spot".

The work is supported by RBRF grant 96-02-18044 and the program "Monitoring of Unique Astrophysical Objects" Russian Ministry of Science.

References

Karitskaya, E.A., Goranskij, V.P. and Grankin, K.N. (1997) Properties of Optical Brightness Variations of Cyg X-1 in 1995– 1996 Including X-ray Hard and Soft States, *JENAM-97*, Thessaloniki, Greece; (in preparation to *Pisma v Astron. Zh.*)

Kemp, J. (1980) On the Allowed Size of an Accretion Disk in Cygnus X-1, *Astrophysical Journal*, **235** pp. 595 – 602

CYGNUS X-1: X-RAY EMISSION MECHANISM AND GEOMETRY

WEI CUI
MIT, Cambridge, MA 02139, USA

S. N. ZHANG
NASA/MSFC, Huntsville, AL 35812, USA

J. H. SWANK, X.-M. HUA AND K. EBISAWA
NASA/GSFC, Greenbelt, MD 20771, USA

AND

T. DOTANI
ISAS, Yoshinodai, Sagamihara, Kanagawa, 229, Japan

1. Introduction

On 1996 May 10, the All-Sky Monitor aboard RXTE revealed that Cyg X–1 started a transition from its hard state to soft state (Cui 1996). Throughout this interesting episode, snapshots were taken with more sensitive detectors on ASCA, RXTE, and CGRO to monitor the temporal and spectral variability of the source over a broad energy range.

The results (both spectral and temporal) from these observations seem to converge toward a self-consistent picture, which not only specifies the X-ray emission geometry but also qualitatively describes how the geometry evolves during the transition.

2. Results and Discussion

The X-ray spectrum of Cyg X–1 extends beyond 100 keV in both states. The spectral shape is rather complicated in details. At low energies, it is dominated by an ultra-soft component, which is thought to be the emission from an optically thick, geometrically thin disk. During the transition, the color temperature varies by a factor of 2 – 3, and the luminosity (of this component) by a factor of 3, indicating that the inner disk edge moves about 3 times closer to the black hole as the soft state is approached (Zhang et al.

1997). At high energies the spectrum can be described by a Comptonized spectrum. It is steeper in the soft state, implying a smaller Comptonizing corona, perhaps due to more efficient cooling (Cui et al. 1997a). Such interpretation is strongly supported by the measured hard X-ray lags: much smaller in the soft state, as well as observed shape of power-density spectra (Cui et al. 1997a,c). The measured coherence function between various energy bands is nearly unity in both states (Vaughan & Nowak 1997; Cui et al. 1997c), but is less during the transition (Cui et al. 1997c), implying that the corona indeed varies during such a period.

A Compton reflection "bump" is apparent on the spectrum, as well as the presence of an iron emission line. In the soft state, the reflecting medium is highly ionized and the solid angle subtended is smaller than in the hard state (Ebisawa et al. 1996; Cui et al. 1997b). This is consistent with the reflection occuring mostly at the inner disk edge that is closer to the black hole (thus hotter) in the soft state. The iron line centers at ~6.4 keV in the hard state, probably due to neutral irons (Ebisawa et al. 1996), while in the soft state it is mostly due to He-like irons (Cui et al. 1997b). This suggests that the line emission also originates in the innermost region of the disk. The broader line profile in the soft state supports this scenario.

Attempts have been made, for the first time, to simultaneously model the observed spectral and timing properties of Cyg X-1 (Hua et al. 1997). The model invokes an extended hot Comptonizing corona with a non-uniform density distribution. It reproduces the observed spectral shape well, and more importantly it is capable of explaining the frequency-dependence of measured hard lags, which has been a "dilema" for Compton models over the past decade (Miyamoto et al. 1988). It now becomes clear that such dependence is sensitive to the density distribution of the corona. Therefore, phase lag measurement may ultimately provide insights into the dynamics of hot corona. In the future, more physics need to be incorporated in the model to account for such features as the reflection bump and iron line.

3. References

Cui, W. 1996, IAUC. 6404
Cui, W., et al. 1997a, ApJ, 474, L57
Cui, W., Ebisawa, K., Dotani, T. & Kubota, A. 1997b, ApJ, submitted
Cui, W., Zhang, S. N.,Focke, W., & Swank, J. 1997c, ApJ, 484, 383
Ebisawa, K. et al. 1996, ApJ, 467, 419
Hua, X.-M., Kazanas, D., & Cui, W. 1997, ApJ, submitted
Miyamoto, S., et al. 1988, Nature, 336, 450
Vaughan, B. A., & Nowak, M. A. 1997, ApJ, 474, L43
Zhang, S. N., et al. 1997, ApJ, 477, L95

X-RAY DETERMINATION OF THE BLACK-HOLE MASS IN CYGNUS X-1

A.KUBOTA AND K.MAKISHIMA
Department of Physics, University of Tokyo
7-3-1, Hongo, Bunkyo-ku, Tokyo 113 JAPAN

T. DOTANI, H. INOUE, K. MITSUDA, F. NAGASE, H. NEGORO AND Y. UEDA
Institute of Space and Astronautical Science, Sagamihara, Kanagawa 229, JAPAN

K. EBISAWA
Universities Space Research Association, NASA/GSFC

S. KITAMOTO
Department of Earth and Space Science, Osaka Univ. Osaka 560, JAPAN

AND

Y. TANAKA
SRON-Utrecht, Sorbonnelaan 2, 3584 CA Utrecht, the Netherlands

About 10 X-ray binaries in our Galaxy and LMC/SMC are considered to contain black hole candidates (BHCs). Among these objects, Cyg X-1 was identified as the first BHC, and it has led BHCs for more than 25 years(Oda 1977, Liang and Nolan 1984). It is a binary system composed of normal blue supergiant star and the X-ray emitting compact object. The orbital kinematics derived from optical observations indicates that the compact object is heavier than ~ 4.8 $M_\odot$ (Herrero 1995), which well exceeds the upper limit mass for a neutron star(Kalogora 1996), where we assume the system consists of only two bodies. This has been the basis for BHC of Cyg X-1.

Black hole binaries exhibits characteristic X-ray properties such as distinctive spectral feature called "hard" or "soft" states. (Tanaka 1995). Although Cyg X-1 has been a prototypical blackhole candidate, it has rarely been found in "soft" states. In 1996 May, the RXTE/ASM reported an X-

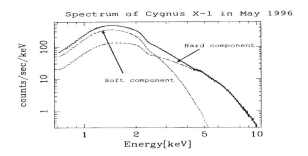

Figure 1. GIS2+GIS3

ray flare up of Cygnus X-1. We observed it with ASCA on 30–31 May for 33 ksec. The observed 0.7–10 keV flux was 1.6×10^{-8} erg/cm^2/sec. The obtained spectrum consists of the "Soft" and "Hard" components (Figure 1). The Soft component of the spectrum was well fitted with the multi-color accretion disk model (MCD model)(Mitsuda 1984) and the general relativistic accretion disk model(GRAD model)(hanawa 1989,ebisawa 1991) as observed in some other BHCs in the Soft states. A color temperature at the disk inner edge is ~ 0.43 keV, and a disk bolometric flux is $\sim 5.60 \times 10^{-8}$ erg/sec/cm^2. Assuming the distance of 2.5 kpc, the inclination of 30°, and the color-to-effective temperature ratio of 1.7, and considering the inner boundary condition, the disk inner radius has been determined to be ~ 90 km.

Assuming the Schwarzshild black hole,the inner edge of the accretion disk is thought to extend down to three times the Schwarzshild radius ($3R_s$),which corresponds to the last stable orbit around the non–rotating black hole. By equationg the disk inner radius to $3R_s$, the mass of central object can be estimated.

Thus,we determined the mass of Cyg X-1 to be 12 ± 2 $M_\odot$. Application of the GRAD model gave a consistent result. These results confirm the existence of the black-hole in the Cygnus X-1 system.

References

Dotani et al. ApJL vol.485 in press
Oda, M. 1977, Space Sci. Rev. 20, 757
Liang, E. P. and Nolan, P. L. 1984, Space Sci. Rev. 38, 353
Herrero, A., Kudritzki, R. P., Gabler, R., Vilchez, J. M., Gabler, A. 1995, AAP 297, 556
Kalogera, V. and Baym, G. 1996, ApJL 470, L61
Tanaka, Y., and Lewin, W. H. G. 1995, *X-Ray Binaries*, eds. W. H. G. Lewin, J. van Paradijs & W. P. J. van den Heuvel, (Cambridge University Press), 126
Makishima, K., Maejima, Y., Mitsuda, K. et al. 1986, PASJ 308, 635
Mitsuda K., Inoue, H., Koyama, K. et al. 1984, PASJ 36, 741
Hanawa T. 1989, ApJ 341, 948
Ebisawa, K., Mitsuda, K., Hanawa, T. 1991, ApJ 367, 213

SPECTRAL EVOLUTION OF THE THERMAL COMPONENT IN CYGNUS X-1 DURING INTENSITY VARIATIONS IN THE SOFT STATE

M. BALUCINSKA-CHURCH AND M. J. CHURCH
University of Birmingham, Birmingham B15 2TT, UK

AND

K. MITSUDA, T. DOTANI, H. INOUE AND F. NAGASE
ISAS, Yoshinodai 3-1-1, Sagamihara, Kanagawa 229, Japan

1. Introduction

Cygnus X-1 was in the Soft State between May and September 1996 for the first time in more than 20 years creating an opportunity to study the thermal component, its spectral evolution, time variability and its relation to the hard component during intensity variations. We will show how the luminosity of the thermal component changes in relation to its temperature and attempt to determine whether the emitting area varied during the variations.

2. Results and Conclusions

Cygnus X-1 was observed with *ASCA* during the Soft State on May 30th, 1996. We present results for spectral analysis of *ASCA* GIS data from this observation lasting $\sim 60,000$ s. The light curve showed that the intensity increased steadily by 30% during the observation. Spectral analysis revealed that the increase was primarily due to brightening of the dominant thermal component. We have selected spectra from 5 parts of the observation where the intensity below 5 keV was stable, i.e without fast changes. Spectral fitting of these with a simple blackbody to model the disc emission plus a power law for the hard component was used to determine the disc luminosity and the colour temperature. Results constrain the relationship $L \propto T_{col}^{\beta}$ to a range for β of 2.3 to 4.8 with 90% confidence. The Figure shows the variation of L with temperature in the Soft State. We have also tested

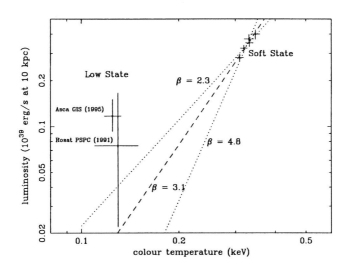

whether the disc inner radius varies with intensity. The actual inner radius R_{in} of the disc is proportional to $f^2 R_*$, where R_* is the apparent inner radius determined by fitting a multi-colour disc, and f is the ratio of T_{col} to effective temperature, determined by fitting the relativistic disc model. Thus we can obtain R_{in}. The data are consistent with a constant R_{in} in the Soft State, although the quality of the data does not allow us to say absolutely that R_{in} does not vary with luminosity. We can also test *whether R_{in} varies between the Low State and the Soft State*. In the Soft State, spectral fitting gives $f \simeq 1.7$. The Low State *ROSAT* and *ASCA* values of disc luminosity (Balucinska-Church et al. 1995, 1997) imply an increase in R_* by ~ 2.5 and 4.0 times respectively compared with the Soft State. The results are consistent with *two explanations*. Firstly, if R_{in} does not change, to explain the change in R_*, f would have to be ~ 1 in the Low State giving a factor of $(1.7)^2 = 2.9$ in R_*. In this case the disc emission in the Low State would have to be simple blackbody at every radius. Alternatively, if $f \simeq 1.7$ in both states, the inner radius decreases by 2.5 - 4.0 times from the Low State to the Soft State in general agreement with the advective disc model (Narayan 1997).

References

Bałucińska-Church M., Belloni T., Church M. J. and Hasinger G., 1995, A&A **302**, L5.
Bałucińska-Church M., Takahashi T., Ueda Y., Church M. J., Dotani T., Mitsuda K., and Inoue H., 1997, ApJ, **480**, L115.
Narayan R., 1997, in: "Accretion Phenomena and Related Outflows", IAU Colloqium no. 163, eds.: D. T. Wickramasinghe, L. Ferrario and G. V. Bicknell.

ASCA OBSERVATIONS OF THE SUPERLUMINAL JET SOURCE GRS1915+105

KEN EBISAWA, T. TAKESHIMA, N. E. WHITE
Laboratory for High Energy Astrophysics, NASA/GSFC
Greenbelt, MD, 20771, USA

T. KOTANI
The Institute of Physical and Chemical Research (RIKEN)
2-1 Hirosawa, Wako, Saitama, 351-01, Japan

T. DOTANI, Y. UEDA
Institute of Space and Astronautical Science
Yoshinodai, Sagamihara, Kanagawa, 229, Japan

B. A. HARMON, C. R. ROBINSON, S. N. ZHANG
NASA/MSFC
Huntsville, AL 35806, USA

W. S. PACIESAS
Department of Physics, University of Alabama in Huntsville
Huntsville, AL 35899, USA

M. TAVANI
Columbia Astrophysics Laboratory, Columbia University
New York, NY 10027, USA

AND

R. FOSTER
Naval Research Laboratory
Washington C. C. 20375, USA

1. Introduction

GRS1915+105 is an extraordinary X-ray transient which exhibits superluminal radio jets. In this paper, ASCA observations of the GRS1915+105 conducted from 1994 to 1997 are reported. Observations are carried out on the following dates each for $\sim$ 20 ksec exposure; Sep 27 1994, April 20 1995, Oct 23 1996 and Apr 25 1997.

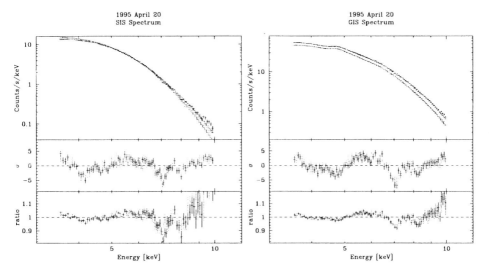

Figure 1. SIS (left) and GIS (right) energy spectra in April 1995 fitted with an absorbed cut-off power-law model. Absorption line features are clearly seen in the residuals.

2. Results

Main results may be summarized as follows:

1. The source is highly variable at various time scales, and in particular characteristic burst/dip structures were observed in October 1996 and April 1997. The flux variation is larger above 3 keV, and the spectral change is most simply described by the reduction of the cut-off energy when the X-ray flux becomes lower.

2. The continuum energy spectra are approximately fitted by an absorbed cut-off power-law with the form $AE^{-\alpha}e^{-E/E_c}$. Overall, there is not a clear correlation between the cut-off energy and the X-ray flux. The hydrogen column densities are always within the range of 3.5 – 4.1 $\times 10^{22}$ cm^{-2}.

3. Characteristic absorption line features are seen in September 1994 and April 1995 at $\sim$ 6.7 keV, 7.0 keV and 8.0 keV (figure 1). The former two may be identified as the $K\alpha$ lines from He-like and H-like irons and the last one may be the $K\beta$ line from He-like iron. These features are vaguely seen in October 1996, and in April 1997, an emission line-like feature is seen at $\sim$ 6.7 keV. These absorption line features are reminiscent of the similar spectral features in the other Galactic superluminal jet source GRO J1655–40 (Ueda et al. 1997, ApJ Letter, in print).

The postscript file of the poster paper presented at the conference can be found at ftp://lheaftp.gsfc.nasa.gov/pub/ebisawa/kyoto.ps.gz.

X-RAY TIMING STUDIES OF GRS 1915+105

with the Indian X-ray Astronomy Experiment (IXAE)

B. PAUL, P. C. AGRAWAL, A. R. RAO, M. N. VAHIA AND
J. S. YADAV
Tata Institute of Fundamental Research
Homi Bhaba road, Mumbai (Bombay), 400 005, India

AND

T. M. K. MARAR, S. SEETHA AND K. KASTURIRANGAN
ISRO Satellite Centre, Bangalore, 560 017, India

1. Introduction

We have made photometric observations of the galactic superluminal jet source GRS 1915+105 in the energy bands of 2-6 and 6-18 keV during 1997 June 12-29 and August 8-10. During our observations, different types of very intense, quasi-regular X-ray bursts have been observed from this source. We present here the light curves and the power density spectra of our observation of this source in its bright state.

2. Observations

Observations were made with the three Pointed Proportional Counters (PPC) of the Indian X-ray Astronomy Experiment (IXAE) onboard the Indian satellite IRS-P3. The PPCs, filled with argon-methane mixture at 800 torr pressure and working in the 2-18 keV energy range, have a total area of 1200 cm^2 and field of view of 2.3° × 2.3°. The satellite was launched from the Shriharikota (SHAR) range, India, on 1996 March 21 and was put into a near circular orbit of height 830 km and 98° inclination. The pointing accuracy of the satellite is better than 0.1°.

Observations were made with a time resolution of 0.1 s or 1 s and more than 55 000 seconds of useful source exposure was obtained.

3. Results

During the 1997 observations, the source was in a bright intensity state. A unique characteristic noted in the light curves is the presence of strong quasi-regular X-ray bursts. Some representative light curves of 500 seconds duration obtained on different days with one of the PPCs are shown in the left column of Figure 1. The power density spectra in the frequency range of 0.001 to 5 Hz, obtained from the

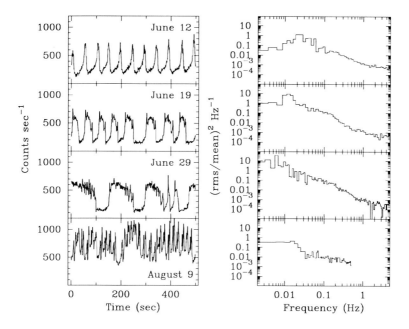

Figure 1. The light curves showing different types of bursts observed in GRS 1915+105 are plotted in the left column along with the day of observation in 1997. The corresponding power density spectra are shown in the right column.

light curves consisting these different type of bursts are shown in the right column of the figure.

GRS 1915+105 is the only black hole source in which regular bursts are observed. The bursts observed in this source also have the unique feature of slow rise and fast decay. The spectrum is found to be harder during the burst decay. These features make the bursts in GRS 1915+105 distinct from both the type I and type II X-ray bursts observed in neutron star sources.

MULTIWAVELENGTH TEMPORAL BEHAVIOR OF GRS 1915+105

D. HANNIKAINEN
Observatory, PO Box 14
00014 University of Helsinki, Finland

AND

PH.DUROUCHOUX
C.E. Saclay, DSM, DAPNIA, Service d'Astrophysique
91191 Gif sur Yvette, Cedex France

GRS 1915+105

The transient X-ray source GRS 1915+105 was discovered in August 1992 with the *GRANAT*/WATCH all-sky monitor (Castro-Tirado et al. 1994). Subsequent VLA observations from March through April 1994 led to the discovery of apparent superluminal motion in a pair of radio condensations moving away from the compact radio core (Mirabel & Rodriguez 1994). These jet-like features are interpreted as a bipolar outflow with bulk velocity $\sim 0.9c$. Although no optical counterpart has been observed, due to the heavy extinction in the Galactic plane, and therefore not enabling measurements of the mass of the compact object, the hard X-ray spectrum and high luminosity ($\sim 10^{39}$ erg s^{-1}), extreme variability in the X-ray light curve and the relativistic jets make GRS 1915+105 a strong black hole candidate.

GRS 1915+105 has been extensively observed in the radio, X-ray and hard X-ray since its discovery. In all wavebands the emission is highly variable and complex, including very large amplitude flaring on timescales of minutes in the X-rays (e.g. Greiner, Morgan & Remillard 1996) and Pooley & Fender (1997) have identified three different types of behavior in the radio flux density following a one and a half years' monitoring at 15 GHz with the Ryle telescope. In Figure 1 we have gathered together a sample of light curves of GRS 1915+105, demonstrating flaring activity and periods of total quiescence in both the radio and X-ray domains. There is no obvious correlation between the radio and X-ray/gamma-ray behavior.

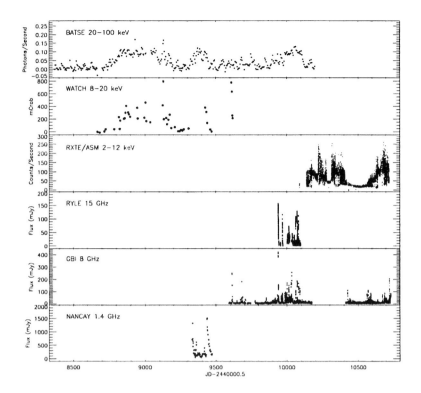

Figure 1. References. BATSE: http://cossc.gsfc.nasa.gov/cossc/batse/hilev/occ.html; WATCH: Finoguenov *et al.* (1994); RXTE: http://space.mit.edu/XTE/ASM_lc.html; RYLE: R. Fender (priv. comm.); GBI: E. Waltman (priv. comm.) and http://www.gb.nrao.edu/gbint/GBINT.html; NANCAY: Rodriguez *et al.* (1995)

Acknowledgements.

The authors would like to thank Robert Fender, Craig Robinson, Elizabeth Waltman, William Heindl and Chris Schrader for their help in gathering the data. Public domain data from the NSF-NRAO-NASA Green Bank Interferometer. DH travelled to Kyoto on a grant from the Chancellor of the University of Helsinki.

References

Castro-Tirado, A. J. *et al.* 1994, ApJS 92, 469
Finoguenov, A. *et al.* 1994, ApJ 424, 940
Greiner, J., Morgan, E.H. & Remillard, R. A. 1996, ApJ 473, L107
Mirabel, I. F. & Rodriguez, L. F. 1994, Nature 371, 46
Pooley, G. & Fender, R. 1997, MNRAS, submitted
Rodriguez, L. F. *et al.* 1995, ApJS 101, 173

RAPID HARD X-RAY VARIABILITY IN GRO J0422+32

F. VAN DER HOOFT, J. VAN PARADIJS AND M. VAN DER KLIS
University of Amsterdam, The Netherlands

C. KOUVELIOTOU, D.J. CRARY AND M.H. FINGER
USRA and NASA/MSFC, USA

B.C. RUBIN
RIKEN Institute, Japan

B.A. HARMON AND G.J. FISHMAN
NASA/MSFC, USA

AND

W.H.G. LEWIN
MIT, USA

1. Introduction

The soft X-ray transient (SXT) GRO J0422+32 (Nova Persei 1992) was detected with the Burst And Transient Source Experiment (BATSE) on board the *CGRO* on 1992 August 5 (Paciesas et al. 1992) (Truncated Julian Day [TJD] 8839). The source intensity of GRO J0422+32 increased rapidly, reaching a flux of $\sim$ 3 Crab (40–230 keV) within days after its first detection (Harmon et al. 1992). Hereafter, the X-ray intensity of the source decreased exponentially with a decay time of $\sim$ 43 days (Vikhlinin et al. 1995). A secondary maximum of the X-ray intensity was reached at TJD 8978, 139 days after the first detection of the source. The daily averaged flux history of GRO J0422+32 in the 40–150 keV energy band is presented in Figure 1.

2. Time series analysis

For our analysis, we considered uninterrupted data segments of 512 successive time bins (of 1.024 s each) on which we performed Fast Fourier Transforms (FFTs) covering the frequency interval 0.002–0.488 Hz. Per day, we obtained typically 35 of such segments while the source was above

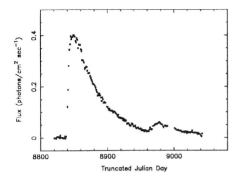 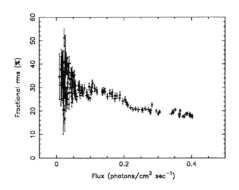

Figure 1. (Left:) Daily averaged flux history of GRO J0422+32 in the 40–150 keV energy band, as determined using the Earth occultation technique. The first detection of the source was on TJD 8839 (1992 August 5). (Right:) The fractional rms amplitudes (20–100 keV) determined in the interval 0.01–0.48 Hz, plotted versus the flux of GRO J0422+32 in the 40–150 keV energy band.

the Earth horizon. For each data segment and for each of the eight detectors separately, we calculated and coherently summed the FFTs of the lowest two energy channels (20–50, 50–100 keV). For those detectors which had the source within 60 degrees of the normal, these FFTs were again coherently summed (weighted by the ratio of the source to the total count rates) and converted to Power Density Spectra (PDSs). The PDSs were normalized such that the power density is given in units of $(\text{rms/mean})^2$ Hz^{-1}. Our analysis fully covers the X-ray outburst of GRO J0422+32.

We determined fractional rms amplitudes by integrating the single-day averaged PDSs of GRO J0422+32 over the frequency interval 0.01–0.48 Hz. During the rise to the first X-ray maximum the fractional rms amplitudes decrease monotonically, while gradually increasing during the exponential decline of the light curve. The largest fractional rms amplitudes were obtained near TJD 8960, shortly before the onset of the secondary maximum in X-rays. At the secondary maximum, the fractional rms amplitudes again reach a local minimum. In Figure 1 we present the fractional rms amplitudes (20–100 keV, 0.01–0.48 Hz), plotted versus the 40–150 keV flux of GRO J0422+32. At low flux levels the fractional rms amplitudes become uncertain and are dominated by detector noise due to unresolved sources in the uncollimated field of view of BATSE. At higher flux levels the fractional rms amplitude and flux appear to be anti-correlated.

References

Harmon, B.A. et al., 1992, IAUC, 5584
Paciesas, W.S. et al. 1992, IAUC, 5580
Vikhlinin, A. et al. 1995, ApJ, 441, 779

THREE-DIMENSIONAL LOCAL MHD SIMULATIONS OF HIGH STATES AND LOW STATES IN MAGNETIC ACCRETION DISKS

T. MATSUZAKI AND R. MATSUMOTO
Department of Physics, Faculty of Science, Chiba University, 1-33 Yayoi-Cho, Inage-Ku, Chiba 263, Japan

T. TAJIMA
Institute for Fusion Studies, the University of Texas at Austin, Austin, TX 78712, USA

AND

K. SHIBATA
National Astronomical Observatory, Mitaka, Tokyo 181, Japan

Black hole candidates sometimes show a transition between the high (or soft) state and the low (or hard) state. In the low state, low frequency time variations are much larger than the high state. A possible mechanism of the large-amplitude, sporadic time variabilities in the low-state is the magnetic energy release in low-β ($\beta = P_{gas}/P_{mag} < 1$) disks (Mineshige, Kusunose & Matsumoto 1995). It had been thought that low-β disks cannot exist because buoyant escape of magnetic flux due to the Parker instability may set the lower limit for β inside the disk. Shibata, Tajima & Matsumoto (1990), however, pointed out that in accretion disks, once a low-β disk is formed, it can stay in low-β state partly because the growth rate of the Parker instability decreases when $\beta < 1$. They suggested that magnetic accretion disks fall into two types; high-β disks and low-β disks.

We carried out local three-dimensional magnetohydrodynamic (MHD) simulations of a gravitationally stratified, isothermal Keplerian disk initially threaded by azimuthal magnetic field. Since both differential rotation and vertical gravity are included, the magnetorotational (or Balbus & Hawley) instability (Balbus & Hawley 1991) couples with the Parker instability when $\beta \sim 1$. Local Cartesian coordinate is used with x, y, z in the radial, azimuthal, and vertical direction, respectively. The vertical gravity is as-

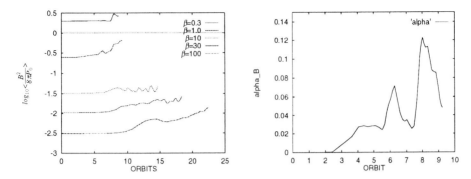

Figure 1. Time evolution of the mean magnetic field strength $\langle B^2/(8\pi P_0(0))\rangle$ for various initial β (left panel) and the angular momentum transport rate $\alpha_B = -\langle B_x B_y/(4\pi P_0(0))\rangle$ for a model with $\beta = 1$ (right panel). The unit of time is the rotation time.

sumed to be $g_z = -GMz/(r_0^2 + z^2)^{3/2}$ where r_0 is the radius from the gravitating center. We assume that β is uniform at the initial state. The size of the simulation box is $(L_x, L_y, L_z) = (1H, 18H, 16H)$, where H is the scale height defined by using the sound speed C_s and Keplerian angular speed Ω_K as $H = C_s/\Omega_K$. The azimuthal boundaries are periodic. The radial boundaries are treated by using the sliding periodic condition.

The left panel of Figure 1 shows the time evolution of the mean magnetic field strength $\langle B^2/(8\pi P_0(0))\rangle$ for various initial plasma β, where $P_0(0)$ is the initial equatorial pressure. Numerical results indicate that in high-β disks, the amplification of magnetic fields due to the Balbus-Hawley instability saturates when $\beta \sim 10 - 30$. The disk approaches to a gas pressure dominated, quasi-steady state. The effective value of the viscosity parameter is $\alpha_B = -\langle B_x B_y/(4\pi P_0(0))\rangle \sim 0.01$. These results are consistent with those reported by Stone et al. (1996). When the initial magnetic energy is comparable to the thermal energy ($\beta \sim 1$), however, the disk stays in the low-β state for time scale longer than the rotation period. In such disks, the amplification of magnetic fields due to the coupling of the Balbus & Hawley instability and the Parker instability overcomes the buoyant loss of magnetic flux. The effective magnetic viscosity in low-β state is the order of 0.1 as shown in the right panel of Figure 1. When the magnetic energy stored in the low-β disk is released, we expect large amplitude sporadic time variations as observed in low-state disks.

References

Balbus, S. A., & Hawley, J. F. (1991), ApJ, 376, 214.
Mineshige, S., Kusunose, M., & Matsumoto, R. (1995), ApJ, 445, L43.
Shibata, K., Tajima, T., & Matsumoto, R. (1990), ApJ, 350, 295.
Stone, J. M., Hawley, J. F., Gammie, C. F., & Balbus, S. A. (1996), ApJ, 463, 656.

ELECTRON-POSITRON PAIR WINDS FROM CENTRAL LUMINOUS ACCRETION DISK WITH RADIATION DRAG

Y. TAJIMA AND J. FUKUE
Astronomical Institute, Osaka Kyoiku University
Asahigaoka, Kashiwara, Osaka 582, Japan

The accretion disks are now supposed to be the main driving source of the active astrophysical phenomena. Even the electron-positron pair plasma will be created at the surface of the sufficiently luminous disk. While the effect of radiation drag which causes in the intense radiation fields around the accretion disk is examined recently. Then, we numerically consider the radiative accelerated pair-winds, which blow off from central luminous accretion disk surrounding a black hole, taking into account radiation drag of the order of v/c.

1. Disk Radiation Fields

For the present purpose, we quantitatively calculate the full components of radiation fields produced by the standard accretion disk around a black hole, considering the Doppler enhancement to the order of v/c (Tajima and Fukue 1997). We emphasize that there appears the azimuthal component of radiative flux, since the disk radiation field has an angular momentum.

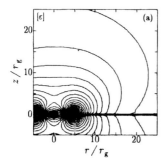 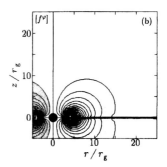

Figure 1. Examples of the contour maps of radiation fields around the accretion disk. (a) The normalized radiation energy density ε. (b) The azimuthal component of normalized radiative flux f^φ.

2. Motions of Pair Winds

Here, we assume that the electron-positron pair plasma is created at an inner region of the disk, and the pair winds blow off explosively from there. Using the components of the radiation fields, we examine the accretion disk winds which are ejected from the central region of the luminous disk. From the comparison between the winds from the less luminous (Fig.2a) and the luminous (Fig.2b) disks, and the non-dragged winds (Fig.2c), it can be understood that the winds, which blow off from luminous disks, are collimated in the vertical direction. Moreover, the radiation drag also collimates the winds. The collimation occurs since the angular momentum of the winds is removed by the radiation drag rapidly (Fig.3). However it is difficult for the streamlines to be collimated to the very close to the z-axis only by the radiation drag, due to the radial component of the radiation flux. Thus, the winds appear the conical streamlines.

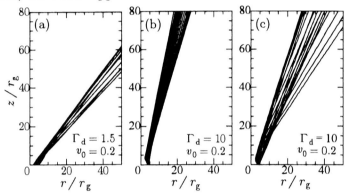

Figure 2. Examples of trajectories of the winds. The parameter Γ_d is the disk luminosity normalized by Eddington luminosity, and v_0 is the initial velocity. (a) Dragged winds from a low luminous disk. (b) Dragged winds from a luminous disk. (c) Non-Dragged winds from a luminous disk.

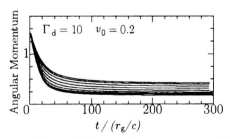

Figure 3. The time variation of the angular momentum of the dragged winds.

References

Tajima Y., Fukue J. (1997) Radiative Disk Winds under Radiation Drag II, *PASJ*, submitted

X-RAY FLUCTUATIONS FROM THE ADVECTION-DOMINATED ACCRETION DISK WITH A CRITICAL BEHAVIOR

M. TAKEUCHI
Kyoto University, Department of Astronomy
Sakyo-ku, Kyoto, 606-01, JAPAN

1. Introduction and Basic Physical Backgrounds

The new model for X-ray fluctuations of Cyg X-1, which is based on the fluid dynamics, is presented. The model is the optically thin and advection-dominated accretion disk model, which has a critical behavior.

The evolution of one-dimensional axisymmetric disk are considered. Basic equations are the same as Takeuchi & Mineshige (1997), which are constructed with mass conservation, momentum conservation, angular momentum conservation, and the energy equation;

$$\frac{\partial}{\partial t}(r\Sigma) + \frac{\partial}{\partial r}(r\Sigma v_r) = 0,$$

$$\frac{\partial}{\partial t}(r\Sigma v_r) + \frac{\partial}{\partial r}(r\Sigma v_r^2) = -r\frac{\partial W}{\partial r} + r^2\Sigma(\Omega^2 - \Omega_K^2) - W\frac{d\ln\Omega_K}{d\ln r},$$

$$\frac{\partial}{\partial t}(r^2\Sigma v_\varphi) + \frac{\partial}{\partial r}(r^2\Sigma v_r v_\varphi) = -\frac{\partial}{\partial r}(r^2\alpha W),$$

$$\frac{\partial}{\partial t}(r\Sigma e) + \frac{\partial}{\partial r}(r\Sigma e v_r) = -\frac{\partial}{\partial r}(rWv_r) - \frac{\partial}{\partial r}(r\alpha W v_\varphi) - rQ\mathrm{rad},$$

where Σ is surface density, W is vertically integrated pressure, Ω and Ω_K are the angular frequency of the gas flow and the Keplerian angular frequency, respectively, M is the mass of the central black hole, α is the viscosity parameter, e is the internal energy of accreting gas, and Q_{rad} is radiative cooling rate per unit surface area due to thermal bremsstrahlung. I assign $M = 10 M_\odot$. Starting from steady solution of these basic equations, I perform time evolutionary calculations using the Lax-Wendroff scheme.

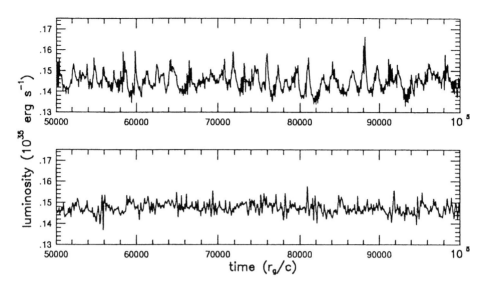

Figure 1. The obtained light curve in cases of $\eta = 0.5$, 1.0, from the top to the bottom, respectively. The unit of time is Schwarzschild radius per the light velocity, r_S/c.

I prescribe the critical value of the surface density, Σ_{crit}, where

$$\Sigma_{\text{crit}}(r) = \Sigma_{\text{steady}}(r)\left(1 + \eta \frac{r - R_{\text{crit}}}{r_{\text{out}} - r_{\text{in}}}\right).$$

Here, $\Sigma_{\text{steady}}(r)$ is surface density of the steady solution, η is a free parameter representing how much Σ_{crit} deviates from $\Sigma_{\text{steady}}(r)$, and the critical radius, R_{crit}, is the radius where $\Sigma_{\text{crit}} = \Sigma_{\text{steady}}$. As the critical behavior, the value of α is switched as following;

$$\alpha = \begin{cases} \alpha_0 & (\Sigma \leq \Sigma_{\text{crit}}) \\ \alpha_{\text{high}} & (\Sigma > \Sigma_{\text{crit}}). \end{cases}$$

I set $R_{\text{crit}} = 30$, $\alpha_0 = 0.1$, and $\alpha_{\text{high}} = 0.12$.

2. Results

In this paper, I introduce the result of $\eta = 0.5$, 1.0 cases. Figure 1 is the obtained light curve in cases of $\eta = 0.5$, 1.0, from the top to the bottom, respectively. I succeed in reproducing X-ray light curve as observed in Cyg X-1. For more details, see Takeuchi & Mineshige (1997).

References

Takeuchi, M., & Mineshige, S. (1997) X-ray Fluctuations from the Advection-Dominated Accretion Disks with A critical Behavior *ApJ*, **486**, 160–168

Session 3: Diagnostics of High Gravity Objects with X- and Gamma Rays

3-3. AGNs

COMPTON REFLECTION IN THE VICINITY OF A ROTATING BLACK HOLE

A. MACIOŁEK-NIEDŹWIECKI
Lódź University, Department of Physics
Pomorska 149/153, 90-236 Lódź, Poland

AND

P. MAGDZIARZ
Jagiellonian University, Astronomical Observatory
Orla 171, 30-244 Kraków, Poland

Abstract. We study the spectra arising from Compton reflection in the innermost parts of the accretion disk. We emphasize that the so far neglected relativistic distortion of the Compton reflection continuum may strongly affect the derived Fe Kα line shapes.

1. Introduction

X-ray irradiation of the cold accretion disk gives rise to two spectral features commonly seen by *Ginga* and *ASCA* in Seyfert 1 galaxies: (1) a broad reflection continuum peaked around 30 keV, and (2) a fluorescent Fe Kα line around 6.4 keV. Current results from the *ASCA* satellite have shown that the Fe Kα line profile is broad and asymmetric in most of Seyfert 1s, indicating strong relativistic effects which distort the reflected spectrum. The analyses of the line profiles show that most of X-ray emission in Seyfert 1s comes from within 10 Schwarzschild radii. Then, since both the Compton reflection and the fluorescence originate from the same reprocessing matter, the reflected continuum is affected by the gravity/Doppler effects in the same manner as the fluorescence line. This effect, neglected in all up-to-date studies, is very likely to affect the fitted line profile, as the low energy tail of the redshifted reflection continuum may significantly contribute to the continuum below 10 keV and the iron edge. To study the importance of this effect, we have recently developed a model treating self-consistently

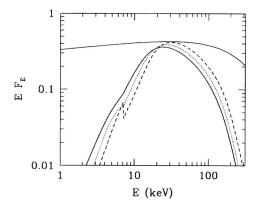

Figure 1. The figure compares the Compton reflected spectra in three different geometries, observed at an inclination angle of $\Theta_{\rm obs} = 30°$. The solid curve corresponds to the Kerr metric for a maximally rotating black hole and the dashed curve corresponds to the Schwarzschild metric. For both cases, we assumed that the bulk of radiation comes from a region within 10 GM/c^2 and that the radial dependence of energy generation follows the emission law $I(r) \propto r^{-3}$. The dotted curve shows the local spectrum of the reflected radiation with no general relativistic transfer effects included. Note the spread of iron absorption edge and strong excess below 10 keV in the reflected spectrum transferred in the Kerr spacetime with respect to the locally generated spectrum. For the Schwarzschild metric the similar effects are less pronounced.

the transfer of the line and the reflection continuum through gravitational field of the black hole with an arbitrary angular momentum (Maciołek-Niedźwiecki, Magdziarz & Zdziarski, in preparation).

2. Results

Figure 1 presents examples of reflected spectra obtained using our model for parameters characteristic for Seyfert 1s (Nandra et al. 1997). For comparison, we also present the spectrum Compton-reflected from a stationary slab (i.e., without taking into account relativistic transfer effects; Magdziarz & Zdziarski 1995). For the Kerr-metric disk relativistic and gravitational shifts of photon energy smear out the bound-free absorption iron edge almost completely, in contrast to the stationary Compton reflection spectrum. The contribution from the innermost radii strongly modifies the shape of the low-energy tail of the reflection continuum integrated over the disk surface. Our results show that the transfer effects have to be taken into account even when modeling broad spectral continuum components.

References

Magdziarz, P., and Zdziarski, A. A. 1995, MNRAS, 273, 837
Nandra K., et al., 1997, ApJ, 477, 602

TWO-TEMPERATURE TRANSONIC ACCRETION DISKS

K.E.NAKAMURA[1], M.KUSUNOSE[2], R.MATSUMOTO[3], S.KATO[1]
[1] *Dept. of Astronomy, Kyoto Univ., Kyoto 606-01*
[2] *Dept. of Physics, Kwansei Gakuin Univ., Nishinomiya 662*
[3] *Dept. of Physics, Chiba Univ., Chiba 263*

1. Introduction

The optically thin, advection-dominated accretion flows are thermally stable against global perturbations. In addition, they have high temperatures because of inefficient radiative cooling. They are thus promising candidates of models to explain the high energy emission of X-ray stars and AGNs. So far, models, however, take no account of the advective heat transport in determining the thermal structure of the electron system. The validy of this neglect, however, must be checked by integrating the electron energy equation globally as well as the ion energy one.

2. Results

In the inner region of the disks, we find that in the ion gas the viscous heating is roughly balanced by advective cooling. That is, the viscously heated ions are transported inward without being cooled. The approximation of advection-dominated flows is good for the ion gas, and only a small fraction of energy is transported to the electron gas by the Coulomb collisions(Λ_{ie}). When the energy balance in the electron gas or the spectrum from the disk are concerned, the following approximation are sometimes adopted: after the ion temperature is calculated by the approximation of advection-dominated flow, the electron temperature is determined under the approximation of $\Lambda_{\mathrm{ie}} = Q^-_{\mathrm{rad}}$. Our results, however, show that the compressive heating of the electron gas is the main ingredient to keep the electron temperature against strong synchrotron cooling of the electron gas. We should take the approximation of $Q_{\mathrm{adv,e}} + Q^-_{\mathrm{rad}} \sim 0$ with a negative Q_{adv} (advective heating), instead of $\Lambda_{\mathrm{ie}} = Q^-_{\mathrm{rad}}$.

ENHANCEMENT OF TURBULENT VISCOSITY BY GLOBAL MAGNETIC FIELDS IN ACCRETION DISKS

YASUSHI NAKAO
Department of Astronomy, Faculty of Science, Kyoto University
Sakyo-ku, Kyoto, 606-01, Japan

1. Summary

A model of magnetohydrodynamic (MHD) turbulence in accretion disks with global magnetic fields is constructed using a second-order closure modeling of turbulence. The transport equations of the Reynolds stress tensor, the Maxwell stress tensor, and the cross-helicity tensor (the correlation of velocity fluctuation and magnetic fluctuation) are closed by second-order quantities using a *two-scale direct interaction approximation* (TSDIA). The quantities appearing these equations are considered to be those averaged in the vertical direction of the disks. The turbulence is assumed to be stationary. We are interested only in the effects of the global magnetic fields on the turbulence in the disks, i.e., no dynamo processes are considered, and the global magnetic fields are supposed to be embedded in the disk *a priori*.

The results show that the presence of a radial global field enhances the turbulent viscosity (and the so-called α-parameter value) in the disk. This suggests that the radial global magnetic field has something to do with advection-dominated disks, in which a rather large viscosity is needed. In advection-dominated disks, the advective motion of the gas is likely to produce a radial global field. On the other hand, both toroidal and vertical global fields diminish the strength of the turbulence. Therefore, the strength of the turbulence in the accretion disk is determined by competition between the above-mentioned two opposite effects of the global fields. Our results show that the enhancement of the turbulence due to a radial field is more effective than the suppression of the turbulence due to a toroidal (or a vertical) field. See Nakao (1997) (Publ. Astron. Soc. Japan **49**, in press) for the details.

ACCRETION-DISK CORONA ADVECTED BY EXTERNAL RADIATION DRAG

Y.WATANABE AND J.FUKUE
Osaka Kyoiku University

1. Motivation

Accretion-disk corona (ADC) is required from observational as well as theoretical reasons. In almost all of traditional studies, however, a stationary corona has been assumed; i.e., the corona gas corotates with the underlying (Keplerian) accretion disk, and the radial motion is ignored. Recently, in the theory of accretion disks a radiative interaction between the gas and the external radiation field has attracted the attention of researchers. In particular the radiation drag between the gas and the external radiation field becomes important from the viewpoint of the angular-momentum removal. We thus examine the effect of radiation drag on the accretion-disk corona above/below the accretion disk (Watanabe, Fukue 1996a, b). We suppose that an accretion disk can be described by the standard disk, and that radiation fields are produced by the central luminous source and the accretion disk, itself. In general an accretion-disk corona under the influence of strong radiation fields dynamically infalls (advected) toward the center.

2. Advection Disk Corona

Let us suppose an axisymmetric disk corona above/below a standard accretion disk, which is steadily rotating around a central object (Fig.1). We ignore the self-gravity of the corona gas and disk gas. The disk corona is assumed to be geometrically thin, and thus, the physical quantities are integrated over the vertical direction. It is also assumed to be effectively optically thin. When the central source becomes luminous, the angular momentum is removed through the action of radiation drag from the central source radiation, and the advection/infall of the corona gas is remarkably enhanced. When the accretion disk is luminous, the corona gas tends to corotate with the disk and the radial infall of the corona gas is suppressed. When the corona gas is hot, the pressure-gradient force supports the radial hydrostatic-equilibrium of the corona and the radial infall of the corona gas is suppressed.

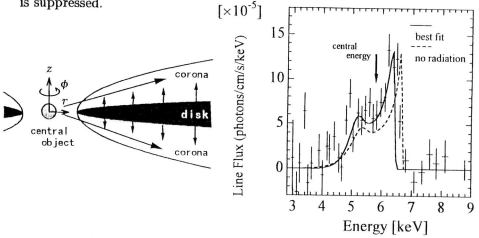

Fig. 1. (Left) Model. Fig. 2. (Right) Best-fit line profile of the present model to the X-ray emission line detected in MCG−6−30−15. The crosses are the observed X-ray lines reproduced from Tanaka et al. (1995).

3. Line Profiles

The line profiles from advection disk coronae generally have double-peaked features. When the central object becomes more luminous, the rotation velocity of the disk corona is smaller, and therefore, the redshift becomes smaller and the separation of two peaks is smaller. The profiles do not depend very much on the disk luminosity. In the relativistic case, the blue peaks become more intense than the red peaks due mainly to the beaming effect, while the centroid energy is smaller than in the non-relativistic case, mainly due to the gravitational redshift.

We fit the X-ray line spectrum of MCG−6−30−15 (Tanaka et al. 1995) by the present model (Fig.2). This means that if the X-ray lines from MCG−6−30−15 are emitted from the corona above the accretion disk, the corona may be strongly affected by the radiation drag, and infall dynamically to the central black hole.

References
Tanaka Y. et al. 1995, Nature 375, 659
Watanabe Y., Fukue J. 1996a, PASJ 48, 841
Watanabe Y., Fukue J. 1996b, PASJ 48, 849

NUMERICAL SIMULATION OF RELATIVISTIC JET FORMATION IN BLACK HOLE MAGNETOSPHERE

SHINJI KOIDE

Faculty of Engineering, Toyama University, Gofuku, Toyama 930, Japan

AND

KAZUNARI SHIBATA AND TAKAHIRO KUDOH

National Astronomical Observatory, Mitaka, Tokyo 181, Japan

1. Introduction

The radio jets ejected from active galactic nuclei (AGNs) sometimes show proper motions with apparent velocity exceeding the speed of light. This phenomenon, called superluminal motion, is explained as relativistic jets propagating in a direction almost toward us, and has been thought to be ejected from the close vicinity of hypothetical supermassive black holes powering AGNs (Rees 1996). The magnetic mechanism has been proposed not only for AGN jets (Lovelace 1976; Blandford & Payne 1983) but also for protostellar jets (Pudritz & Norman 1986; Uchida & Shibata 1985; Shibata & Uchida 1986), although no one has yet performed nonsteady general relativistic magnetohydrodynamic (GRMHD) numerical simulations on the formation of jets from the accretion disk around a black hole.

2. Numerical Results

We use 3+1 formalism of general relativistic conservation laws of particle number, momentum, and energy and Maxwell equations with the infinite electric conductivity (Thorne *et al.* 1986; Koide *et al.* 1996; Koide 1997) to simulate the relativistic magnetohydrodynamic jet formation. The Schwarzschild metric, which provides the space-time around the black hole at rest, is used in the calculation.

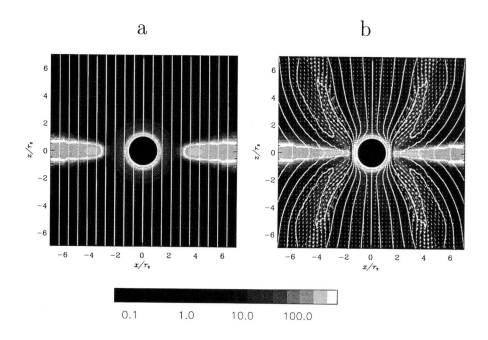

Figure 1. The jet formation in a black hole magnetosphere. (a) The initial condition. (b) $t = 93\tau_S$.

Figure 1 shows the initial and final states of the relativistic jet formation in the black hole magnetosphere. The proper mass density (gray-scale), velocity (vector), and magnetic field (solid lines) are shown. The black circles at the centers represent the black hole inside the event horizon at Schwarzschild radius r_S. Initially, the plasma of the corona is in hydrostatic equilibrium. The accretion disk rotates around the black hole with Keplerian velocity. The density of the disk is 400 times that of the background corona. The uniform magnetic field crosses the disk perpendicularly. Figure 1b shows the final stage of this simulation at $t = 91\tau_S$. The maximum poloidal component of the jet velocity is 0.88c, which corresponds to a Lorentz factor of 2.1 at $x \sim 3r_S$, $z \sim 3r_S$. The jet has two-layered structure consisting of the inner, fast (low-density) jet and the outer, slow (high-density) jet. The former is ejected from the corona, while the latter is from the disk.

We thank M. Inda-Koide for her discussion and important comments for this study. We also thank M. Takahashi, T. Yokoyama, K. Hirotani, K.-I. Nishikawa, J.-I. Sakai, and R. L. Mutel for their discussion and encouragement.

RADIATION FROM ADVECTION-DOMINATED FLOWS

Application to Sgr A*

T. MANMOTO
Department of Astronomy, Faculty of Science, Kyoto University
Sakyo-ku, Kyoto 606-01, Japan

1. Introduction

Advection-dominated accretion flow (hereafter ADAF) is the only self-consistent solution to describe the optically thin accretion flows around compact objects. The main feature of ADAF is that the dynamics of the flow is dominated by accretion process rather than radiation process. As a result of advection domination, the luminosity of ADAFs is very low. Coupled with the existence of the event horizon, ADAF has been successfully applied to the dim accretion black holes such as central core of our Galaxy :Sgr A*. In this issue, we calculate the spectrum radiated from the optically thin ADAFs and show that the observed spectrum of Sgr A* is explained with the accretion massive black hole.

2. Accretion Flow Model

We solve the basic equations for the optically thin accretion gas around a black hole globally. We assume that the gas is two-temperature plasma with randomly oriented magnetic field. The radiation processes taken into account are synchrotron emission, bremsstrahlung, and Comptonization. We also include the effect of redshift due to bulk motion and gravity. The basic equations used in our calculations are described in detail in Manmoto, Mineshige and Kusunose 1997. We show in the figure the observed spectrum of Sgr A* and our model (shown by solid line). The observational data in the figure are ones formerly compiled and tabulated in Narayan et al. 1997.

3. Results and Discussions

We fit the observational data with the standard values of physical parameters such as $\alpha = 0.1$, $\beta = 0.5$ (case of equipartition), $\dot{m} = 2.5 \times 10^6$. The

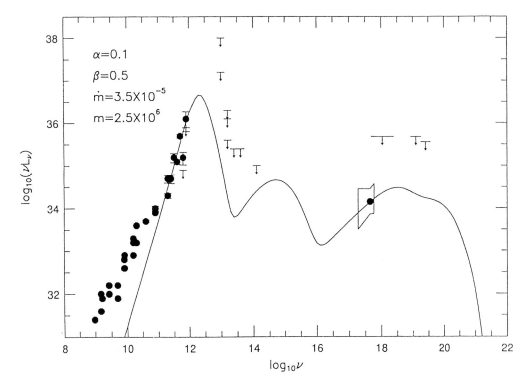

Figure 1. Spectrum of Sgr A*

parameters we assigned are very close to those assigned by Narayan et al. 1997. In Manmoto et al 1997, we concluded that the several X-ray data which were asserted to be from Sgr A* are in fact not from Sgr A* itself, if we set $\alpha \sim 0.1$ which is considered to be the standard value for ADAFs. The newly added infrared upper limit (Menten et al. 1997) makes it improbable that the X-ray data more luminous than the upper limits shown in the figure previously obtained (see Narayan et al. 1995) are from Sgr A* itself only.

4. References

Manmoto, T., Mineshige, S., & Kusunose, M. 1997, ApJ, in press
Menten, K. M., Reid, M. J., Eckart, A., & Genzel, R. 1997, ApJL, 475, L111
Narayan, R., Mahadevan, R., Grindlay, J. E., Popham, G., & Gammie, C. 1997, ApJ, in press
Narayan, R., Yi, I., & Mahadevan, R., 1995, Nature 374, 623

BLACK-HOLE ACCRETION CORONA IMMERSED IN THE DISK RADIATION FIELDS

T. MIWA, Y. WATANABE, J. FUKUE
*Astronomical Institute, Osaka Kyoiku University,
Asahigaoka, Kashiwara, Osaka 582, Japan*

1. Accretion Corona

We examined an accretion-disk corona around a black hole immersed in the disk radiation fields (cf. Watanabe, Fukue 1996a, b). The corona is supposed to be initially at rest far from the center. During infall above and below the disk, the corona is suffered from the disk radiation fields. As a disk model, we adopted the standard α-disk, and in order to mimic the general relativisitic effects, we use the pseudo-Newtonian force proposed by Artemova et al. (1996). Moreover, we assume that the corona is geometrically thin and optically thin, and ignored any motion such as wind. We consider the cold case, where the pressure-gradient force is ignored. Under these assumptions, we calculated the motion of the corona gas and found that the infall of corona is supressed due to disk radiation fields.

2. Results

The results are shown in the following figures. If the disk luminosity increases, the radial infall velocity of the corona gas deviates from the free-fall one and becomes small, since the corona gas gains the angular momentum from the underlying disk via radiation drag of disk radiation fields. When the disk is sufficiently luminous, the infalling of the gas is stopped (in such a case, the corona gas may be blown off from the disk).

These properties are summarized in the Γ_d-a plane, where a is the spin parameter of the hole and Γ_d is the normalized disk luminosity ($= L_d/L_{\rm Edd}$). The corona gas can infall when Γ_d is smaller than some critical values, and this critical Γ_d becomes small, as a increases.

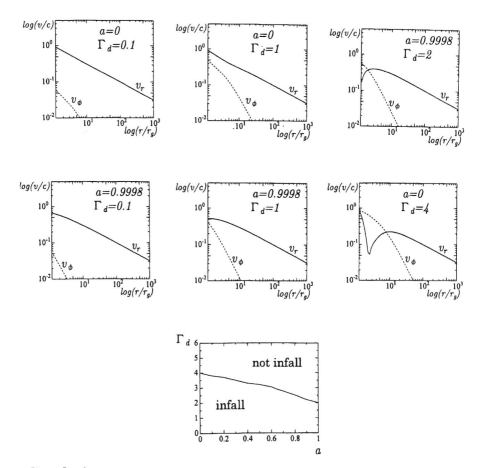

3. Conclusion

When we consider the normal (proton-electron) plasmas, the corona gas can always infall, since Γ_d does not exceed unity. The infall velocity, however, deviates from the free-fall one due to the radiation drag of disk radiation fields. When we consider electron-positron plasmas, however, the infall will be remarkably modified, since Γ_d can exceed unity ($\Gamma_d \sim 1836$). In this case where Γ_d becomes so large, the corona gas will escape as *radiatively driven winds* (Tajima, Fukue 1996).

Reference

Artemova I. V., Björnsson G., Novikov I. D. 1996, ApJ 461, 565
Tajima Y., Fukue J. 1996, PASJ 48, 529
Watanabe Y., Fukue J. 1996a, PASJ 48, 841
Watanabe Y., Fukue J. 1996b, PASJ 48, 849

DETERMINATION OF NGC 4151 NUCLEUS MASS FROM PARAMETERS OF NARROW SATELLITES OF BROAD LINES

N.G.BOCHKAREV
Sternberg Astronomical Institute, Moscow, 119899, Russia

The paper considers the possibility of AGN mass estimation from observations of spectral lines emitted by jets. Jets are a typical for AGNs.

Similar to SS433 the inner parts of AGN jets may emit spectral lines. SS433 spectra show that jet lines are week in comparison with those of slowly outflowing gas. Consequently the proper time for AGNs jet lines observations is the deep brightness minima (so-called transition from Sy1 type to Sy2 type of spectra). It is possibly the only chance for jet lines observations, because on the Sy1 stage gas surrounding the central object can be optically thick in the broad lines wings or at least light up the week jet lines. The only well studied case of AGN transition from Sy1 to Sy2 with evidence of jets is NGC 4151 when this galaxy was in a deep minimum of brightness during the 1984-1987.

Analysis of NGC 4151 spectra observed by IUE during 1984 reveals evidence of variations in narrow satellites of CIV1550 Å line (Ulrich et al. 1985), (Clavel et al. 1985). Later, when NGC 4151 returned to the bright state the CIV1550 Å satellites were seen no more. Bochkarev et al. (1991) have got some evidence of possible existence of week narrow satellites of hydrogen lines during 1986-1987. They estimated the mass of the nucleus using the satellites radial velocity. However as observations (Bochkarev et al. 1991) have been done with low signal to noise (S/N) ratio, they are not sure of the results. Meanwhile, the method described by Bochkarev et al. (1991) could be used for another Seyfert type transitions.

References

Ulrich, M.-H., Altamore, A., Boksenberg, A, et al. (1985) *Nature* v. 313, p. 474
Clavel, J., Altamore, A., Boksenberg, A., et al. (1987) *Astrophys.J* v. 321, p.215
Bochkarev, N.G., Shapovalova, A.I., and Zhekov, S.A. (1991) *Astron.J.* v.102, p.1278

THE X-RAY SPECTRUM AND VARIABILITY OF NGC 4151

K. M. LEIGHLY
Columbia Astrophysics Laboratory, Columbia University
538 West 120th Street, New York, NY 10027
leighly@ulisse.phys.columbia.edu

AND

M. MATSUOKA, M. CAPPI, T. MIHARA
Cosmic Radiation Laboratory, RIKEN
Hirosawa 2-1, Wako-shi, Saitama 351, Japan

We report investigation of the iron Kα line in a long (100 ks) *ASCA* observation of NGC 4151. This observation offers unprecedented good statistics; however, the situation is complicated by the fact that the absorption in NGC 4151 is complex and therefore it is difficult to deconvolve a broad iron line from the power law strongly curved by the absorption. Preliminary spectral fitting with a dual absorber model, using updated abundances and response matrices, and also allowing for iron overabundance, revealed significant spectral residuals around 5 keV which could be modeled with a broad Gaussian. This profile resembles the line characteristic of emission from a relativistic accretion disk; however, that model fit the spectra poorly. Since the energy of the narrow core is nearly 6.4 keV, the orientation of the accretion disk should be nearly face-on, because if the inclination were higher, the blue horn should be shifted to higher energies. If the orientation is face-on, there should be no emission blueward of 6.4 keV; however, a small blue wing as well as a long red wing are present in the residuals.

Atomic physics requires that an iron Kβ line should be produced with about 13.5% of the Kα line flux at 7.06 keV. If we include an additional Gaussian to model the Kβ line with energy, width and flux scaled appropriately to the Kα line parameters, the model fit becomes as good as the broad+narrow line description with no change in number of free parameters. The disk line then models the red wing and, with ~ 100 eV equivalent width, contributes only a small fraction of the total line emission.

The Kβ interpretation of the blue wing implicitly assumes that the narrow core of the line near 6.4 keV is intrinsically narrow and therefore arises

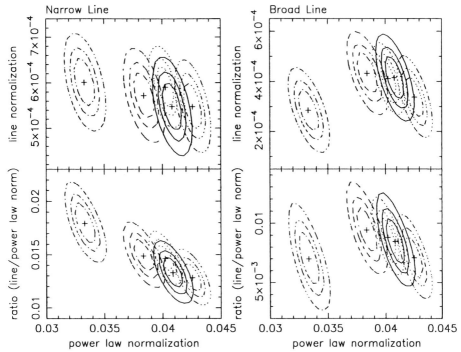

Figure 1. χ^2 contours from fits to five time-sliced spectra from the long ASCA observation of NGC 4151. The left and right sides show the results for the narrow $K\alpha$ (including the $K\beta$ line scaled to 13.5% of the flux of the $K\alpha$ line), and broad disk line respectively. The upper and lower panels shows the line flux, and the ratio of the line to power law flux, respectively.

in regions where the velocity of the emitting material is relatively small, such as the broad-line emission clouds, the molecular torus, or the material responsible for the continuum absorption. Therefore, the line flux should vary less than the continuum flux on short time scales. During the observation, variability in total flux with amplitude about 40% was observed. The light curve was divided into 5 time slices corresponding roughly to the slow changes in flux observed. We fit the time-sliced spectra with the complex line model described above and found indeed that the flux of the narrow line component, composed of the $K\alpha$ and $K\beta$ lines, did not vary significantly. However, the ratio of the line flux to the power law flux did vary (Figure 1). Since the photon index was fixed to best fit value, this ratio approximates the equivalent width. Neither the flux nor the equivalent width of the broad component varied significantly. Detection of such variability was hampered because the equivalent width is low, and also because observed continuum spectral variability attributed to changes in the absorption column required us to allow the absorption columns to vary independently while fitting each time-sliced spectrum.

REVERBERATION MAPPING ANALYSIS OF THE BROAD-LINE REGION IN SEYFERT GALAXY NGC 4151

S.J. XUE

The Institute for Physical and Chemical Research (RIKEN) Hirosawa 2-1, Wako, Saitama 351-01, Japan

Shaanxi Astronomical Observatory, Academia Sinica Lintong 710600, China

AND

F.Z. CHENG

The University of Science & Technology of China (USTC) Heifei 230026, China

1. Introduction

One of the primary goals of AGN variability studies has been to determine the size of broad-line region (BLR) through the *reverberation mapping* technique. In a recent international multiwavelength spectroscopic monitoring campaign, NGC 4151 has been observed intensively by ground-based telescopes for a period of over 2 months, with a typical temporal resolution of 1 day. The main result from this optical campaign is that finding the variation in the emission line flux ($H\beta$ or $H\alpha$) lagging the continuum by 0–3 days (1993 campaign: Kaspi et al. 1996). This is in contrast to the past results in which a time lag of 9±2 days was found for the same emission line (1988 campaign: Maoz et al. 1991). Such a BLR "size problem" may be caused by a different variability timescale of the ionizing continuum or a real change in BLR gas distribution in the 5.5 yr interval between the two watch campaigns. In order to clarify which of the two possibilities is most likely the real case, we performed further reverberation analysis on both optical datasets.

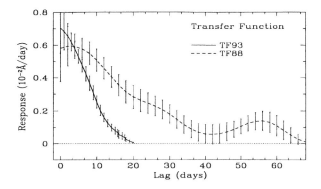

Figure 1. Transfer functions recovered from two optical watch campaigns.

2. Monte-Carlo Simulation

We first evaluated the uncertainties associated with the lag measurement caused by any "ionizing" continuum behaviors via Monte-Carlo simulations. The results show that, for the 1993's continuum behavior, if the BLR structure keeps stationary as it was in 1988, we would get the BLR size of 10.5±6.3 days in the 95 % error range. This is in sharp contrast to the practical result of 0–3 days. We thus concluded that the hypothesis of the stationary BLR structure between the 5.5 yr interval of the two observation campaigns is highly excluded.

3. Comparison of Transfer Functions

We then worked out BLR transfer functions TF93 and TF88 for both datasets using the *Regularized Linear Inversion* method (Krolik & Done, 1995). Figure 1 shows the significant difference between the shapes of the two. In contrast to TF88, TF93 reveals a much "condense" BLR, which is most likely a shell with 10 day thickness and the inner radius of $\lesssim 1$ day.

The dramatic change of the transfer function indicates the real difference in BLR gas distribution between 1988 and 1993. This observed BLR evolution not only naturally explains the present "size problem", but also provides an exciting new dimension in AGN variability studies.

References

Kaspi, S., et al. 1996, ApJ, 470, 336
Krolik, J.H. & Done, C. 1995, ApJ, 440, 166
Maoz, D., Netzer, H., Leibowitz, E. et al., 1991, ApJ, 421, 34

X-RAY PROPERTIES OF HIGH-Z RADIO-QUIET QUASARS: *ASCA* OBSERVATIONS

M. CAPPI[1,2], M. MATSUOKA[1], C. VIGNALI[3], A. COMASTRI[4] AND
T. MIHARA[1], C. OTANI[1], G.G.C. PALUMBO[3,2], S.J. XUE[1]
[1] *RIKEN, Hirosawa 2-1, Wako, Saitama 351-01, Japan*
[2] *ITeSRE/CNR, Via Gobetti 101, 40129, Bologna, Italy*
[3] *Dpt. of Astronomy, Bologna University, Via Zamboni 33, 40126, Italy*
[4] *Osservatorio Astronomico di Bologna, Via Zamboni 33, 40126, Italy*

Abstract. The first-ever high energy ($\sim$ 1-30 keV) X-ray spectra of three z $\sim$ 2 radio-quiet quasars (RQQs) are presented. If confirmed, the most interesting result is the marginal, but consistent, evidence for FeK$_\alpha$ emission lines in at least two of the sources. Further *ASCA* observations of high-z RQQs are needed for firm conclusions.

1. *ASCA* Preliminary Results

To date, most of the available X-ray data from high-z objects concerns radio-loud quasars (RLQs) while spectral information for radio-quiet quasars (RQQs) are quite scarce. Recent *ROSAT* and *ASCA* results allow us to derive a fairly well defined characteristic X-ray spectrum of high-z RLQs (Cappi et al. 1997). However, most spectral properties of high-z RQQs have been derived using color techniques based on *ROSAT* PSPC hardness ratios and assuming no intrinsic absorption (Bechtold et al. 1994, Fiore et al. 1997). Given these general considerations, we have started a collaborative effort between Japanese and Italian researchers with the aim of obtaining *ASCA* X-ray spectra of a representative sample of high-z RQQs, thus filling this gap. Preliminary results obtained from 3 out of 8 quasars observed (see Cappi 1998 and Vignali et al. in preparation for a more detailed study) of these quasars are shown here.

Best-fit spectral parameters are given in Table 1 for three z$\simeq$ 2 radio-quiet quasars. The average $\sim$ 1-30 keV rest-frame spectrum has $<\Gamma>$ $\simeq 1.75 \pm 0.2$ and there is no indication of excess-absorption. A Compton reflection component (as commonly observed in Seyfert 1 galaxies) or other

TABLE 1. Best-fit models with an absorbed power law, with and without an emission line

name	z	N_H (10^{20} cm^{-2})	Γ	E_{line}^a (keV)	EW^a (eV)	χ^2/dof	logL^c (2-10 keV)
0300-4342	2.3	<7	$1.78^{+0.21}_{-0.14}$	....	....	73/72	46.4
....		1.83 fix	$1.83^{+0.15}_{-0.14}$	....	....	73/73	....
....		1.83 fix	$1.76^{+0.16}_{-0.15}$	$3.76^{+0.03}_{-0.04}$	307^{+180}_{-168}	64/71	....
1101-264	2.15	<34.1	$2.08^{+0.56}_{-0.45}$	....	....	32.3/25	46.3
....		5.68 fix.	$1.90^{+0.24}_{-0.23}$	....	....	32.8/26	....
....		5.68 fix.	$1.93^{+0.27}_{-0.24}$	$6.50^{+0.18}_{-0.17}$	728^{+605}_{-577}	30.2/24	....
1352-2242	2.0	<11	$1.60^{+0.25}_{-0.14}$	....	....	60/74	46.4
....		1.66 fix	$1.63^{+0.13}_{-0.14}$	....	....	60/75	....
....		1.66 fix	$1.64^{+0.15}_{-0.13}$	$6.54^{+0.12}_{-0.12}$	276^{+216}_{-201}	54/73	....

^{a} Energy and EW of the line in the source rest-frame; ^{b} in units of 10^{-13} erg cm^{-2}s^{-1}; ^{c} in units of erg s^{-1}

complex continuum emission models are not required by the data (see Fig. 1). If confirmed, the most significant result of this study is the marginal, but consistent, detection of an iron emission line. In two cases (1352-2242 and 1101-264), the line energy is consistent with a rest energy of 6.4 keV, thus can be interpreted as Fe K$_\alpha$ emission from neutral or mildly ionized matter. In the third quasar (0300-4342), the origin of the line is puzzling because its energy does not correspond to any expected instrumental or source feature.

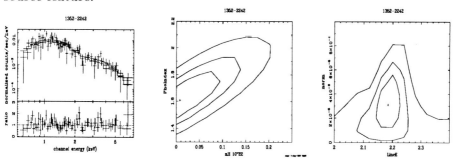

Figure 1. Best-fit spectrum, residuals and χ^2 contour plots (N_H-Γ and Line E-Intensity) for 1352-2242 (z=2), as an example.

2. References

Bechtold, J., et al., 1994, AJ, 108, 374
Cappi, M., 1998, PhD Thesis, submitted
Cappi, M., et al., 1997, ApJ, 478, 492
Dickey, J.M. & Lockman, F.J., 1990, ARA&A, 28, 215
Fiore, F., et al., 1997, ApJ, in press (astroph/9708050)

ASCA OBSERVATIONS OF NARROW LINE SEYFERT 1 GALAXIES

K. HAYASHIDA

Department of Earth & Space Science, Osaka University
1-1, Machikaneyama, Toyonaka, Osaka, Japan
hayasida@ess.sci.osaka-u.ac.jp

Abstract. Study of nine Narrow Line Seyfert 1 galaxies (NLS1) observed with ASCA is presented. The X-ray spectra are fitted with a two component model. The soft component can be represented by a blackbody model with $kT \sim 100$eV. To make the observed luminosity below the Eddington limit, we have to assume the emission region has a size much smaller than $3R_s$. The X-ray variabilities of the NLS1s have systematically shorter time scale than the typical (broad line) Seyferts 1 (BLS1), which suggests lower black hole masses in the NLS1s.

We also study the the X-ray spectral structure around 1.1 keV, which has been reported for some NLS1s. Although the structure has been interpreted as an absorption feature, we propose an alternative model employing emission lines. One possible origin of the emission lines is preprocessing from a highly ionized accretion disk.

1. X-ray Spectra: Soft Component

The NLS1s we have analyzed are IRAS13224, H0707, PG1404, Mkn478, RE1034, PG1211, Mrk766, IZw1, and PG1244 (Note that results of some sources have already been published by authors). The ASCA spectra (0.4-10keV) of these NLS1s were fitted with a two component model (a soft component and a hard power law component), though inclusion of the soft component is not necessary for some sources (*e.g.* IZw1). We employed a blackbody model to fit the soft component. One important consequence is that the best fit blackbody temperatures concentrate around $kT \sim 100$eV.

With the blackbody model fit, the size of the emission region was derived for each sources. If we assume it corresponds to $3R_s$, we can estimate the black hole mass, ranging from 2×10^4 to 10^6 M_o. Employing those mass values, however, leads to the super-Eddington luminosity. Using a smaller radius, e.g. 0.5 Rs, the corresponding Eddington limit becomes larger by a factor of 6 and the observed soft component luminosity go in the Eddignton limit. This result may imply the Kerr metric for these sources if the soft component come from an optically thick accretion disk.

2. X-ray Variability

Rapid and large amplitude X-ray variability is one of the characteristics of the NLS1s. We employed the analysis method previously applied to BLS1s (Hayashida et al., 1997a). From the variability time scale, we estimated the central black hole masses of nine sources, ranging from 2×10^4 to 10^7 M_o. These mass values of the NLS1s are systematically lower than those for the BLS1s. We speculate that the lower black hole masses in the NLS1 is is one of the main causes for their peculiar properties among the Seyferts.

3. X-ray Spectral Feature around 1.1 keV: Emission Line Model

In some NLS1s, a peculiar spectral structure was reported around 1.1 keV so far(IRAS13224: Otani et al., 1996, PG1404:Comastri et al., 1997, H0707: Hayashida, 1997b). Although the feature has been considered as absorption feature (e.g. highly blue-shifted O edge, see Leighly et al., 1997), we propose an alternative interpretation, i.e. emission lines from ionized materials. For example, Gaussian emission lines with energies of $0.84(\pm 0.01)$, $0.99(\pm 0.01)$, and $1.66(\pm 0.03)$ keV improve the fit for IRAS13224. In the case of H0707, the energies are $0.68(\pm 0.02)$, $0.87(\pm 0.02)$, and $1.41(\pm 0.02)$ keV. One possible cause of the emission lines is fluorescent reprocessed trom highly ionized matter on the surface of an accretion disks, such as indicated by Ross and Fabian (1993). Although detailed study is needed for the model fit, if this model is appropriate it means that we have a new probe to observe the status and the abundance of the matter very close to the central massive black hole.

References

Hayashida, K. et al. (1997a), *ApJ*, accepted.
Otani, C., Kii, T. & Miya, K. (1996) *MPE report*, p.263.
Comastori, A., Molendi, S. & Ulrich, M.H. (1997) *X-ray Imaging and Spectroscopyof Cosmic Hot Plasmas*, p.279.
Hayashida, K. (1997b), *Emission Line in Active Galaxies (IAU colloquium 159)*, p.40.
Leghily, K.M. et al. (1997) *ApJ*, in press.
Ross, R.R. & Fabian, A.C. (1993), *MNRAS*, **261**, p.74.

A MODEL OF THE BROAD-BAND CONTINUUM OF NGC 5548

P. MAGDZIARZ

Astronomical Observatory, Jagiellonian University
Orla 171, 30-244 Cracow, Poland

AND

O. BLAES

Department of Physics, University of California
Santa Barbara, CA 93106, USA

Abstract. We discuss a model of the central source in Seyfert 1 galaxy NGC 5548. The model assumes a three phase disk structure consisting of a cold outer disk, a hot central disk constituting a Comptonizing X/γ source, and an intermediate unstable and complex phase emitting a soft excess component. The model qualitatively explains broad-band spectrum and variability behavior assuming that the soft excess contributes significantly to the continuum emission and drives variability by geometrical changes of the intermediate disk zone.

1. Introduction

Broad-band spectral analysis of NGC 5548 shows that a soft X-ray excess dominating below 1 keV is related to the disk and may contribute significantly to the source energetics (Magdziarz et al. 1997). Since an optical/UV continuum requires the disk temperature of order a few eV, the soft excess component needs an additional continuum emitting phase. The standard opticly thick accretion disk (Shakura & Sunyaev 1973) breaks down at high accretion rate and, thus it puffs up producing the central, hot ($\sim$100 keV) region (e.g. Shapiro, Lightman & Eardley 1976) which constitutes the Comptonizing X/γ source (e.g. Zdziarski et al. 1997). In a such model the third intermediate phase of the disk (with characteristic temperature $\sim$100 eV) naturally appears as an effect of an instability.

2. Physics of the central source

Correlated variability of the spectral index, amount of reflection, and the total flux emitted in the X/γ continuum (Magdziarz et al. 1997) suggests that a number of seed photons controls the Compton cooling of the central X/γ source. This explains that the source is soft in a bright state and hard (i.e. photon starved) in a faint state. The opposite relation in the overall optical to soft excess continuum suggests that the main contributor of seed photons for Comptonization is the soft excess. Lack of ionized Compton reflection and lack of emission lines in EUV range requires complex structure of the soft excess. If the part of emitting matter is Thomson thin and dense it may emit dominant part of energy in a form of EUV pseudocontinuum (e.g. Kuncic, Celotti & Rees 1997).

3. Phenomenological model

The model assumes that the observed variability is driven by geometrical changes of the unstable inner edge of the cold disk on thermal time scales (cf. Kaastra & Barr 1989; Loska & Czerny 1997). This edge couples to the hot plasma by energy reprocessing modulated with the solid angle of cold matter seen from the central hot disk. The instability of the intermediate zone between the cold and the hot phase puffs up the inner cold disk and leads to its fragmentation which may be responsible for the observed self-organized critically behavior of variability (e.g. Leighly 1997). The unstable zone has a complex structure of clouds or filaments which explains the observed continuum of cold Comptonization in the soft excess range (Magdziarz et al. 1997). The positive feedback between the number of seed photons emitted from the unstable inner edge of the cold disk and the energy radiated out from the hot central disk produces substantial nonlinearity in the variability and naturally explains the correlation of the total flux emitted in X/γ continuum with the spectral index and amount of reflection. In such a physical picture, the soft excess dominantes energetics of the broad-band spectrum and drives variability of the central source.

References

Kaastra, J. S., & Barr (1989) *AA* **226**, 59
Kuncic, Z., Celotti, A., & Rees, M. J. (1997) *MNRAS* **284**, 717
Leighly, K. M. (1997) *this proceedings*
Loska, Z., & Czerny, B. (1997) *MNRAS* **284**, 946
Magdziarz, P., et al. (1997) *Proceedings of the 4-th Compton Symposium*, in press
Shakura, N. I., & Sunyaev, R. A. (1973) *AA* **24** 337
Shapiro, S. L., Lightman, A. P., & Eardley, D. M. (1976) *ApJ* **204** 187
Zdziarski, A. A., et al. (1997) *Proceedings of the 2nd INTEGRAL Workshop*, ESA SP-382

HOW UNBIASED IS A [OIII]λ5007-BRIGHT SAMPLE?

X-ray observations of Type-2 Seyfert galaxies with ASCA

SHIRO UENO AND J. DUNCAN LAW-GREEN
*Department of Physics & Astronomy, University of Leicester
Leicester LE1 7RH, UK*

AND

HISAMITSU AWAKI AND KATSUJI KOYAMA
*Department of Physics, Faculty of Science, Kyoto University
Sakyo-ku, Kyoto 606-01, JAPAN*

1. Introduction

We have observed a dozen bright Seyfert 2's selected by their [OIII]λ5007 emission line flux (Ueno et al 1996, Awaki et al 1996). We chose this to be a flux-limited sample free from X-ray selection effects such as intrinsic absorbing columns (N_H) and differences in viewing angle (Mulchaey et al 1994). Is it genuinely free from these selection effects? To investigate this, we compare the [OIII] luminosity of various Seyfert 2 galaxies. For this paper, we assume the Hubble constant to be $H_0 = 50 \text{km s}^{-1} \text{ Mpc}^{-1}$.

2. Histogram of Absorbing Column Thickness

It is difficult to get an unbiased histogram of column density thickness, but important to the origin of the cosmic X-ray background (Awaki 1991, Comastri et al 1995), and determination of the spatial distribution of absorbing material.

We made a histogram of the absorbing column thickness of Seyfert 2's (Ueno et al 1997). One remarkable result is the concentration of the column densities between $10^{23} - 10^{24} \text{cm}^{-2}$ for our [OIII]-selected sample. There are only a few objects with $N_H > 10^{24} \text{cm}^{-2}$ in our sample. However, the number of such sources could be an underestimate if extremely X-ray absorbed sources also have small [OIII] luminosities, possibly due to obscuration. Thus, we compare [OIII] luminosities in a larger Seyfert 2 sample, consisting of our [OIII]-selected sample, plus additional data to increase the number of completely X-ray absorbed sources.

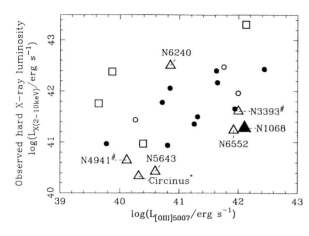

Figure 1. Plot of Hard X-ray vs. [OIII]λ5007 emission

3. Plot of Hard X-ray vs. [OIII]λ5007 emission

Figure 1 is a plot of hard X-ray vs. [OIII]λ5007 luminosity for Type-2 AGN. We distinguish between less absorbed ($N_H < 10^{23}$ cm^{-2}), moderately absorbed ($10^{23} < N_H < 10^{24}$ cm^{-2}) and completely blocked ($N_H > 10^{24}$ cm^{-2}) AGN.

We found no significant correlation between [OIII]λ5007 luminosity and absorbing column thickness. This implies that an [OIII]λ5007 flux-limited sample should be reasonably free from X-ray selection effects such as intrinsic N_H and differences in viewing angle.

Most of the completely blocked sources have low observed hard X-ray luminosity when compared with other Seyfert 2s having the same [OIII] luminosity. This is consistent with a model in which only the scattered X-ray emission is visible. The exceptions so far are NGC 6240 and Circinus. The large X-ray/[OIII] ratios in NGC 6240 may also be due to a reduction in the observed [OIII]λ5007 line flux due to reddening, as reported in Circinus (Oliva et al 1994).

4. References

Awaki H. 1991, *Ph.D. thesis, Nagoya University*
Awaki H., Ueno S., Koyama K. 1996, *X-ray Imaging and Spectroscopy of Cosmic Hot Plasma*, F.Makino and K.Mitsuda(Eds), p.271
Comastri A., Setti G., Zamorani G., Hasinger G. 1995, *A&A*, **296**, 1
Mulchaey J.S., Koratkar A., Ward M.J., Wilson A.S. et al. 1994 *ApJ*, **436**, 586
Oliva E., Salvati M., Moorwood A.F.M., Marconi A. 1994 *A&A*, **288**, 457
Salvati M., Bassani L., Dellaceca R., Maiolino R. et al. 1997 *A&A*, **323**, L1
Ueno S. et al. 1996, *X-ray Imaging and Spectroscopy of Cosmic Hot Plasma*, p.243
Ueno S. et al. 1997, in preparation

BEPPOSAX OBSERVATION OF THE QUASAR 3C273

T. MINEO, G. CUSUMANO
*Istituto di Fisica Cosmica ed Applicazioni all'Informatica, CNR
via U. La Malfa 153, I-90146 Palermo, Italy*

M. GUAINAZZI
*BeppoSAX Science Data Center, ASI
via Corcolle 19, I-00131 Roma, Italy*

AND

P. GRANDI
*Istituto di Astrofisica Spaziale, CNR
via E. Fermi 21, I-00044 Frascati (Roma), Italy*

The quasar 3C273 was observed from 18th to 21st July 1996 with the Narrow Field Instruments (NFIs) onboard BeppoSAX satellite (Boella et al 1997). The total exposure time was 12 ksec for LECS, 131 ksec for MECS and 64 ksec for HPGSPC and PDS.

LECS and MECS data have been accumulated on circular region of 8' and 4' respectively, and their relative background has been extracted from blank fields observations. The canonical procedure has been applied to obtain PDS and HPGSPC data: the off-source spectrum is subtracted from the on-source one.

Figure 1 presents the observed light curves in three energy ranges: 0.1–2 keV (LECS), 2–10 keV (MECS) and 13–200 keV(PDS). Each point corresponds to an integration time of 40 ksec. MECS light curve shows a significant decrease of 17% of the total count rate. A smaller variation is detected in LECS data (12%), that can be justified by the presence of some soft excess. PSD data do not show any significant variation, being the errors too large to detect variations of small percentages.

The four NFIs spectra were fitted simultaneously with a single power law plus low energy absorption. A detailed description of the analysis can be found in Grandi et al 97. The fit gives for the spectral index $\Gamma = 1.568 \pm 0.01$ and for the unabsorbed flux in the 2–10 keV energy band $F_{2-10} = (7.1 \pm 0.2) \times 10^{-11}$ erg cm^{-2} s^{-1}, being the low energy absorption constrained to the galactic value.

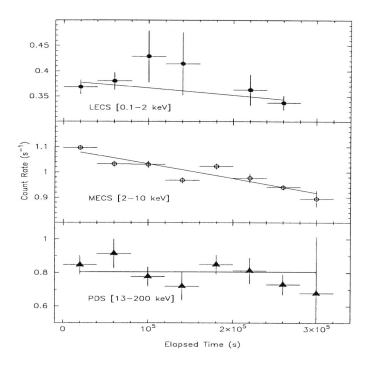

Figure 1. Observed count rate in the energy ranges 0.1–2 keV (LECS), 2–10 keV (MECS) and 13–200 keV (PDS). Data are fitted with a line in order to evaluate the variation with time.

Some distorsions from the single power law are moreover evident. An emission line at E=6.22±0.12 keV (source rest frame) and EW=30±12 eV is present in MECS spectrum. The residuals of the fit of LECS data with a single power law presents an excess below 0.5 keV and a depression around 0.6 keV, that can be taken into account fitting with a broken power law plus an absorbing feature.

The spectral analysis has been performed using public available matrices as December 31st 1996, however, the values of all spectral parameters are confirmed using the last release of the matrixes (August 31st 1997).

Reference

Boella, G. Butler R.C., Perola G.C. et al. 1997, A&AS, 122, 299
Grandi, P., Guainazzi, M., Mineo, T., et al 1997 A&AS in press

CHANGE IN THE WARM ABSORBER IN MR 2251−178

C. OTANI[1], T. KII[2], & K. MIYA[2]
[1] *The Institute of Physical and Chemical Research (RIKEN)*
[2] *The Institute of Space and Astronautical Science (ISAS)*

Abstract. We report the results of the search for the change in the warm absorber of the quasar MR 2251-178 with *ASCA*. We detected its change along with the X-ray intensity change. In the bright phase in 1993, its X-ray spectra showed the absorption edge features by O VII and O VIII edges. In contrast, in the medium intensity (in 1996), the absorption apparently increased and the edge energy was shifted to the lower energy. These results are well explained by the ionization change of the same absorber. The timescale of the change requires moderately high density and the location to be nearer to the center than the Narrow Line Region. We briefly discuss the stability of the warm absorber.

1. Results

Six *ASCA* observations of MR 2251−178 were performed during 2 months in 1993 and four during 6 months in 1996. The source was in the bright phase in 1993 and in the medium phase in 1996. The absorption feature of the mean spectrum in 1993 was well reproduced by the combination of O VII and O VIII edges in the quasar's rest frame ($z=0.064$). The significant change of the feature from 1993 is detected in 1996. The edge energy was shifted from 0.74 to 0.67 keV, and the bulk absorption increased (Figure). Both spectra are well reproduced by a single warm absorber model with the ionization parameter of $\log \xi = 1.35$ and 0.96 in 1993 and 1996, respectively. The total column density was almost constant ($\log N_W = 21.6-21.7$). This means that the observed change of the absorption feature is explained by the change of the single warm absorber. The warm absorber fits of the individual spectra show that ξ correlates with the flux and that N_W stays almost constant (see Figure again).

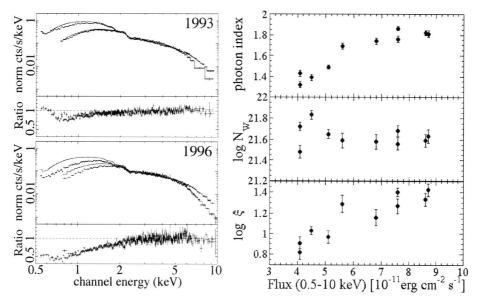

Figures. (left) Mean spectra and the residuals from the power-law model determined above 3 keV in 1993 (upper) and 1996 (lower). (right) The correlation of the incident flux in 0.5–10 keV with Γ, $N_{\rm W}$, and ξ in the individual observations.

2. Discussion

The smooth correlation between the flux and the ionization parameter implies that the recombination timescale in the absorber is shorter than about months to a year. This gives the lower limit of the density to be $\sim 10^{4-5}$ cm^{-3} and the distance from the center to be smaller than ~ 10 pc. Since the quasar's narrow line region (NLR) is expect to locate around 1-10 kpc, the warm matter should locate closer to the center than NLR.

We explored the thermal stability of the warm absorber in terms of the "two-phase" model (Krolik, McKee, Tarter 1981). Adopting the observed flux-index(Γ) relation, we calculated the $\Xi - T$ equilibrium curves for $\Gamma = 1.1 - 1.9$ using CLOUDY (Ferland 1996). We found that, though the curves changes and the ionization state of the absorber in 1993 moves as Γ changes, its track seems keep stable. Although this result is still preliminary, such a behavior may explain the stable presence of the warm absorber in MR 2251−178.

References

Ferland G.J. 1996, 'Hazy, a Brief Introduction to Cloudy', Univ. of Kentucky, Dept. of Physics and Astronomy

Krolik J.H., McKee C.F., Tarter C.B. 1981, ApJ 371, 541

ASCA OBSERVATIONS OF THE QUASAR CONCENTRATION 1338+27

T. YAMADA
Astronomical Institute Tohoku University,
Aoba-ku, Sendai 980-77, Japan

Y. UEDA, T.TAKAHASHI
Institute of Space and Astronautical Science,
3-1-1, Yoshinodani, Sagamihara, Kanagawa 229, Japan

T. MIHARA, N, KAWAI
Institute of Physical and Chemical Research (RIKEN),
Wako, Saitoma, 351-01, Japan

AND

Y. ISHISAKI
Department of Physics, Tokyo Metropolitan University,
Hachioji, Tokyo 192-03, Japan

1. Quasar Concentration 1338+27

There are several regions where a group of quasars are significantly clustered in the physical space. In the "CFHT grens survey" conducted by Crampton et al. (Crampton et al. 1989 and references therein), the 23 quasars between z=1.036 and 1.185 were found to be clustered over $\sim 2° \times 2°$ in the region denoted as 1338+27 At the mean redshift $z_{ave} = 1.113$, the angular extent 6000 arcsec (CHH89) of this cluster corresponds to 60 h^{-1}Mpc($q_0 = 0.5$) and the dispergion of the redshift $\Delta z = 0.044$ to 45 h^{-1}Mpc.

2. ASCA Observations

We observed two regions in the quasar-concentration 1338+27 where surface density of quasars are high. Our main purpose is to search for type-2 obscured quasars in the field. If the clustering of the quasar has physical reasons and not just a statistical fluctuation of the observation, we may

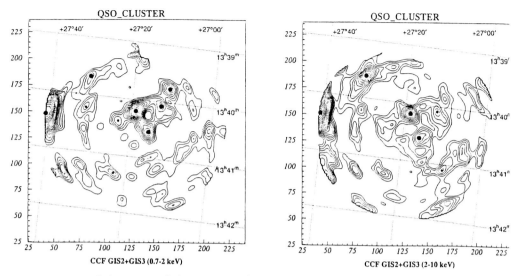

Figure 1. GIS images of the 1338+27 Southern Peak below and above 2 keV. Large filled circles indicate 5σ sources and small ones 4σ

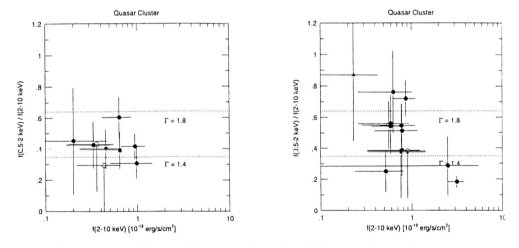

Figure 2. Distribution of flux ratio f(2-10 keV)/f(0.5-2 keV) for SIS and GIS

expect that type-2 quasars are also frequent in the region of the quasar concentration. Figure 1 shows the GIS images of the southern peak, nearly certered at the quasar xxx, and Figure 2 shows the distiribution of hardness (hard-to-soft flux ratio) of the detected sources. It is quite interesting that there are many sources whose have much harder X-ray spectra than those of ordinary quasars.

Search for 10 TeV Gamma-Rays from the Nearby AGNs with the Tibet Air Shower Array

L.K. Ding[1], T. Kobayashi[2], K. Mizutani[2], A. Shiomi[2], Y.H. Tan[1], T. Yuda[3] and the other members of the Tibet ASγ Collaboration
[1]*Institute for High Energy Physics, Academia Senica, Beijing, China,*
[2]*Saitama University, Shimo-Okubo 255, Urawa 338, Japan*
[3]*Institute for Cosmic Ray Research, University of Tokyo,*

1. Introduction

The detection by EGRET[1] of gamma-rays from more than 50 active galactic nuclei (AGNs) allowed us to expect these objects to be the sources of extragalactic cosmic rays at very high energy. The TeV gamma-rays from nearby BL Lac objects of the AGNs examined were detected by the Whipple Observatory collaboration[2]. In this paper, we present the results given by the Tibet air shower array on the search for 10 TeV gamma-ray emission from 18 relatively nearby AGNs with redshifts of $z < 0.07$.

2. Experiment

The Tibet air shower array is located at Yangbajing (4300 m a.s.l., 606 g/cm^2, $90.52°E$, $30.11°N$) in Tibet, China[3]. This array allows us night and day to detect small air showers initiated by cosmic rays with the energy around 10 TeV. Each arrival direction is determined with an accuracy of about $1°$ in the energy around 10 TeV by using a fast timing method.

3. Analysis

We selected the data from the original data set by imposing suitable conditions, to obtain the showers with energy more than about 8 TeV[3]. An equi-zenithal scan method was used to search for a gamma-ray emission from each object. That is, a comparison of the number of events between into the source and into the background windows at the same zenith angle was made to examine an excess of signals from the source direction.

4. Results and Discussion

We used the data taken during the period from October 1995 through August 1996. The target AGNs and the results on the search for continuous emission are listed in Table 1. No excellent excess is found for steady emission of 10 TeV gamma-rays from any object. Flux upper limits at the

90% confidence level for 10 TeV emission were estimated to be 0.32 for NGC 315, 0.35 for NGC 1275, 0.67 for Mkn 421, 1.42 for 3C 120.0, 0.85 for Mkn 501, 2.03 for Mkn 180 and 0.24 for 1Zw 187, respectively, in the unit of 10^{-12} cm^{-2} s^{-1}, by using the simulation calculation in which a differential power-law spectrum with the form of E^{-2} was assumed for each of AGNs.

Since November 1996, we are operating also the high-density detector-array. The statistics will be much improved in the near future.

Table 1. The target AGNs and the results.

Source	z	$\geq$ 10 TeV		$\geq$ 30 TeV	
		N_b	σ	N_b	σ
0055+300 (NGC 315)	0.017	171314	-0.79	6605	1.32
0316+413 (NGC 1275)	0.017	191790	-0.41	6990	0.03
2201+044	0.028	104540	-1.33	3946	0.04
1101+384 (Mkn 421)	0.031	199570	0.69	7185	1.02
0430+052 (3C 120.0)	0.033	109829	2.31	4135	-1.09
1652+398 (Mkn 501)	0.034	194437	1.73	7049	-0.06
2344+514	0.044	156851	-0.07	5885	1.41
1133+704 (Mkn 180)	0.046	56721	2.41	2378	0.14
1959+650	0.047	83340	0.41	3378	-0.18
1807+698 (3C371.0)	0.051	57139	-0.82	2378	1.11
1514+004	0.052	82455	-0.66	3183	0.70
0402+379	0.055	197554	-0.63	7122	0.75
1727+502 (1Zw 187)	0.055	162801	-0.70	6085	-0.11
2321+419	0.059	189433	1.61	6884	-0.43
0116+319	0.059	175649	-1.55	7040	0.23
0802+243 (3C 192.0)	0.060	189989	0.88	6845	-0.09
1214+381	0.062	200168	0.68	7281	-1.40
2200+420 (BL Lac)	0.069	188779	-1.23	6865	-0.65

This work is supported in part by Grants–in–Aid for Scientific Research and also for International Scientific Research from the Ministry of Education, Science and Culture, in Japan and the Committee of National Nature Science Foundation in China.

References

1. D.J. Thompson et al., Astrophys. J. Suppl., 101, 259 (1995)
2. J. Quinn et al., Astrophys. J., 456, L83 (1996)
3. M. Amenomori et al., Phys. Rev. Letters, 69, 2468 (1992)

MICROVARIABILITY OF S5 0716+714, A γ-RAY BLAZAR

R. NESCI, E, MASSARO, M. MAESANO AND F. MONTAGNI
Istituto Astronomico, Universita' "La Sapienza", Roma, ITALY

F. D'ALESSIO
Osservatorio Astronomico Collurania, Teramo, ITALY

AND

G. TOSTI AND M. LUCIANI
*Osservatorio Astronomico, Universita' di Perugia, Perugia
ITALY*

1. Introduction

Like many radio-selected blazars, 0716+714 shows a high level of variability on different time scales, as short as a few days (e.g. Wagner et al 1996, Ghisellini et al. 1997). The mechanism of the emission in the optical band, in the general scheme of a relativistic plasma jet highy collimated toward the observer, is generally believed to be synchrotron radiation from electrons in a strong magnetic field. Our monitoring of S5 0716+714 is aimed to clarify whether the flux variations are chromatic or achromatic. In the first case, one could guess that variations in the spectrum of the injected electrons are responsible for the flux variations; in the second case, geometrical effects like small changes in the angle between the jet and the line of sight could be more likely (Wagner et al. 1993).

2. Observations and data analysis

Photometry was carried out with several telescopes: the 0.4m automatic telescope of the Perugia University, the 0.5m telescope at Vallinfreda (Roma), the 0.7m telescope of the Astronomical Institute and IAS-CNR at Monte Porzio, the 0.7m telescope of the Collurania-Teramo Observatory. In our search for intranight variations we generally used different telescopes, each one observing in a single filter, in order to sample each band in the most tight manner. Besides the standard Johnson-Cousins B,V,R,I filters, also

TABLE 1. Observations log

JD	Day	MP	VA	TE	PG	R mag.	Δmag(max)
2,450,369.5	Oct 12	-	R	99	-	13.6	0.00
2,450,396.4	Nov 08	80	B	-	-	13.3	0.10
2,450,400.4	Nov 12	80	B	-	V	13.5	0.17
2,450,401.4	Nov 13	-	-	86	V,R,I	13.7	0.16
2,450,424.4	Dic 06	86	-	B	-	14.1	0.25
2,450,425.4	Dic 07	86	-	B	-	14.3	0.00
2,450,465.4	Gen 16	-	I	B	-	13.6	0.09
2,450,511.4	Mar 03	-	B,I	-	I	13.9	0.00
2,450,512.4	Mar 04	-	B,I	-	I	13.9	0.17
2,450,520.4	Mar 12	-	B,I	-	-	13.9	0.13

the 80, 86 and 99 narrow-band filters of the Arizona system (Johnson and Mitchell 1975) were used.

The main observational results of the intranight monitoring may be summarized as follows:

a) the luminosity of 0716+714 was generally variable with an amplitude of at least several hundredths of magnitude on time scales of a few hours.

b) whenever the flux variation was substantially greater than our measuring accuracy (typically 0.02 mag), the slope of the light-curve was steeper at shorter wavelengths.

c) when we could observe both the raising and the falling branches of a local maximum we found that there was no plateau and the raising and falling times were substantially equal.

The dependence of the spectral index on the flux variations suggests that the most likely mechanism at work is a localized injections of energy in the form of fresh and rapidly cooling relativistic particles, superimposed to a steadier emission, possibly coming from larger regions. The lack of a plateau in the local maxima may be used to put constraints on the structure of the emitting region in the framework of SSC models, which will be discussed elsewhere.

References

Ghisellini G., Villata M., Raiteri C.M., et al. 1997, A&A in press.
Johnson H.L., Mitchell R.I., 1975, Re. Mexicana Astron. Astrophys. 1, 299
Wagner, S.J., Witzel, A., Krichbaum, T.P., et al. 1993, A&A 271,344.
Wagner, S.J., Witzel, A., Heidt, J., et al., 1996, AJ, 111, 2187

X-RAY PROPERTIES OF LINERS AND LOW LUMINOSITY SEYFERT GALAXIES OBSERVED WITH ASCA

Y.TERASHIMA, H.KUNIEDA
Department of Physics, Nagoya University, Chikusa-ku, Nagoya, Japan

AND

P.J.SERLEMITSOS, A.PTAK
Code 662, NASA/Goddard Space Flight Center, Greenbelt, MD 20771, USA

Abstract.

We present *ASCA* spectra of low luminosity AGNs with X-ray luminosities of 10^{40-41} ergs s^{-1}. Their X-ray continua are very similar to Seyfert galaxies. Although iron K emissin centered at 6.4 keV or 6.7 keV is detected from a few objects, iron K emission is weak on average. Possible reasons for variety of iron line properties are discussed.

1. Observations and Results

We compiled LINERs (Low Ionization Nuclear Emission-line Regions) and low luminosity Seyfert galaxies from which X-ray emission is considered to be dominated by low luminosity AGNs (LLAGNs): NGC 1097, NGC 3031, NGC 3998, NGC 4450, NGC 4579, NGC 4594, and NGC 5033. These galaxies have a X-ray nucleus, larger X-ray to B band luminosity ratio than normal galaxies, and similar X-ray to Hα luminosity ratio to Seyfert galaxies.

Their X-ray continuum shape are very similar to Seyfert 1 galaxies; photon index $\sim$1.8 with small absorption ($< 10^{21}$ cm^{-2}). Iron K emission is detected from NGC 3031 [1][4], NGC 4579 [5], and NGC 5033. The line center energy is 6.7 keV for NGC 3031 and NGC 4579, and 6.4 keV for NGC 5033. Other galaxies (NGC 1097, NGC 3998, NGC 4450, NGC 4594)

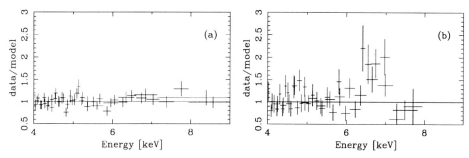

Figure 1. Data / continuum model around iron emission lines (a) averaged spectrum of NGC 1097, NGC 3998, NGC 4450, and NGC 4594 (b) NGC 4579

show no significant iron line (upper limit of an equivalent width EW = 200–300 eV at 90% confidence for one parameter). Since these LLAGNs have very similar X-ray characteristics, we summed up their X-ray spectra and made a composite spectrum. No significant iron emission line is seen in the composite spectrum and an upper limit of an equivalent width is ~ 100 eV for a narrow line at 6.4 keV or 6.7 keV.

2. Discussion

The line center energies of NGC 3031 and NGC 4579 (6.7 keV) are consistent with He-like iron and significantly higher than usual Seyfert 1s. An upper limit on the iron line equivalent width of the composite spectrum is weaker than Seyfert 1s. If iron lines are emitted from the accretion disk around the central blackhole, possible explanations of variety of the iron line properties are difference of inclination angle and/or ionization state of the accretion disk. If inclination angle becomes large, the line becomes very broad and hard to be detected. The blue shifted component from inclined disk may also explain higher line center energy. Alternatively, if the disk is ionized, the line center energy increases. When the ionization state of iron lies in FeXVII–FeXXIII, iron emission is suppressed since effective fluorescence yield is small due to resonant trapping [3] [6]. A possible problem is to maintain high ionization parameter under very low luminosity.

References

1. Ishisaki et al. 1996, PASJ, 48, 237
2. Makishima et al. 1994, PASJ, 46, L77
3. Ross & Fabian 1993, MNRAS, 261, 74
4. Serlemitsos, Ptak, & Yaqoob 1996, in Physics of LINERs in view of Recent Observations, eds. Eracleous et al.
5. Terashima et al. 1997, in preparation
6. Życki & Czerny 1994, MNRAS, 266, 653

EVIDENCE FOR A DRAMATIC ACTIVITY DECLINE IN THE NUCLEUS OF THE RADIO GALAXY FORNAX A

N. IYOMOTO, K. MAKISHIMA, M. TASHIRO, K. MATSUSHITA,
Y. FUKAZAWA, H. KANEDA AND S. OSONE
Department of Physics, University of Tokyo,
7-3-1 Hongo, Bunkyo-ku, Tokyo 113, Japan

Fornax A (NGC 1316) is a radio galaxy with prototypical double lobes. Feigelson et al. (1995 ApJ 449, L149) and Kaneda et al. (1995 ApJ 454, L13) detected inverse Compton X-rays for the first time from its radio lobes, and unambiguously derived the lobe magnetic field intensity of 2–4 μG. Accordingly, the radio-emitting electrons in the lobes are inferred to have a Lorentz factor of 10^4, and hence a synchrotron life time of $\sim 10^8$ yr. This means that Fornax A was highly active at least 10^8 years ago.

In contrast, the Fornax A nucleus is very faint in radio (Slee et al. 1994 MNRAS 269, 928), optical and UV (Fabbiano et al. 1994 ApJ 434, 67), although it remains uncertain whether the nucleus has switched off or is heavily absorbed. We re-analized the data obtained with ASCA on 1994 January 11. The obtained spectra are dominated by thin thermal emission from hot inter-stellar medium, while the 2–10 keV luminosity of hard excess, which may include AGN emission, is less than 1.2×10^{40} erg s^{-1}. This is at least three orders of magnitude lower than those of typical radio galaxy nuclei. Since fluorescence Fe-K line emission, a reliable index of absorption, is not statistically significant in ASCA spectra, X-ray emission from the Fornax A nucleus dose not suffer from heavy absorption; consequently, the nucleus of Fornax A is really inactive at present.

From these two results, we conclude that Fornax A has become dormant during the last 10^8 years, due presumably to a dramatic decrease in the mass accretion rate. This conclusion suggests the possible abundance of "dormant" quasars in nearby galaxies.

ROSAT X-RAY OBSERVATIONS OF THE RADIO GALAXY 4C46.09

D.A. LEAHY
University of Calgary
Calgary, Alberta, Canada T2N 1N4

1. Introduction

4C46.09 is the radio source that shows up as a point-like x-ray source inside the supernova remnant HB9 (Leahy, 1987). Leahy, 1987 found a 0.2-4 keV Einstein IPC flux of approximately $1.5 \times 10^{-12} erg\ cm^{-2} s^{-1}$ and a significantly higher hardness ratio than the rest of HB9. Too few counts were available for any spectral analysis. Seward et al, 1991, found 4C46.09 to be a large radio galaxy at redshift 0.195 and distance 1280 Mpc ($H_o = 50\ km\ s^{-1}\ Mpc^{-1}$). 4C46.09 is of further interest due to the observation of a high energy component in the spectrum of HB9 observed by GINGA (Yamauchi and Koyama, 1993). Whether this was due to HB9 or to 4C46.09 could not be determined.

4C46.09 is the only 4C source inside HB9. and is clearly seen in the radio maps of HB9 of Leahy and Roger, 1991, i.e. at 408 MHz (their Fig. 1) and 1420 MHz (their Fig. 2). The radio flux densities at 408 and 1420 MHz (Leahy and Roger 1995) are 1480 mJy and 248 mJy, respectively. Its position is (1950) R.A. $4^h 54^m 45.61^s$, decl. $46°19'58.0''$ (Seward et al, 1991).

The current work has purpose to clarify whether the hard tail observed by GINGA is due to 4C46.09 or to HB9, and also to learn more about 4C46.09. The ROSAT All-Sky-Survey data of the HB9 region is re-examined here with an analysis of the soft x-ray spectrum of 4C46.09.

2. Observations and Data Analysis

The data was obtained from the ROSAT All Sky Survey data base at MPE, Garching and analyzed at MPE using the EXSAS software. The ROSAT map of the region is dominated by HB9 (Leahy and Aschenbach, 1995), and 4C46.09 is not easily seen within HB9. However there is a bright region associated with the known coordinates of 4C46.09. This bright region coincides with northwest "arm" of the bright cross-shaped region in the western half of HB9. 4C46.09 shows up as a high hardness ratio region inside the western boundary of HB9 in the the hardness ratio (0.4-0.9 keV/ 0.9-2.4 keV) map of HB9.

Spectral fitting to the background subtracted source spectrum was carried out for a Raymond-Smith model. The best fit values are kT= 1.7 keV and $N_H = 6 \times 10^{20} cm^{-2}$. The emission measure for the best fit is $3 \times 10^{42} cm^{-5} \times D^2$, with D in kpc. The 0.1-2.4 keV flux from 4C46.09 is $1 \times 10^{-10} erg\ cm^{-2}\ s^{-1}$.

3. Results and Discussion

The hard x-ray emission from 4C46.09 and HB9 has been previously discussed by Yamauchi and Koyama (1993) in non-imaging observations of HB9 with the LAC detectors on GINGA. The GINGA 2-10 keV flux of $6 \times 10^{-12} erg cm^{-2} s^{-1}$ for the 4C46.09 plus HB9 (the LAC field of view is just larger than HB9, see their Figure 1). Based on the spectral fits here, this flux is consistent with an origin entirely from 4C46.09.

The 0.1-2.4 keV x-ray luminosity of 4C46.09 for a distance of 1280 Mpc is $6 \times 10^{46}\ erg\ s^{-1}$, which makes it rather bright for an AGN. The optical luminosity of 4C46.09 (Seward et al 1991) is $5 \times 10^{11}\ L_{sun}$ and the optical spectrum is dominated by starlight. Since an AGN nucleus does not dominate the optical light, the x-ray luminosity is not likely to be from a bright AGN but rather is probably due to diffuse emission from the cluster. The x-ray emission measure gives $n_e n_H V/D^2$. For a cluster of 1 Mpc radius, one finds $n_e n_H = 3 \times 10^{-7} cm^{-6}$, which is characteristic of diffuse cluster x-ray emission.

In, summary, it is found that the x-ray emission from 4C46.09 may be due to diffuse emission from the cluster, and that it can account for the hard x-ray emission observed from the HB9/4C46.09 region by GINGA.

References
Leahy, D. 1987, ApJ, 322 917
Leahy, D., and Roger, R. 1991, AJ, 102, 2047
Leahy, D., and Roger, R. 1996, A&A Supp., 115, 345
Leahy, D. and Aschenbach, B. 1995, A&A, 293, 853
Seward, F. et al 1991, AJ, 102, 2047
Yamauchi, S. and Koyama, K. 1993, PASJ 45, 545

THE BLACK HOLE GRAZER

A NEW BINARY BLACK HOLE ENGINE IN ACTIVE GALACTIC NUCLEI

Y. TANIGUCHI & O. KABURAKI
Astronomical Institute, Tohoku University
Aramaki, Aoba, Sendai 980-77, JAPAN

1. THE BLACK HOLE GRAZER

We propose an alternative model for the powering of active galactic nuclei (AGN), based on the assumption that all AGN have experienced mergers. In our model (Kaburaki and Taniguchi 1996; Taniguchi and Kaburaki 1996), a close pair of super-massive black holes (the black hole grazer) orbit one another in a plane roughly perpendicular to the galactic center magnetic field. The orbital motion induces surface charges on the black holes which produce an electric field. This field is strong enough to cause pair creation so that the Roche lobe of the binary system is filled with pair plasmas. Rigid-body rotation of the Roche-lobe magnetosphere drives electrodynamically a powerful synchrotron jet emanating from the center of mass of the binary. Furthermore, a pair of equatorial jets flow from the outer Lagrangian points of the binary system. Although these jets are not so collimated, they interact with the accreting gas ring formed around the orbital plane of the binary, causing broad line regions or H_2O maser emission regions (Taniguchi et al. 1996). In addition to the primary jet, two secondary jets are also driven by local accretion disks around the two black holes. The interaction among the primary and the secondary jets may explain detailed jet morphology observed by VLBI facilities.

References

Kaburaki, O., and Taniguchi, Y. (1996) Binary Black-Hole Engines in Active Galactic Nuclei, *Physics of Accretion Disks: Advection, Radiation, and Magnetic Field*, edited by S. Kato et al. (Gordon and Breach Science Publishers), 327

Taniguchi, Y., and Kaburaki, O. (1996) A New Unified Model of Active Galactic Nuclei, *The Physics of LINERs in View of Recent Observations*, ASP Conf. Ser. **103**, 227

Taniguchi, Y., Kaburaki, O., Ohyama, Y., and Murayama, T. (1996b) A Supermassive Black Hole Binary in NGC 4258, *The Physics of LINERs in View of Recent Observations*, ASP Conf. Ser. **103**, 221

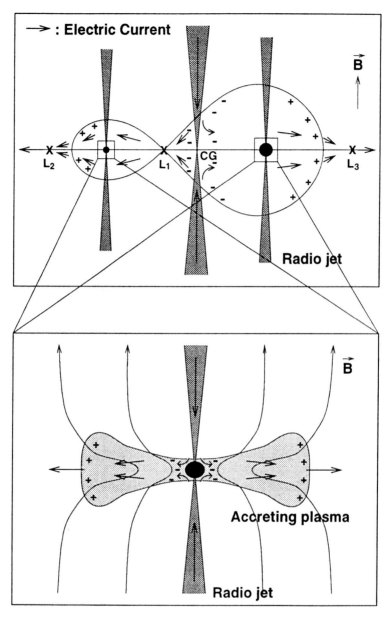

Figure 1. The grazing model of the super massive black-hole binary system. [1] (upper panel) The primary jets emanating from the center of gravity (CG) of the binary system. When the two super massive black holes graze near the pericenter, the two black holes emit intense electric dipole emission. The dipole emission introduces a large-scale electric field around the black holes, resulting electron-positron pair creation. The rotation of the plasma filling the Roche lobe drives an outward electric current from CG. [2] (lower panel) The two secondary jets emanating from the two black holes. The rotation of accreting plasma disks formed around the two black holes drives similar systems of electric current. The physical mechanism of the jets is the same as that of the primary jets. Each vertical pair of return current along the magnetic field grows into a highly collimated, bipolar radio jets.

OPTICAL VARIABILITY IN AGNS; DISK INSTABILITY OR STARBURSTS?

T. KAWAGUCHI AND S. MINESHIGE

Department of Astronomy, Kyoto University
Sakyo-ku, Kyoto 606-01, Japan

M. UMEMURA

Center for Computational Physics, University of Tsukuba
Tsukuba, Ibaraki 305, Japan

AND

E. L. TURNER

Princeton University Observatory
Peyton Hall, Princeton, NJ 08544

1. Introduction

Aperiodic optical variability is a common property of Active Galactic Nuclei (AGN), though its physical origin is still open to question. We have compared light curves among the following two models and observation of quasar 0957+561A,B (Kundić et al. 1997) in terms of structure function analysis (§2).

Starburst (SB) model: Variability is due to random superposition of supernova light curves in the nuclear starburst region. In this study, we calculated a fluctuation light curve, following Aretxaga et al. (1997).

Disk-instability (DI) model: Variability is caused by some instabilities in the accretion disk around a supermassive black hole. We followed Mineshige et al. (1994) for the calculation methods, assuming that optical variability simply reflects X-ray variability.

2. Structure Function Analysis

When time series of magnitude $[\,m(t_i),\ i=1,2,\ldots;\ t_i < t_j\,]$ is given, the first-order structure function, $V(\tau)$, is defined as

$$V(\tau) \equiv \frac{1}{N(\tau)} \sum_{i<j} [\,m(t_i) - m(t_j)\,]^2. \quad (1)$$

Summation is made over all pairs in which $t_j - t_i = \tau$, and $N(\tau)$ denotes a number of such pairs. Generally, $V(\tau)$ is described by a power-law below the typical timescale of shots;

$$[V(\tau)]^{1/2} \propto \tau^\beta. \quad (2)$$

We especially focus our attention to the logarithmic slope (β), since this distinguishes between the SB model and the DI model.

In addition, to evaluate the time-asymmetry of the light curve quantitatively, we separate $V(\tau)$ into two parts, $V_+(\tau)$ and $V_-(\tau)$, depending on the sign of $m(t_i) - m(t_j)$;

$$V_\pm(\tau) \equiv \frac{1}{N_\pm(\tau)} \sum_{i<j} [\,m(t_i) - m(t_j)\,]^2 \quad \text{for} \quad m(t_i) - m(t_j) \gtrless 0, \quad (3)$$

where the summations in the expressions of $V_+(\tau)$ and $V_-(\tau)$ are made, respectively, only for pairs which have plus and minus signs of $[m(t_i) - m(t_j)]$, and $N_+(\tau)$ and $N_-(\tau)$ are the numbers of such pairs. A significant difference between $V_+(\tau)$ and $V_-(\tau)$, if it exists, indicates a deviation from time-symmetry of light curve. The results are as follows:

SB model; We find $\beta \sim 0.75-0.90$. There is a trend that $V_+(\tau) \gtrsim V_-(\tau)$, which means rapid rise and gradual decay in light curve.

DI model; We estimate $\beta \sim 0.41 - 0.49$. There is a slight tendency that $V_+(\tau) \lesssim V_-(\tau)$, which seems to reflect our neglect of hydrodynamical effects.

Observations (0957+561A,B); We have $\beta \sim 0.35$. A little time-asymmetry is shown $[V_+(\tau) \lesssim V_-(\tau)]$, but it is not clear whether the deviation is real or not. Longer observational data will clarify this issue.

In conclusion, the DI model is favored over the SB model to account for the statistical properties of the AGN optical light curve.

The details will be presented elsewhere (e.g., Kawaguchi et al. 1997).

References

Aretxaga, I., Cid Fernandes, R., & Terlevich R. J. 1997, MNRAS, 286, 271
Kawaguchi, T., Mineshige, S., Umemura, M., & Turner, E. L. 1997, in preparation
Kundić, T. et al. 1997, ApJ, 482, 75
Mineshige, S., Takeuchi, M., & Nishimori, H. 1994, ApJ, 435, L125

TESTING THE CENTRAL ENGINES IN AGNS

N. YAMAZAKI, Y. TANIGUCHI, AND O. KABURAKI
Astronomical Institute, Tohoku University
Aramaki, Aoba, Sendai 980-77, JAPAN

1. Theoretical Predictions

It is very important to estimate the relationship between the central black-hole mass $M_{\rm BH}$ and bolometric luminosity $L_{\rm bol}$ of AGNs, because such a relation may be useful to judge the validity of various models for AGNs (Koratkar & Gaskell 1991, hereafter KG91). The predictions of various models are summerized in Table 1, where B is the magnetic field around the central object.

TABLE 1. The model dependence of $M_{\rm BH}$ - $L_{\rm bol}$ relationship

Model	$M_{\rm BH} - L_{\rm bol}$ relationship	Reference
Radiation-driven jet model	$L_{\rm bol} \propto M_{\rm BH}$	Jaroszyński et al. 1980
Accretion-driven jet model	$L_{\rm bol} \approx$ independent of $M_{\rm BH}$	Pringle 1981, Kaburaki 1986
Spinning black-hole model	$L_{\rm bol} \propto B^2 M_{\rm BH}^2$	Blandford & Znajek 1977
Grazer model	$L_{\rm bol} \propto B^2 M_{\rm BH}^2$	Kaburaki & Taniguchi 1996
		Taniguchi & Kaburaki 1996

2. Data

Using the data obtained by Kaspi et al., we have newly derived the masses ($M_{\rm BH}$) of central objects, PG0804+761 and PG0953+414 according to the method used in KG91. Adding these results to the 10 data of KG91 and omitting, instead, 5 original data points with very large errors, we obtain the mass-luminosity relation for AGNs. We find that $L_{\rm bol}$ is proportional to $M_{\rm BH}^2$ rather than to $M_{\rm BH}^1$ (figure1).

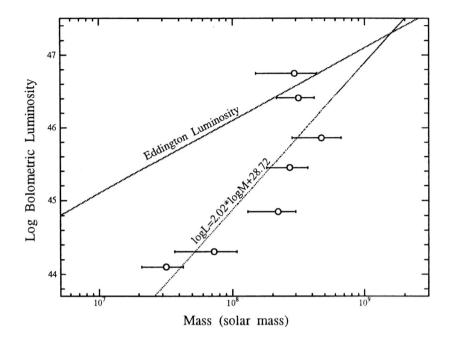

Figure 1. The relation between M_{BH} and L_{bol}

3. Conclusion

Our tentative conclusion is that the most appropriate models of AGN engines are the spinning black hole model and the grazer model (Yamazaki et al. 1997). Radiation-driven jet model is not so appropriate but can not be rejected. The normal accretion-driven jet model is inappropriate.

References

Blandford, R. D., & Znajek, R. L. (1977) *MNRAS*, **179**, 433
Jaroszyński, M., Abramowicz, M. A., & Paczyński, B. (1980) *Acta Astronomica*, **30**, 1
Kaburaki, O. (1986) *MNRAS*, **220**, 331
Kaburaki, O., and Taniguchi, Y. (1996) in *Physics of Accretion Disks: Advection, Radiation, and Magnetic Field*, edited by S. Kato et al. (Gordon and Breach Science Publishers), 327
Kaspi, S., Smith, P. S., Maoz, D., Netzer, H., & Jannuzi, B. T. (1996) *ApJ*, **471**, L75
Koratkar, A. P. & Gaskell, C. M. (1991) *ApJ*, **370**, L61 (KG91)
Padovani, P., & Rafanelli, P. (1988) *A&A*, **205**, 53
Pringle, J. E. (1981) *ARA&A*, **19**, 137
Taniguchi, Y., and Kaburaki, O. (1996) in *The Physics of LINERs in View of Recent Observations, ASP Conf. Ser.* **103**, 227
Yamazaki, N., Taniguchi, Y., & Kaburaki, O. (1997) *ApJ*, submitted

DYNAMICAL EVOLUTION OF ACCRETION FLOW ONTO A BLACK HOLE

M. YOKOSAWA
Department of Physics
Ibaraki University, Mito, Japan

1. Introduction

Active galactic nuclei(AGN) produce many type of active phenomena, powerful X-ray emission, UV hump, narrow beam ejection, gamma-ray emission. Energy of these phenomena is thought to be brought out binding energy between a black hole and surrounding matter. What condition around a black hole produces many type of active phenomena ? We investigated dynamical evolution of accretion flow onto a black hole by using a general-relativistic, hydrodynamic code which contains a viscosity based on the alpha-model. We find three types of flow's pattern, depending on thickness of accretion disk. In a case of the thin disk with a thickness less than the radius of the event horizon at the vicinity of a marginally stable orbit, the accreting flow through a surface of the marginally stable orbit becomes thinner due to additional cooling caused by a general-relativistic Roche-lobe overflow and horizontal advection of heat. An accretion disk with a middle thickness, $2r_\mathrm{h} \leq h \leq 3r_\mathrm{h}$, divides into two flows: the upper region of the accreting flow expands into the atmosphere of the black hole, and the inner region of the flow becomes thinner, smoothly accreting onto the black hole. The expansion of the flow generates a dynamically violent structure around the event horizon. The kinetic energy of the violent motion becomes equivalent to the thermal energy of the accreting disk. The shock heating due to violent motion produces a thermally driven wind which flows through the atmosphere above the accretion disk. A very thick disk, $4r_\mathrm{h} \leq h$, forms a narrow beam whose energy is largely supplied from hot region generated by shock wave. The accretion flowing through the thick disk, $h \geq 2r_\mathrm{h}$, cannot only form a single, laminar flow falling into the black hole, but also produces turbulent-like structure above the event horizon. The middle disk may possibly emit the X-ray radiation observed in active galactic nuclei.

The thin disk may produce UV hump of Seyfert galaxy. Thick disk may produce a jet observed in radio galaxy. The thickness of the disk is determined by accretion rate, such as $h \approx \kappa_{es}/c\dot{M}f(r) \approx 10 r_h \dot{m} f(r)$, at the inner region of the disk where the radiation pressure dominates over the gas pressure. Here, $\dot{M}$ is the accretion rate and $\dot{m}$ is the normarized one by the critical-mass flux of the Eddington limit. κ_{es} and c are the opacity by electron scattering and the velocity of light. $f(r)$ is a function with a value of unity far from the hole.

2. Numerical Calculation of an Accretion Disk in Curved Spacetime

We consider the structure and evolution of the innermost parts of the accretion disk based on a numerical simulation. We used a numerical technique of flux-corrected transport, which was developed from numerical codes used for special relativistic hydrodynamics and for general-relativistically magnetohydrodynamic accretion. In order to examine a standard accretion disk, we include the viscosity in the transport of angular momentum and in the heat equation. The heat generated by viscosity, $-t_{\alpha\beta}\sigma^{\alpha\beta}$, is transported by the radiation. The transport of radiation was solved by the diffusion approximation.

In the case of a thin disk, $h_0 = 0.1 - 1 r_h$, the accreting fluid is smoothly swallowed by a hole. Within the critical surface, $r = r_{ms}$ the fluid rapidly falls, and then the gradient of the velocity becomes very large. This causes an additional cooling due to general-relativistic Roche lobe overflow and horizontal advection of heat. The rate of work due to the pressure force in the radial direction, $\dot{E}_{P\nabla V_r} = -P/\sqrt{\gamma}\,\partial_r(W\sqrt{\gamma}v^r)$, is negative there. On the other hand, its rate becomes positive in the vicinity of the horizon. Its sign is determined by the divergent $\dot{E}_{P\nabla V_r} \propto -\partial_r(W\sqrt{\gamma}v^r) = -\partial_r(\sqrt{g}u^r) \propto -\partial_r(r^2 u^r)$. When $\partial_r(u^r)/u^r < -2/r$, then $\dot{E}_{P\nabla V_r} < 0$. If the fluid freely falls into a hole, $u^r = -\sqrt{2/r}$, then $\dot{E}_{P\nabla V_r} > 0$. The rapidly falling fluid through the critical surface enters a state of free fall near to the horizon.

In the case of the medium thickness of the disk, $h_0 = 2 - 3r_h$, some portion of the fluid in the disk can not be swallowed by a hole, and then begins to blow out into the atmosphere. The initial density at the vicinity of the critical surface, $r = r_{ms}$ distributes along the lines of iso-density, $z(r)_\rho \propto r^n$, $n > 1$. These lines bend strongly at $r \approx r_{ms}$. The fluid in the neighbourhood of the equatorial plane is attracted by the hole and cooled by the horizontal advection of heat. Although the upper part of the disk is also attracted by a hole, its direction of force is radial. Then, the advection of heat, $-P/\sqrt{\gamma}\partial_r(W\sqrt{\gamma}v^r)$, becomes positive. Work due to movement of fluid heats the flow, thus causes an expansion of the flow.

Session 3: Diagnostics of High Gravity Objects with X- and Gamma Rays

3-4. Gamma-Ray Bursts

POSSIBLE X-RAY COUNTERPARTS TO GAMMA-RAY BURSTS, GRB930131 AND GRB940217

R. SHIBATA, T. MURAKAMI AND Y. UEDA
The Institute of Space and Astronautical Science,
3-1-1, Yoshinodai, Sagamihara, Kanagawa 229 Japan

A. YOSHIDA, F. TOKANAI, C. OTANI AND N. KAWAI
The Institute of Physical and Chemical Research,
2-1 Hirosawa, Wako, Saitama 351 Japan

AND

K. HURLEY
UC Berkeley Space Sciences Laboratory,
Berkeley, CA 94720-7450

1. Abstract

We made a search of quiescent X-ray counterparts of two Gamma-Ray Bursts (GRBs), GRB930131 and GRB940217. These GRBs were detected with BATSE, EGRET, COMPTEL on board *CGRO* together with the GRB detector on *Ulysses* spacecraft, then they were localized in small error regions. These observations showed that the bursts were remarkably bright accompanying delayed high energy gamma-rays. *ASCA* observations have found a single X-ray source for each GRB on the possible location determined with the above instruments.

2. Introduction

GRB930131 was the brightest GRB recorded by BATSE in the 3rd catalog [1]. It was a rare GRB in which very high energy (GeV) photons were detected [2] and was also detected by several other gamma-ray detectors, including EGRET, COMPTEL on board *CGRO*, as well as the GRB detector on board *Ulysses*. GRB940217 was also detected with the above instruments [3][4][5] and was the most peculiar burst observed so far. It was one of the GRBs which showed the strongest fluence, and had the longest duration among the GRBs detected with *CGRO*. COMPTEL observed six separate emission peaks during this burst.

3. Observations and Results

3.1. GRB930131

The searched sky area was determined to cover the 2.5 σ EGRET error region with *ASCA*-GIS along the IPN annulus of 43″ wide (90% confidence). There is only one source found in the combined IPN/EGRET error region taking account of the *ASCA* location uncertainty of 1′. The detection significance of the source is 4.5 σ_D(SIS)/4.8 σ_D(GIS), and the source is localized to be at R.A. = $12^h15^m11^s$, Dec. = $-10°18'21''$(J2000) with a 90% confidence error radius of 1′. This location coincides with an X-ray source detected with the *ROSAT* All-Sky Survey, which is claimed to be associated with HR4657, an F-type star of V=6.1 mag at D=34pc [6]. If the extremely intense GRB930131 came from the normal star HR4657, how was the burst energy produced? Even if the burst energy is produced by flares, the luminosity of this X-ray source cannot reach this burst luminosity. This result is likely that the X-rays from the counterpart were too weak to detect with *ASCA*. More detailed discussions have been published in [7].

3.2. GRB940217

ASCA observations were made to cover the combined IPN/EGRET-3σ error region. The analysis has revealed one X-ray source on the combined IPN/EGRET (95%) error region with the detection significance of 6 σ_D at R.A. = $2^h00^m6^s.48$, Dec. = $4°12'46''.8$ (J2000). Archived optical plates were studied with the APM system at Royal Greenwich Observatory, and we found two objects within radius of 1′ around the *ASCA* source. One of them was located inside the 3σ IPN annulus. This optical source being 37″ apart from the *ASCA* location shows blue color, B-R = 0.27, and is likely an AGN. More detailed discussions have been published in [8].

References
[1] Meegan, C. A., et al., ApJS **106**, 65, 1996
[2] Sommer, M., et al., ApJ **422**, L63, 1994
[3] Hurley, K., et al., Nature **372**, 652, 1994
[4] Winkler, C., et al., A&A **302**, 765, 1995
[5] Jones, B. B., et al., ApJ **463**, 565, 1996
[6] Schaefer, B. E., et al., ApJ **422**, L71, 1994
[7] Shibata, R., et al., ApJ **486**, 938, 1997
[8] Tokanai, F., et al., PASJ **49**, 207, 1997

IMPORTANCE OF THE HIGH ENERGY CHANNEL FOR THE GAMMA-RAY BURST DATA

A. MÉSZÁROS
Dpt. Astron., Charles Univ., Prague 5, Švédská 8, Czech Rep. and Konkoly Obs., Budapest, Box 67, H-1525, Hungary

Z. BAGOLY
Eötvös Univ., Lab. Inform. Technol., Múzeum krt.6-8, H-1088 Budapest, Hungary

L. G. BALÁZS
Konkoly Obs., Budapest, Box 67, H-1525, Hungary

AND

I. HORVÁTH, P. MÉSZÁROS
Dpt. Astron., Penn. State Univ., 525 Davey Lab., University Park, PA 16802, USA

1. Principal Component Analysis

Extensive data bases on Gamma-Ray Burst (GRB) properties such as the BATSE 3B catalog (Meegan et al. 1996) contain a wealth of statistical information. The nine entries of the 3B database for each GRB consist of two durations, T_{50}, T_{90}, which contain 50% and 90% of the burst energy, respectively; four fluences (time-integrated energy fluxes) $F1$, $F2$, $F3$, $F4$, defined over different energy channels; and three measures of the peak flux (each summed over the four energy channels), measured over three different resolution timescales (64 ms, 256 ms and 1024 ms). Thus the initial number of variables is $n = 9$. There is, of course, some incompleteness in the catalog. There are 625 GRBs having all 9 non-zero quantities, and only they are considered here.

The Principal Component Analysis (PCA) is done on these. This statistical procedure is explained in (Jolliffe 1986) and (Kendall & Stuart 1976). It can determine the number m of important quantities ($m < n$).

2. Results

The results of this procedure are the following. The first PC is roughly given by the sum of all the quantities, with some extra weight on the first three fluences. Because of the different dimensions involved, it has only a formal meaning. The second PC, accounting for 26.5% of the variation, is clearly important, this value being far above Jolliffe's $70/n = 7.8\%$ cut-off level (Jolliffe 1972). This PC is roughly given by the formal difference of the logarithmic peak fluxes and durations. Hence, the duration, peak flux and total fluence are undoubtedly important quantities, but only two of them are independent. This means that in the roughest approximation, it is enough to consider, e.g., a total fluence and a duration, and these two represent 91.6% of the information content in the 3B catalog. The third PC, practically defined by $F4$ alone, accounts for 5.1% of the variation and is already below Jolliffe's level. Hence $m = 2$. Because it is just below Jolliffe's level, this third PC might however be of some importance. The fourth PC, with its 1.5 %, is far below Jolliffe's limit, and is unimportant.

Some of the results are unexpected. For instance, the analysis indicates that an allowable approximation would be to combine (add) the fluences in the first three channels, and consider them in conjunction with the fluence $F4$ in the fourth channel as the basis vectors for the fluence space. This singling out of $F4$ based on its information content appears to be new.

The details of this contribution will be presented elsewhere (Bagoly et al. 1997).

This research was supported by Domus Hungarica Scientiarium Artiumque Foundation and Charles University grant 101-10/736 (A.M.), NASA grants NAG5-2362, NAG5-2857 (P.M., I.H.), OTKA grants F14324, T14304, F26666 (I.H.), T024027 (L.G.B.), KOSEF grant and Széchenyi Foundation (I.H.). I.H. and P.M. are grateful to E.D. Feigelson and E.E. Fenimore for discussions, and A.M. acknowledges the kind hospitality of Konkoly Observatory.

References

Bagoly, Z., et al. (1997) A Principal Component Analysis of the 3B Gamma-Ray Burst Data, *Astrophys. J.*, submitted for publication.
Jolliffe, I. T. (1972) Discarding Variables in a Principal Component Analysis, I: Artificial data, *Appl. Statist.*, **21**, 160-173.
Jolliffe, I. T. (1986) *Principal Component Analysis*. Springer, New York.
Kendall, M. and Stuart, A. (1976) *The Advanced Theory of Statistics*. Griffin, London.
Meegan, C. A., et al. (1996) The Third BATSE 3B Gamma-Ray Catalog, *Astrophys. J. Suppl.*, **106**, 65-110.

Session 4: Large Scale Hot Plasmas and Their Relation with Dark Matter

OPTICAL FOLLOW-UP OBSERVATIONS OF THE *ASCA* LARGE SKY SURVEY

M. AKIYAMA AND K. OHTA
Department of Astronomy, Kyoto University
Kyoto 606-01, Japan

T.YAMADA
Astronomical Institute, Tohoku University, Japan

Y.UEDA AND T.TAKAHASHI
Institute of Space and Astronautical Science, Japan

M.SAKANO AND T.TSURU
Department of Physics, Kyoto University, Japan

AND

B.BOYLE
Anglo-Australian Observatory, Australia

1. The *ASCA* Large Sky Survey

To reveal the origin of the cosmic X-ray background (CXB) in the hard band, we are now conducting a wide (~ 7 deg^2) and deep ($\sim 1 \times 10^{-13}$ erg sec^{-1} cm^{-2} in the 2–10 keV band) survey with the *ASCA* (the *ASCA* Large Sky Survey, hereafter LSS). We have detected 83 sources above 4 sigma level in the 0.7–10 keV band with the GIS and resolved $\sim 30\%$ of the CXB in the 2–10 keV band into discrete sources (Ueda 1996). AGNs (type 1 and type 2) and clusters of galaxies are expected to be major contributors to these X-ray sources.

2. Optical Observations and First Results

We have made optical imaging observations of almost all the LSS region with the KISO 1m Schmidt telescope. Typical magnitude limit is $R \sim 21$. To compensate R band data of 21 X-ray sources and B magnitude data, we also used APM catalog (McMahon & Irwin 1992, limits are $R \sim 20$ and

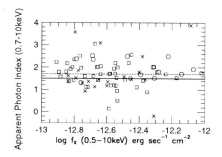

Figure 1. Distribution of X-ray sources in X-ray flux vs. apparent photon index diagram. Circles show candidates of clusters of galaxies, rectangles candidates of type 1 AGNs and crosses remaining sources. The dashed and solid line represent the average photon index of candidates of type 1 AGNs and photon index of the CXB in the LSS region (1.5, Ishisaki 1996), respectively.

$B \sim 22$). Compiling these data, we have made catalogs of optical objects within $1'$ from the center of each X-ray source.

At first, we picked up candidates for clusters of galaxies using number of objects within each error circle in the R band. We identified X-ray sources which have more than 16 optical objects as candidates of clusters. We found 13 candidates in the LSS field. For 6 of these candidates, we have made deep imaging observations with UH 2.2m telescope and confirmed the existence of galaxy concentration. Next, we picked up candidates for type 1 AGNs (type 1 Seyferts and QSOs) in each error circle using the criteria of $B - R \leq 1$ and $-2.5\log(f_x) - 14.2 \leq R \leq -2.5\log(f_x) - 12.2$ (f_x in erg sec^{-1} cm^{-2} in the 0.5–2 keV band). We picked up 48 sources each of which have a optical object meeting the criteria.

In figure 1, we show the distribution of the X-ray sources in flux vs. apparent photon index diagram. Photon indices of the candidates of clusters (circles) are consistent with thermal (1–10 keV) X-ray spectrum. The distribution of photon indices of the candidates of type 1 AGNs (squares) is also consistent with that of type 1 AGNs in the 0.5–10 keV band in the past observations. Many remaining sources (crosses) distribute around the photon index of the CXB. This result may indicate an existence of a population of objects which have hard X-ray spectrum and different optical characteristics from type 1 AGNs.

References

Ishisaki Y., 1996, Ph.D thesis, University of Tokyo
McMahon R., Irwin M., 1992, in Digitized Optical Sky Survey, ed. H.T.MacGillivray and E.B.Thomson (Kluwer, Dordrecht). p.417
Ueda Y., 1996, Ph.D thesis, University of Tokyo

RECENT REPORT ON THE ASCA GIS SOURCE CATALOG PROJECT

Y. ISHISAKI AND T. OHASHI
Department of Physics, Tokyo Metropolitan University
1-1 Minami-Osawa, Hachioji, Tokyo 192-03, Japan

Y. UEDA AND T. TAKAHASHI
The Institute of Space and Astronautical Science
3-1-1 Yoshinodai, Sagamihara, Kanagawa 229, Japan

K. MAKISHIMA
Department of Physics, University of Tokyo
7-3-1 Hongo, Bunkyo-ku, Tokyo 113, Japan

AND

THE GIS TEAM

1. Introduction

We are constructing the ASCA GIS source catalog from the ASCA public archive, mainly for extra-galactic sky. The large field of view and the low-background characteristics of the GIS make it suitable for a search for serendipitous sources in a wide energy band of 0.7–10 keV. Sources to be detected by the project will provide valuable information on the $\log N$-$\log S$ relation over the entire sensitivity band, which has never been available before. About this project, also refer to Ishisaki *et al.* (1995), Ueda *et al.* (1997) and Takahashi *et al.* (1997). There is the SIS source catalog project, too (Gotthelf *et al.*, 1996). These catalogs are going to appear on the WEB.

2. Field Selection and Present Progress

The automatic source-finding procedure has been applied to all the archival data (1993–1996) which satisfy the following selection criteria; the Galactic latitude $|b|$ is higher than 10 degrees, the net exposure is longer than 10 ks, and the primary target is less bright than 10 c/s/GIS in the total count rate. Data out of 481 pointings which amount to 262 deg^2 have been analyzed so far. After running the automatic process, we further rejected the field where the results of the 2-dimensional fitting turned out to be unacceptable. With these filters, we detected 992 sources above 5σ detection in 0.7–7 and/or 2–10 keV band including the main targets.

3. Derived log$N$-log$S$ Relation

The detected source fluxes distribute in a wide range from 10^{-14}–10^{-10} erg s^{-1} cm^{-2} (2–10 keV). To convert flux distribution to the log N-log S relation, we must carefully estimate the survey area at this detection limit, as well as the influence of the main target. To reduce the influence of the main targets, we further selected 27 fields observed in 1994 with severer conditions that $|b| > 50°$ and the primary target is less bright than 0.25 c/s/GIS. To estimate the detection significance for a given flux, we utilized a simplified simulation at every position (4 × 4 arcmin cell in practice). Figure 1 shows the derived log N-log S relation from the 2–10 keV band survey of the selected 27 sample fields. Although the contributions of the main targets are not excluded, our result lies on the extrapolation from the results of *Ginga* fluctuation analysis (Stewart 1992) and the source counts by HEAO-1 A2 (Piccinotti *et al.* 1982), which is consistent with the ASCA Large Sky Survey (LSS) results (Ueda 1996).

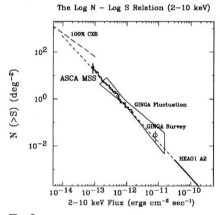

Figure 1. The 2–10 keV log N-log S relation (labeled "ASCA MSS") derived from the selected 27 sample fields. Main targets are also included. 4σ significance detection level are chosen for this plot.

References

Gotthelf, E. and White, N. G. (1996), in "X-ray Imaging and Spectroscopy of Cosmic Hot Plasmas" ed F. Makino and K. Mitsuda (Universal Academy Press, Tokyo)
Ishisaki, Y., Takahashi, T., Kubo, H., Ueda, Y., Makishima, K.,and the GIS team (1995), ASCA News No. 3, 19.
Piccinotti, G. *et al.* (1982), *Astrophys. J.*, **269**, 423.
Stewart, G.C. 1992, in *The X-Ray Background*, eds. X. Barcons and A.C. Fabian (Cambridge University Press: Cambridge), p187.
Takahashi, T., Ueda, Y., Ishisaki, Y., Makishima, K., Ohashi, T. (1997), in "X-Ray Surveys" workshop at Potsdam, Germany
Ueda, Y. (1996), PhD. thesis, The University of Tokyo
Ueda, Y., Takahashi, Y., Ishisaki, Y., Makishima, K., Ohashi, T. (1997), in "All-Sky X-Ray Observations in the Next Decade" workshop at RIKEN Saitama Japan

ASCA DEEP SKY SURVEY

The Log N–Log S relation of faint extragalactic objects and their contribution to the Cosmic X-Ray Background

Y. OGASAKA
NASA Goddard Space Flight Center

T. KII, Y. UEDA, T. TAKAHASHI, H. INOUE
The Institute of Space and Astronautical Science

Y. ISHISAKI
Department of Physics, Tokyo Metropolitan University

K. OHTA
Department of Astronomy, Kyoto University

T. YAMADA
Astronomical Institute, Tohoku University

K. MAKISHIMA
Department of Physics, University of Tokyo

AND

T. MIYAJI, G. HASINGER
Astrophysikalisches Institut Potsdam

1. Introduction

ASCA DSS was intended to carry out unbiased surveys in wide energy range of 0.5–10 keV. The strategy of this project is to survey small sky region with extremely high sensitivity reaching to the source confusion limit of *ASCA* XRT, in contrast to the Large Sky Survey project (Ueda 1996) which covers much larger sky area with relatively shallow exposure.

The regions surveyed and so far studied include Lynx Field (75 ksec), Lockman Hole (55 ksec) and Selected Area 57 (SA57, 250 ksec). These fields are all known to be less extincted by galactic absorption and have deep previous survey observations in optical and other wavebands.

2. Log N–Log S relation

From the number density distribution of faint sources detected from DSS, we derived 2–10 keV $\log N$–$\log S$ relation at the limiting flux of 3.80×10^{-14}

erg sec^{-1} cm^{-2}. This is consistent with the extrapolation of Log$N$–Log$S$ relations from previous experiments(e.g., Hayashida 1991; Piccinotti et al. 1982). At this flux limit about 40% of the CXB intensity in the 2–10 keV band is resolved into discrete sources.

We also derived 0.5–2 keV Log$N$–Log$S$ relation at the flux limit of 9.23×10^{-15} erg sec^{-1} cm^{-2}. The number density obtained seems to be slightly larger than *ROSAT* values(e.g., Hainger et al. (1993)), although it is consistent within the statistical errors and systematic errors due to uncertainties of assumption of spectral index.

3. Source spectrum

Averaged spectrum of detected sources was obtained by hardness-ratio study in 2–10 keV band, as $\alpha = 0.3 \pm 0.4$ which is consistent with the index of the CXB ($\alpha = 0.4$).

At brighter flux regions, *Ginga* and *HEAO-1* experiments obtained averaged indices of 0.8 ± 0.1 and 0.9 ± 0.1, respectively. In this flux region, main contributor to the CXB is considered to be Seyfert 1 Galaxies whose mean index is about 0.7. The hardening of the spectrum we observed can be interpreted as a contribution from hard X-ray dominant sources like absorbed AGNs.

The comparison of the source spectrum for quasar and non-quasar candidates was obtained from cross-identification and spectral analysis of sources detected in SA57 with KKC optical quasar catalog (Koo, Kron and Cudworth 1986; Koo and Kron 1988). The averaged 0.5–10 keV band spectral index for samples with KKC identifications is $\alpha = 0.63 \pm 0.31$. This is consistent with averaged spectrum of high-z quasars observed with *ASCA*, which is $\alpha = 0.61 \pm 0.04$ (0.5–10 keV, Cappi et al. 1997). On the other hand, averaged spectrum of sources with no KKC identifications is hard. If we express this spectrum by absorbed power law, absorption column is estimated as $N_H = (9 \pm 4) \times 10^{21}$ cm^{-2} with fixed power law index of $\alpha = 0.7$.

Acknowledgements. YO acknowledges the support of the Postdoctoral Fellowships for Research Abroad of the Japan Society for the Promotion of the Science.

References

Cappi, M. et al. 1997, *in the proceedings of "X-ray Imaging and Spectroscopy of Cosmic Hot Plasmas", Tokyo, March 1996*
Koo, D. C., Kron, R. G., and Cudworth, K. M. 1986, *Publ. Astron. Soc. Pacific*, **98**, 285.
Hasinger, G. et al., 1993, *Astron. Astrophys.*, **275**, 1.
Hayashida, K. 1989, *Ph. D dissertation of Univ. of Tokyo, ISAS RN 466.*
Ogasaka, Y. 1996, *Ph. D dissertation of Gakushuin University*
Piccinotti, G. et al 1982, *Astrophys. J.*, **253**, 485.
Ueda, Y. 1996, *Ph. D dissertation of University of Tokyo*

OPTICAL FOLLOW-UP OBSERVATIONS OF ASCA LYNX DEEP SURVEY

K. OHTA, M. AKIYAMA, K. NAKANISHI
Dept. Astronomy, Kyoto Univ., Kyoto 606-01, Japan

T. YAMADA
Astronomical Institute, Tohoku Univ., Sendai 980-77, Japan

K. HAYASHIDA
Dept. Earth and Space Science, Osaka Univ., Osaka 560, Japan

T. KII
Institute of Space and Astronautical Science, Kanagawa 229, Japan

AND

Y. OGASAKA
NASA/GSFC, Greenbelt, MD 20771, USA

1. Introduction

Since the bulk of the energy density of the Cosmic X-ray Background (CXB) resides in the harder energy band than that of the ROSAT band (0.5-2 keV) and since the X-ray sources identified in the ROSAT band have X-ray spectra softer than that of the CXB, investigation of nature of the X-ray sources at the harder energy band is indispensable to solve the origin of the CXB. However, only 2-3% of the CXB in the hard band (2-10 keV) had been resolved into discrete sources (Piccinotti et al. 1982, ApJ 253, 485). We present our preliminary results of optical follow-up observations of the ASCA Lynx deep survey.

2. X-ray observations

The X-ray observations were made with the ASCA with an exposure time of about 80 ksec. The flux limit in the 2-10 keV band is more than 100 times deeper than that of the previous survey. The observed field is Lynx

3A and the field of view of SIS is about 20' by 20'. Details of the X-ray observations are presented in Ogasaka (1997, in this proceedings). We define an X-ray source sample containing 8 sources by selecting X-ray sources detected above 5.5 σ confidence. The fluxes of the sources range from 3×10^{-14} to 12×10^{-14} erg s^{-1} cm^{-2} in 0.7 - 7 keV band and from 5×10^{-14} to 10×10^{-14} erg s^{-1} cm^{-2} in 2 - 7 keV band.

3. Optical follow-up observations

Optical imaging observations were made in R-band with the Kiso Schmidt (1.05m) in Japan and in I-band with the University of Hawaii 2.2m and 0.6m telescopes. We picked up candidates of optical counterpart in an error circle (radius $\sim 30''$) of an ASCA source. In this process, we also used the ROSAT SRC catalogue, in which X-ray sources are cataloged with a typical error radius of about $10''$. Note that all the sources identified below are also ROSAT sources. Optical spectroscopic observations were carried out for these candidates using the KPNO 2.1m with the Gold Camera and the KPNO 4m with the Cryo Cam.

4. Results

Two sources are identified with type-1 (broad emission line) AGNs with redshifts of 0.46 and 0.57. One type 1.5 AGN is also identified at $z = 0.56$. These objects have X-ray luminosities of $(1 - 7) \times 10^{44}$ erg s^{-1} (2-10 keV). We have two more type-1 AGN counterparts, but only one broad emission line is seen in the spectra; possible redshifts of these objects are 4.2, 3.2, and 1.3, if the emission is Ly α, CIV, and MgII, respectively. Other 3 X-ray sources have not yet optically identified. However since one of them shows an excess number density of faint galaxies, it may be a cluster of galaxies.

Our 5.5 σ sample has total flux of 3.95×10^{-13} erg s^{-1} cm^{-2} in the 2 - 7 keV band which corresponds to about 27 % of the CXB in this band. Type-1 AGNs/QSOs including type-1.5 and the objects without the certain redshift contribute 19 % of the CXB in this band. The possible cluster has about 4 % contribution.

It should be worth noting that in course of the follow-up program, we obtained optical spectra for less significant X-ray sources. One of which is AX J08494+4454. This object was identified with a type-2 QSO at $z = 0.9$ (Ohta et al. 1996, ApJ 458, L57) and contributes about 7 % of the CXB in the 2 -7 keV band.

THE NEP ROSAT SURVEY

C.R. MULLIS, I.M. GIOIA[1] AND J.P. HENRY
Institute for Astronomy, University of Hawaii
2680 Woodlawn Drive, Honolulu, HI 96822, USA
[1]*Istituto di Radioastronomia del CNR*
Via Gobetti 101, 40129, Bologna, ITALY

1. Introduction

The Rosat All-Sky Survey (Trümper 1991, Adv. Spce Res., 2, 241) has its largest exposure times, approaching 10 ks, at the ecliptic poles where the scan circles overlap. The North Ecliptic Pole (NEP) region covers a $9° \times 9°$ field, and contains a total of 465 X-ray sources detected at $> 4\sigma$ in the 0.1−2.4 keV. We are identifying all sources in the field. The principal derivative is a statistically complete sample of galaxy clusters appropriate for more fully characterizing X-ray Luminosity Function (XLF) evolution. We report preliminary results for two subregions which are identified to the 95% level. These subregions are observed to approximately the median NEP survey exposure. They enclose 16% of the survey area and contain 96 X-ray sources representing 21% of the total NEP. The typical mix of sources is consistent with the Einstein Extended Medium Sensisitivity Survey (EMSS; Gioia et al., 1990a, ApJS, 72, 567). In particular there are 56% AGN/ELG (59% in the EMSS), 25% stars (27% in the EMSS) and 14% clusters against 13% clusters in the EMSS.

2. The Sample

There are 13 clusters in the sample drawn from the completely identified subregions of the NEP. The observed 0.1−2.4 keV count rates are converted to 0.5−2.0 keV band fluxes using a 6 keV thermal bremmstrahlung model including Galactic absorption and K-corrections. Compensations for flux falling outside the detect cell are based on the convolution of the RASS PSF (De Grandi private communication) and a King profile.

3. Evolution

X-ray luminosities of clusters span three orders of magnitude from 10^{42} to 10^{45} erg s^{-1}, and any study of their statistical properties must take this into account. In the context of cluster evolution, the relevant quantity is the X-ray luminosity function. Gioia et al. (1990a, ApJ, 365, L35) and Henry et al. (1992, ApJ, 386, 408) analyzed the XLF for the EMSS cluster sample at z~0.17 and z~0.33. Their analysis suggests a deficit of high luminosity clusters at high redshifts by a factor of three. We reproduce the EMSS XLFs in Figure 1. Two models XLFs are overplotted. H92 from Henry et al. (1992) is the best-fitting theoretical function for the EMSS data. BCS is the recent determination of the low redshift XLF from the ROSAT Brightest Cluster Sample (Ebeling et al. 1997, ApJ, 479, L101). As a first step towards addressing the evolution issue, we add two preliminary NEP XLF determinations at high redshift. To date, 59% of the ROSAT NEP targets have been identified. The complete survey will yield roughly 65 clusters, comparable in size to the EMSS sample. This will provide a significant and unique extension to previous attempts to confirm evolution.

This work received partial support from NSF grant AST95-00515 and NASA grants NAG5-2594 and NAG 5-2914.

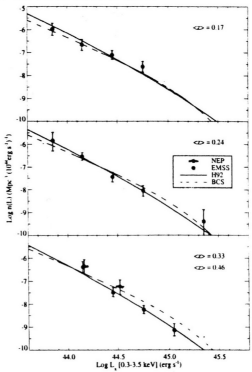

Figure 1. The XLF for clusters of galaxies in redshift shells.